Studies in Systems, Decision and Control

Volume 367

The series "Studies in Systems, Decision and Control" (SSDC) covers both new developments and advances, as well as the state of the art, in the various areas of broadly perceived systems, decision making and control–quickly, up to date and with a high quality. The intent is to cover the theory, applications, and perspectives on the state of the art and future developments relevant to systems, decision making, control, complex processes and related areas, as embedded in the fields of engineering, computer science, physics, economics, social and life sciences, as well as the paradigms and methodologies behind them. The series contains monographs, textbooks, lecture notes and edited volumes in systems, decision making and control spanning the areas of Cyber-Physical Systems, Autonomous Systems, Sensor Networks, Control Systems, Energy Systems, Automotive Systems, Biological Systems, Vehicular Networking and Connected Vehicles, Aerospace Systems, Automation, Manufacturing, Smart Grids, Nonlinear Systems, Power Systems, Robotics, Social Systems, Economic Systems and other. Of particular value to both the contributors and the readership are the short publication timeframe and the world-wide distribution and exposure which enable both a wide and rapid dissemination of research output.

Indexed by SCOPUS, DBLP, WTI Frankfurt eG, zbMATH, SCImago.

All books published in the series are submitted for consideration in Web of Science.

More information about this series at http://www.springer.com/series/13304

Tadeusz Kaczorek · Kamil Borawski

Descriptor Systems of Integer and Fractional Orders

Springer

Tadeusz Kaczorek
Faculty of Electrical Engineering
Bialystok University of Technology
Białystok, Poland

Kamil Borawski
Faculty of Electrical Engineering
Bialystok University of Technology
Białystok, Poland

ISSN 2198-4182 ISSN 2198-4190 (electronic)
Studies in Systems, Decision and Control
ISBN 978-3-030-72479-5 ISBN 978-3-030-72480-1 (eBook)
https://doi.org/10.1007/978-3-030-72480-1

This Springer imprint is published by the registered company Springer Nature Switzerland AG
The registered company address is: Gewerbestrasse 11, 6330 Cham, Switzerland

Preface

This monograph covers some selected problems of the descriptor integer- and fractional-order positive continuous-time and discrete-time systems.

The monograph consists of three chapters, four appendices and the list of references.

Chapter 1 is devoted to descriptor integer-order continuous-time and discrete-time linear systems. The state equations and their solutions of the linear systems are presented. Necessary and sufficient conditions for the positivity, stability and superstability of the systems are established. The considerations are based on the Laurent series expansion method, Drazin inverse matrix method and Weierstrass–Kronecker decomposition method. In addition, for each of the methods and for the assumed approximation of the derivative (Euler method) formulae linking together, the matrices of the continuous-time model and its discrete-time counterpart are given and the influence of the discretization period on the properties of descriptor systems are examined.

In Chapter 2, descriptor fractional-order continuous-time and discrete-time linear systems are considered. The Euler gamma function, Mittag–Leffler functions, Caputo derivative integral and Grünwald–Letnikov fractional-order difference and their basic properties are recalled. The state equations and their solutions of continuous-time and discrete-time linear systems are presented. The positivity, stability and superstability of the factional continuous-time and discrete-time linear systems are investigated. Similarly, to Chapter 1, the considerations are based on the Laurent series expansion method, Drazin inverse matrix method and Weierstrass–Kronecker decomposition method.

Chapter 3 is devoted to the stability of descriptor continuous-time and discrete-time systems of integer and fractional orders. Some stability tests for positive descriptor linear systems are considered. Necessary and sufficient stability conditions for positive descriptor interval linear systems are given. New sufficient conditions for the global stability of nonlinear feedback systems with positive descriptor linear parts are established. These considerations are limited to the Drazin inverse matrix method and Weierstrass–Kronecker decomposition method.

In Appendix A, extensions of the Caley–Hamilton theorem for descriptor linear systems are given. Some methods for computation of the Drazin inverse are presented in Appendix B. In Appendix C, some basic definitions and theorems on Laplace transforms and $\mathcal{Z}$-transforms are given. Some properties of the nilpotent matrices are given in Appendix D.

The monograph contains some original results of the authors. These considerations are an extension of their research presented in articles and the Ph.D. thesis [14]. It is dedicated to scientists and Ph.D. students from the field of control systems theory.

January 2021

Taedusz Kaczorek
Kamil Borawski

Acknowledgment

This research was supported by National Science Centre in Poland under work No. 2017/27/B/ST7/02443.

Contents

List of Symbols

$\mathbb{R}$	the set of real numbers
$\mathbb{R}_+$	the set of real nonnegative numbers
$\mathbb{Z}$	the set of integers
$\mathbb{Z}_+$	the set of nonnegative integers
$\mathbb{C}$	the set of complex numbers
$\mathbb{N}$	the set of natural numbers
$\mathbb{R}^{n\times m}$	the set of $n \times m$ real matrices
$\mathbb{R}_+^{n\times m}$	the set of $n \times m$ real matrices with nonnegative entries
$\mathbb{M}_n$	the set of $n \times n$ Metzler matrices (real matrices with nonnegative off-diagonal entries)
$\mathbb{I}_n$	the $n \times n$ identity matrix
$\delta^{(i)}(t)$	i-th distributional derivative of the Dirac impulse
$\det A$	determinant of the matrix A
$\mathrm{rank} A$	rank of the matrix A
A^{-1}	inverse of the matrix A
A^T	transpose of the matrix A
A^D	Drazin inverse of the matrix A
$\mathrm{Im} A$	Image of the matrix A
$\mathrm{Ker} A$	kernel of the matrix A
$\|\|x\|\|$	norm of the vector x
$\|\|A\|\|$	norm of the matrix A
$\deg p(s)$	degree of the polynomial $p(s)$
$\mathrm{diag}[a_1, \ldots, a_n]$	diagonal matrix with $a_1, \ldots, a_n$ on the diagonal
$n!$	factorial of the natural number n
$\Gamma(x)$	Euler gamma function
$E_\alpha(z)$	one parameter Mittag-Leffler function of z
$E_{\alpha,\beta}(z)$	two parameters Mittag-Leffler function of z
${}_0I_t^\alpha f(t)$	fractional (α-order) integral of the function $f(t)$
${}_0D_t^\alpha f(t)$	fractional (α-order) derivative of the function $f(t)$
${}_0\Delta_k^n x_i$	n-order backward difference of x_i on the interval $[0, k]$

$\mathcal{L}[f(t)] = F(s)$	Laplace transform of the function $f(t)$
$\mathcal{Z}[f_i] = F(z)$	$\mathcal{Z}$-transform of the function f_i
$f(t) * f(t)$	convolution of functions $f(t)$ and $g(t)$
$\dot{f}(t) = \frac{df(t)}{dt}$	first derivative of the function $f(t)$
$f^{(n)}(t) = \frac{d^n f(t)}{dt^n}$	n-th derivative of the function $f(t)$

Chapter 1
Descriptor Linear Systems

In this chapter descriptor continuous-time and discrete-time linear systems will be analyzed. In the subsequent sections solutions of the state equation and analytical conditions of the positivity, stability and superstability will be established considering the Laurent series expansion method, Drazin inverse matrix method and Weierstrass-Kronecker decomposition method. In addition, for each of the methods and for the assumed approximation of the derivative (Euler method) formulae linking together the matrices of the continuous-time model and its discrete-time counterpart will be given and the influence of the discretization period on the properties of descriptor systems will be examined.

1.1 Continuous-Time Systems

1.1.1 State Equations of Descriptor Continuous-Time Linear Systems

Consider the continuous-time linear system described by state equations

$$E\dot{x}(t) = Ax(t) + Bu(t), \tag{1.1a}$$

$$y(t) = Cx(t) + Du(t), \tag{1.1b}$$

where $x(t) \in \mathbb{R}^n$, $u(t) \in \mathbb{R}^m$, $y(t) \in \mathbb{R}^p$ are state, input and output vectors and $E, A \in \mathbb{R}^{n\times n}$, $B \in \mathbb{R}^{n\times m}$, $C \in \mathbb{R}^{p\times n}$, $D \in \mathbb{R}^{p\times m}$. We assume that $\det E = 0$ and

$$\det[Es - A] \neq 0 \text{ for some } s \in \mathbb{C}. \tag{1.2}$$

T. Kaczorek and K. Borawski, *Descriptor Systems of Integer and Fractional Orders*, Studies in Systems, Decision and Control 367,
https://doi.org/10.1007/978-3-030-72480-1_1

If the matrix E is nonsingular ($\det E \neq 0$), then we can transform (1.1a) to the state equation of a standard continuous-time linear system through premultiplication of (1.1a) by E^{-1}. The system (1.1) with the singular matrix E ($\det E = 0$) will be called a descriptor continuous-time linear system.

1.1.2 Solution to the State Equation of Descriptor Continuous-Time Linear Systems

Let U be the set of admissible input vectors $u(t) \in U \subset \mathbb{R}^m$ and $X_0 \subset \mathbb{R}^n$ the set of consistent initial conditions $x(0) = x_0 \in X_0$ for which the Eq. (1.1a) has the solution $x(t)$ with $u(t) \in U$.

1.1.2.1 Laurent Series Expansion Method

In this section the solution to the Eq. (1.1a) will be given using the Laurent series expansion method [26, 47, 49, 111, 147]. Two forms of the solution will be considered: the one that contain Dirac impulses and their derivatives and the one without these components.

Theorem 1.1. Solution to the Eq. (1.1a) for $x_0 \in X_0$ and $u(t) \in U$ is given by the formula

$$x(t) = \sum_{k=0}^{\infty} \left(\frac{T_k E t^k}{k!} x_0 + \int_0^t \frac{T_k (t-\tau)^k}{k!} B u(\tau) d\tau \right) + \sum_{k=1}^{n_0 - 1} T_{-k} E x_0 \delta^{(k-1)}(t) + \sum_{k=1}^{n_0} T_{-k} B u^{(k-1)}(t), \tag{1.3a}$$

where the matrices $T_k \in \mathbb{R}^{n \times n}$ are determined by

$$\begin{cases} ET_k - AT_{k-1} = \mathbb{I}_n & \text{for } k = 0, \\ ET_k - AT_{k-1} = 0 & \text{for } k \neq 0, \\ ET_k = 0 & \text{for } k \leq -n_0, \end{cases} \tag{1.3b}$$

$\delta^{(k)}(t)$ is the k-th distributional derivative of the Dirac impulse, $u^{(k)}(t)$ is the k-th time derivative of the input vector and $n_0 \in \mathbb{Z}_+$ is called the index of the system.

Proof. Using the Laplace transform on (1.1a) we obtain

$$[Es - A]X(s) = Ex_0 + BU(s). \tag{1.4}$$

If the condition (1.2) holds, then we have

$$X(s) = [Es - A]^{-1}[Ex_0 + BU(s)]. \tag{1.5}$$

Expanding the matrix $[Es - A]^{-1}$ in the Laurent series we obtain

$$[Es - A]^{-1} = \sum_{k=-n_0}^{\infty} T_k s^{-(k+1)}. \tag{1.6}$$

If $\det E \neq 0$, then $n_0 = 0$. From definition of the inverse matrix

$$[Es - A]^{-1}[Es - A] = [Es - A][Es - A]^{-1} = \mathbb{I}_n \tag{1.7}$$

and (1.6) we have

$$\sum_{k=-n_0}^{\infty} T_k[Es - A]s^{-(k+1)} = \sum_{k=-n_0}^{\infty} [Es - A]T_k s^{-(k+1)} = \mathbb{I}_n. \tag{1.8}$$

Comparison of the coefficients at the same powers of s in (1.8) yields (1.3b). Substituting (1.6) into (1.5) we obtain

$$X(s) = \sum_{k=-n_0}^{\infty} \left[T_k E x_0 s^{-(k+1)} + T_k B s^{-(k+1)} U(s)\right]. \tag{1.9}$$

The transform (1.9) can be written as a sum

$$X(s) = X_1(s) + X_2(s), \tag{1.10a}$$

where

$$X_1(s) = \sum_{k=0}^{\infty} \left[T_k E x_0 s^{-(k+1)} + T_k B s^{-(k+1)} U(s)\right], \tag{1.10b}$$

$$\begin{aligned} X_2(s) &= \sum_{k=-n_0}^{-1} \left[T_k E x_0 s^{-(k+1)} + T_k B s^{-(k+1)} U(s)\right] \\ &= \sum_{k=1}^{n_0} \left[T_{-k} E x_0 s^{k-1} + T_{-k} B s^{k-1} U(s)\right]. \end{aligned} \tag{1.10c}$$

Applying to (1.10b) the inverse Laplace transform we obtain the first two components of the solution (1.3) since $\mathcal{L}\{t^k\} = \frac{k!}{s^{k+1}}$, $k \in Z_+$. Applying to (1.10c) the inverse Laplace transform and taking into account that $T_{-n_0}E = 0$ we obtain the third and fourth component of the solution (1.3) since $\mathcal{L}\{\delta^{(k)}(t)\} = s^k$, $k = 0, 1, \ldots$. □

Remark 1.1. If $E = \mathbb{I}_n$, then $n_0 = 0$ and $T_k = A^k$, $k = 0, 1, \ldots$. In this case from (1.3) we have

$$
\begin{aligned}
x(t) &= \sum_{k=0}^{\infty}\left(\frac{A^k t^k}{k!}x_0 + \int_0^t \frac{A^k (t-\tau)^k}{k!} Bu(\tau)d\tau\right) \\
&= e^{At}x_0 + \int_0^t e^{A(t-\tau)}Bu(\tau)d\tau,
\end{aligned}
\tag{1.11}
$$

since $e^{At} = \sum_{k=0}^{\infty}\frac{A^k t^k}{k!}$. The formula (1.11) is the solution to the state equation of a standard continuous-time linear system [47].

Note that the solution (1.3a) contains the component with Dirac impulse and its derivatives, which will not appear in physically realizable systems. It should be taken into account that this component affects the state vector only for $t = 0$ and it does not have impact on the trajectory of state vector for $t > 0$. Thus we can neglect this part of the solution and assume that for $t = 0$ the initial condition $x_0 \in X_0$. We will show that the formula (1.3a) without the third component, which will be called the non-impulse solution to the state equation, also satisfies the equality (1.1a).

Theorem 1.2. Non-impulse solution to the Eq. (1.1a) for $x \in X_0$ and $u(t) \in U$ has the form

$$
x(t) = \sum_{k=0}^{\infty}\left(\frac{T_k E t^k}{k!}x_0 + \int_0^t \frac{T_k (t-\tau)^k}{k!} Bu(\tau)d\tau\right) + \sum_{k=1}^{n_0} T_{-k}Bu^{(k-1)}(t), \tag{1.12}
$$

and the consistent initial conditions of the system (1.1) are given by the formula

$$
x_0 = T_0 E v + \sum_{k=1}^{n_0} T_{-k}Bu^{(k-1)}(0), \tag{1.13}
$$

where the matrices $T_k \in \mathbb{R}^{n\times n}$ can be calculated from (1.3b) or (1.6), $v \in \mathbb{R}^n$ is an arbitrary vector, $u^{(k)}(t)$ is the k-th time derivative of the input vector and $n_0 \in \mathbb{Z}_+$ is called the index of the system.

Proof. From Eqs. (1.1a) and (1.12) we have

$$
\begin{aligned}
E\dot{x}(t) &= \sum_{k=1}^{\infty}\left(\frac{ET_k t^{k-1}}{(k-1)!}Ex_0 + \int_0^t \frac{ET_k (t-\tau)^{k-1}}{(k-1)!} Bu(\tau)d\tau\right) \\
&\quad + ET_0Bu(t) + \sum_{k=1}^{n_0} ET_{-k}Bu^{(k)}(t)
\end{aligned}
\tag{1.14}
$$

and

$$Ax(t) = \sum_{k=0}^{\infty}\left(\frac{AT_k t^k}{k!}Ex_0 + \int_0^t \frac{AT_k(t-\tau)^k}{k!}Bu(\tau)d\tau\right) + \sum_{k=1}^{n_0} AT_{-k}Bu^{(k-1)}(t) \tag{1.15}$$

Taking into account that $ET_k - AT_{k-1} = 0$ for $k \neq 0$ we obtain

$$E\dot{x}(t) - Ax(t) = ET_0Bu(t) + \sum_{k=1}^{n_0} ET_{-k}Bu^{(k)}(t) - \sum_{k=1}^{n_0} AT_{-k}Bu^{(k-1)}(t) = Bu(t), \tag{1.16}$$

since $ET_{n_0} = 0$ i $ET_0 - AT_{-1} = \mathbb{I}_n$. The solution (1.12) satisfies the Eq. (1.1a). We obtain the formula (1.13) assuming in (1.12) $t = 0$. □

Example 1.1. Consider the descriptor continuous-time linear system (1.1a) with

$$E = \begin{bmatrix} -0.4 & 0 & -0.5 & 0 \\ -0.2 & 0 & 0 & 0 \\ 0.4 & 1 & 0.5 & 0 \\ 0.2 & 0 & 0 & 0 \end{bmatrix}, \quad A = \begin{bmatrix} -0.2 & 1.8 & 0.5 & 0 \\ 0.4 & 0.4 & 0 & 0 \\ 0.2 & -1.8 & -0.5 & 0.5 \\ -0.4 & 0.6 & 0 & 0 \end{bmatrix}$$
$$B = \begin{bmatrix} -1 & -3.6 \\ 0 & -0.8 \\ -1 & 2.6 \\ 0 & -0.2 \end{bmatrix}. \tag{1.17}$$

The matrix pencil (E, A) of (1.17) is regular since

$$\det[Es - A] = -0.05(s+1)(s+2) \neq 0. \tag{1.18}$$

The matrix $[Es - A]^{-1}$ has the form

$$[Es - A]^{-1} = \begin{bmatrix} -0.4s + 0.2 & -1.8 & -0.5s - 0.5 & 0 \\ -0.2s - 0.4 & -0.4 & 0 & 0 \\ 0.4s - 0.2 & s + 1.8 & 0.5s + 0.5 & -0.5 \\ 0.2s + 0.4 & -0.6 & 0 & 0 \end{bmatrix}$$
$$= \begin{bmatrix} 0 & -\frac{3}{s+2} & 0 & \frac{2}{s+2} \\ 0 & -1 & 0 & -1 \\ -\frac{2}{s+1} & \frac{6}{s+2} & 0 & \frac{2(s+4)}{(s+1)(s+2)} \\ -2 & -2s & -2 & -2s \end{bmatrix}. \tag{1.19}$$

Using (1.6) for $n_0 = 2$, the matrix (1.19) can be written in the form

$$[Es - A]^{-1} = \sum_{k=-2}^{\infty} T_k s^{(k+1)}, \tag{1.20a}$$

where

$$T_k = \begin{cases} \begin{bmatrix} 0 & 0 & 0 & 0 \\ 0 & 0 & 0 & 0 \\ 0 & 0 & 0 & 0 \\ 0 & -2 & 0 & -2 \end{bmatrix} & \text{for } k = -2, \\ \begin{bmatrix} 0 & 0 & 0 & 0 \\ 0 & -1 & 0 & -1 \\ 0 & 0 & 0 & 0 \\ -2 & 0 & -2 & 0 \end{bmatrix} & \text{for } k = -1, \\ \begin{bmatrix} 0 & -3(-2)^k & 0 & 2(-2)^k \\ 0 & 0 & 0 & 0 \\ -2(-1)^k & 6(-2)^k & 0 & 6(-1)^k - 4(-2)^k \\ 0 & 0 & 0 & 0 \end{bmatrix} & \text{for } k = 0, 1, 2, \ldots . \end{cases} \tag{1.20b}$$

Using (1.17) and (1.20b) we have

$$T_k E = \begin{cases} \begin{bmatrix} 0 & 0 & 0 & 0 \\ 0 & 0 & 0 & 0 \\ 0 & 0 & 0 & 0 \\ 0 & -2 & 0 & 0 \end{bmatrix} & \text{for } k = -1, \\ \begin{bmatrix} (-2)^k & 0 & 0 & 0 \\ 0 & 0 & 0 & 0 \\ 2(-1)^k - 2(-2)^k & 0 & (-1)^k & 0 \\ 0 & 0 & 0 & 0 \end{bmatrix} & \text{for } k = 0, 1, 2, \ldots, \end{cases} \tag{1.21a}$$

$$T_k B = \begin{cases} \begin{bmatrix} 0 & 0 \\ 0 & 0 \\ 0 & 0 \\ 0 & 2 \end{bmatrix} & \text{for } k = -2, \\ \begin{bmatrix} 0 & 0 \\ 0 & 1 \\ 0 & 0 \\ 4 & 2 \end{bmatrix} & \text{for } k = -1, \\ \begin{bmatrix} 0 & 2(-2)^k \\ 0 & 0 \\ 2(-1)^k & 6(-1)^k - 4(-2)^k \\ 0 & 0 \end{bmatrix} & \text{for } k = 0, 1, 2, \ldots . \end{cases} \tag{1.21b}$$

Next, we compute

$$\sum_{k=0}^{\infty}\frac{T_k t^k}{k!}=\begin{bmatrix}0 & -3e^{-2t} & 0 & 2e^{-2t}\\ 0 & 0 & 0 & 0\\ -2e^{-t} & 6e^{-2t} & 0 & 2e^{-2t}(3e^t-2)\\ 0 & 0 & 0 & 0\end{bmatrix}. \tag{1.22}$$

Using (1.3a), (1.17), (1.21) and (1.22) we obtain

$$\begin{aligned}
x(t) &= \sum_{k=0}^{\infty}\left(\frac{T_k E t^k}{k!}x_0 + \int_0^t \frac{T_k(t-\tau)^k}{k!}Bu(\tau)d\tau\right)\\
&\quad + T_{-1}Ex_0\delta(t) + T_{-1}Bu(t) + T_{-2}Bu^{(1)}(t)\\
&= \begin{bmatrix} e^{-2t} & 0 & 0 & 0\\ 0 & 0 & 0 & 0\\ 2e^{-2t}(e^t-1) & 0 & e^{-t} & 0\\ 0 & 0 & 0 & 0\end{bmatrix}x_0\\
&\quad + \int_0^t \begin{bmatrix} 0 & 2e^{-2(t-\tau)}\\ 0 & 0\\ 2e^{-(t-\tau)} & 2e^{-2(t-\tau)}(3e^{t-\tau}-2)\\ 0 & 0\end{bmatrix}u(\tau)d\tau\\
&\quad + \begin{bmatrix} 0 & 0 & 0 & 0\\ 0 & 0 & 0 & 0\\ 0 & 0 & 0 & 0\\ 0 & -2 & 0 & 0\end{bmatrix}x_0\delta(t) + \begin{bmatrix} 0 & 0\\ 0 & 1\\ 0 & 0\\ 4 & 2\end{bmatrix}u(t) + \begin{bmatrix} 0 & 0\\ 0 & 0\\ 0 & 0\\ 0 & 2\end{bmatrix}u^{(1)}(t).
\end{aligned} \tag{1.23}$$

The equation above shows the solution (1.3a) of the state equation of the descriptor linear system (1.1a) with (1.17). In further considerations we will use the solution (1.12), which would have been obtained neglecting the third component of (1.23).

1.1.2.2 Drazin Inverse Matrix Method

In this section the solution to the Eq. (1.1a) will be given using the Drazin inverse matrix method [26, 27, 34, 47]. Two forms of the solution will be considered: the one that contain Dirac impulses and their derivatives and the one without these components.

Consider the Eq. (1.1a) and assume that the condition (1.2) is met. There exists a number $c \in \mathbb{C}$ such that $\det[Ec - A] \neq 0$. Premultiplying the Eq. (1.1a) by the matrix $[Ec - A]^{-1}$ we obtain

$$\bar{E}\dot{x}(t) = \bar{A}x(t) + \bar{B}u(t), \tag{1.24a}$$

where

$$\bar{E} = [Ec - A]^{-1}E,\ \ \bar{A} = [Ec - A]^{-1}A,\ \ \bar{B} = [Ec - A]^{-1}B. \tag{1.24b}$$

Note that Eqs. (1.1a) and (1.24a) have the same solution $x(t)$.

Definition 1.1. [47] The smallest nonnegative integer q satisfying

$$\operatorname{rank}\bar{E}^q = \operatorname{rank}\bar{E}^{q+1} \tag{1.25}$$

is called the index of the matrix $\bar{E} \in \mathbb{C}^{n\times n}$.

Definition 1.2. [47] A matrix $\bar{E}^D$ is called the Drazin inverse of $\bar{E}$ if it satisfies the conditions:

$$\bar{E}\bar{E}^D = \bar{E}^D\bar{E}, \tag{1.26a}$$

$$\bar{E}^D\bar{E}\bar{E}^D = \bar{E}^D, \tag{1.26b}$$

$$\bar{E}^D\bar{E}^{q+1} = \bar{E}^q. \tag{1.26c}$$

The Drazin inverse of a square matrix always exists and is unique. If $\det\bar{E} \neq 0$ then $\bar{E}^D = \bar{E}^{-1}$. Some methods for computation of the Drazin inverse are given in [47, 68, 69].

Lemma 1.1. [47] The matrices $\bar{E}$ and $\bar{A}$ defined by (1.24b) have the following properties:

$$\bar{A}\bar{E} = \bar{E}\bar{A},\ \ \bar{A}^D\bar{E} = \bar{E}\bar{A}^D,\ \ \bar{E}^D\bar{A} = \bar{A}\bar{E}^D,\ \ \bar{A}^D\bar{E}^D = \bar{E}^D\bar{A}^D, \tag{1.27a}$$

$$\ker\bar{A} \cap \ker\bar{E} = \{0\}, \tag{1.27b}$$

$$(\mathbb{I}_n - \bar{E}\bar{E}^D)\bar{A}\bar{A}^D = \mathbb{I}_n - \bar{E}\bar{E}^D,\ \ (\mathbb{I}_n - \bar{E}\bar{E}^D)(\bar{E}\bar{A}^D)^q = 0. \tag{1.27c}$$

Also, the matrices $\bar{E}$, $\bar{E}^D$ and $\bar{A}$ can be written in the form

$$\bar{E} = T\begin{bmatrix} J & 0 \\ 0 & N \end{bmatrix}T^{-1},\ \ \bar{A} = T\begin{bmatrix} A_1 & 0 \\ 0 & A_2 \end{bmatrix}T^{-1},\ \ \bar{E}^D = T\begin{bmatrix} J^{-1} & 0 \\ 0 & 0 \end{bmatrix}T^{-1}, \tag{1.27d}$$

where $T \in \mathbb{R}^{n\times n}$, $\det T \neq 0$, $J \in \mathbb{R}^{n_1\times n_1}$ is a nonsingular matrix, $N \in \mathbb{R}^{n_2\times n_2}$ is a nilpotent matrix with the nilpotency index q, i.e. $N^{q-1} \neq 0$ i $N^q = 0$, $A_1 \in \mathbb{R}^{n_1\times n_1}$, $A_2 \in \mathbb{R}^{n_2\times n_2}$ and $n_1 + n_2 = n$.

Lemma 1.2. If

$$x_1(t) = \bar{E}\bar{E}^D x(t),\ \ x_2(t) = (\mathbb{I}_n - \bar{E}\bar{E}^D)x(t), \tag{1.28a}$$

$$x_1(t) + x_2(t) = x(t), \tag{1.28b}$$

then the Eq. (1.24a) is equivalent to the following equations:

$$\dot{x}_1(t) = \bar{E}^D\bar{A}x_1(t) + \bar{E}^D\bar{B}u(t), \tag{1.29a}$$

$$\bar{A}^D\bar{E}(\mathbb{I}_n - \bar{E}\bar{E}^D)\dot{x}_2(t) = x_2(t) + (\mathbb{I}_n - \bar{E}\bar{E}^D)\bar{A}^D\bar{B}u(t). \tag{1.29b}$$

Proof. Premultilpying (1.24a) by $\bar{E}\bar{E}^D$ we obtain

$$\bar{E}^2\bar{E}^D\dot{x}(t) = \bar{A}\bar{E}\bar{E}^D x(t) + \bar{E}\bar{E}^D\bar{B}u(t), \tag{1.30}$$

which is equivalent to

$$\bar{E}\dot{x}_1(t) = \bar{A}x_1(t) + \bar{E}\bar{E}^D\bar{B}u(t). \tag{1.31}$$

Next, multiplying (1.31) by $\bar{E}^D$ we get

$$\bar{E}\bar{E}^D\dot{x}_1(t) = \bar{E}^D\bar{A}x_1(t) + \bar{E}^D\bar{E}\bar{E}^D\bar{B}u(t) \tag{1.32}$$

and

$$\dot{x}_1(t) = \bar{E}^D\bar{A}x_1(t) + \bar{E}^D\bar{B}u(t), \tag{1.33}$$

since $\bar{E}\bar{E}^D x_1(t) = \bar{E}\bar{E}^D\bar{E}\bar{E}^D x(t) = \bar{E}\bar{E}^D x(t) = x_1(t)$. Then, subtracting (1.30) from (1.24a) we have

$$\bar{E}(\mathbb{I}_n - \bar{E}\bar{E}^D)\dot{x}(t) = \bar{A}(\mathbb{I}_n - \bar{E}\bar{E}^D)x(t) + (\mathbb{I}_n - \bar{E}\bar{E}^D)\bar{B}u(t) \tag{1.34}$$

and

$$\bar{E}\dot{x}_2(t) = \bar{A}x_2(t) + (\mathbb{I}_n - \bar{E}\bar{E}^D)\bar{B}u(t). \tag{1.35}$$

Premultiplying (1.35) by $\bar{A}^D(\mathbb{I}_n - \bar{E}\bar{E}^D)$ we obtain

$$\bar{A}^D\bar{E}(\mathbb{I}_n - \bar{E}\bar{E}^D)\dot{x}_2(t) = \bar{A}\bar{A}^D(\mathbb{I}_n - \bar{E}\bar{E}^D)x_2(t) + (\mathbb{I}_n - \bar{E}\bar{E}^D)^2\bar{A}^D\bar{B}u(t). \tag{1.36}$$

Using (1.27c) and the fact that $(\mathbb{I}_n - \bar{E}\bar{E}^D)^k = \mathbb{I}_n - \bar{E}\bar{E}^D$ for $k = 1, 2, \ldots$ we have

$$\bar{A}^D\bar{E}(\mathbb{I}_n - \bar{E}\bar{E}^D)\dot{x}_2(t) = x_2(t) + (\mathbb{I}_n - \bar{E}\bar{E}^D)\bar{A}^D\bar{B}u(t), \tag{1.37}$$

since $(\mathbb{I}_n - \bar{E}\bar{E}^D)x_2(t) = (\mathbb{I}_n - \bar{E}\bar{E}^D)^2 x(t) = (\mathbb{I}_n - \bar{E}\bar{E}^D)x(t) = x_2(t)$. □

Theorem 1.3. Solution to the Eq. (1.24a) for $x_0 \in X_0$ and $u(t) \in U$ has the form

$$\begin{aligned} x(t) = e^{\bar{E}^D\bar{A}t}\bar{E}\bar{E}^D x_0 + \int_0^t e^{\bar{E}^D\bar{A}(t-\tau)}\bar{E}^D\bar{B}u(\tau)d\tau \\ + (\bar{E}\bar{E}^D - \mathbb{I}_n)\left[\sum_{k=1}^{q-1}\delta^{(k-1)}(\bar{A}^D\bar{E})^k x_0 + \sum_{k=0}^{q-1}(\bar{A}^D\bar{E})^k\bar{A}^D\bar{B}u^{(k)}(t)\right], \end{aligned} \tag{1.38}$$

where $\delta^{(k)}$ is the k-th distributional derivative of the Dirac impulse, $u^{(k)}(t)$ is the k-th time derivative of the input vector and q is the index of $\bar{E}$.

Proof. The equality (1.29a) is the state equation of a standard continuous-time system, hence its solution has the form [47]

$$\begin{aligned} x_1(t) &= e^{\bar{E}^D\bar{A}t}x_{10} + \int_0^t e^{\bar{E}^D\bar{A}(t-\tau)}\bar{E}^D\bar{B}u(\tau)d\tau \\ &= e^{\bar{E}^D\bar{A}t}\bar{E}\bar{E}^D x_0 + \int_0^t e^{\bar{E}^D\bar{A}(t-\tau)}\bar{E}^D\bar{B}u(\tau)d\tau, \end{aligned} \tag{1.39}$$

where $x_{10} = \bar{E}\bar{E}^D x_0$.

Next, taking into account that the matrix $\bar{E}(\mathbb{I}_n - \bar{E}\bar{E}^D)$ is a nilpotent matrix with the nilpotency index q, i.e. $\bar{E}^q(\mathbb{I}_n - \bar{E}\bar{E}^D) = 0$, since according to (1.27d) we have

$$\begin{aligned} \bar{E}(\mathbb{I}_n - \bar{E}\bar{E}^D) &= T\begin{bmatrix} J & 0 \\ 0 & N \end{bmatrix}\left(\mathbb{I}_n - \begin{bmatrix} J & 0 \\ 0 & N \end{bmatrix}\begin{bmatrix} J^{-1} & 0 \\ 0 & 0 \end{bmatrix}\right)T^{-1} \\ &= T\begin{bmatrix} 0 & 0 \\ 0 & N \end{bmatrix}T^{-1}. \end{aligned} \tag{1.40}$$

Using the Laplace transform on (1.29b) we obtain

$$\begin{aligned} X_2(s) = &-[\mathbb{I}_n - \bar{A}^D\bar{E}(\mathbb{I}_n - \bar{E}\bar{E}^D)s]^{-1} \\ &\times [\bar{A}^D\bar{E}(\mathbb{I}_n - \bar{E}\bar{E}^D)x_{20} + (\mathbb{I}_n - \bar{E}\bar{E}^D)\bar{A}^D\bar{B}U(s)], \end{aligned} \tag{1.41}$$

where $x_{20} = (\mathbb{I}_n - \bar{E}\bar{E}^D)x_0$. Using the equality [47]

$$[\mathbb{I}_n - \bar{A}^D\bar{E}(\mathbb{I}_n - \bar{E}\bar{E}^D)s]^{-1} = \sum_{k=0}^{q-1}[\bar{A}^D\bar{E}(\mathbb{I}_n - \bar{E}\bar{E}^D)]^k s^k \tag{1.42}$$

and (1.41) we get

$$\begin{aligned} X_2(s) = &-\sum_{k=1}^{q-1}[\bar{A}^D\bar{E}(\mathbb{I}_n - \bar{E}\bar{E}^D)]^k s^{k-1}x_{20} \\ &-\sum_{k=0}^{q-1}[\bar{A}^D\bar{E}(\mathbb{I}_n - \bar{E}\bar{E}^D)]^k(\mathbb{I}_n - \bar{E}\bar{E}^D)\bar{A}^D\bar{B}s^kU(s) \\ &= (\bar{E}\bar{E}^D - \mathbb{I}_n)\left[\sum_{k=1}^{q-1}(\bar{A}^D\bar{E})^k s^{k-1}x_{20} + \sum_{k=0}^{q-1}(\bar{A}^D\bar{E})^k\bar{A}^D\bar{B}s^kU(s)\right]. \end{aligned} \tag{1.43}$$

Applying to (1.43) the inverse Laplace transform and using (1.28) we obtain

$$x_2(t) = (\bar{E}\bar{E}^D - \mathbb{I}_n)\left[\sum_{k=1}^{q-1} \delta^{(k-1)}(\bar{A}^D\bar{E})^k x_0 + \sum_{k=0}^{q-1}(\bar{A}^D\bar{E})^k\bar{A}^D\bar{B}u^{(k)}(t)\right]. \tag{1.44}$$

Taking into account that $x_1(t) + x_2(t) = x(t)$, from (1.39) and (1.44) we get (1.38). □

Note that the solution (1.38) contains the component with Dirac impulse and its derivatives. Similarly to the Laurent series expansion method, we assume that for $t = 0$ the initial condition $x_0 \in X_0$ and neglect the third component of (1.38). This form of the solution to the Eq. (1.24a) have been considered in [26, 47] and it will be called the non-impulse solution to the state Eq. (1.24a).

Theorem 1.4. [26, 47] Non-impulse solution to the Eq. (1.24a) for $x_0 \in X_0$ and $u(t) \in U$ has the form

$$\begin{aligned} x(t) &= e^{\bar{E}^D\bar{A}t}\bar{E}\bar{E}^D x_0 + \bar{E}^D\int_0^t e^{\bar{E}^D\bar{A}(t-\tau)}\bar{B}u(\tau)d\tau \\ &+ (\bar{E}\bar{E}^D - \mathbb{I}_n)\sum_{k=0}^{q-1}(\bar{E}\bar{A}^D)^k\bar{A}^D\bar{B}u^{(k)}(t), \end{aligned} \tag{1.45}$$

where $u^{(k)}(t)$ is the k-th time derivative of the input vector and q is the index of $\bar{E}$.

From (1.45) for $t = 0$ we have

$$x_0 = \bar{E}\bar{E}^D v + (\bar{E}\bar{E}^D - \mathbb{I}_n)\sum_{k=0}^{q-1}(\bar{E}\bar{A}^D)^k\bar{A}^D\bar{B}u^{(k)}(0), \tag{1.46}$$

where $v \in \mathbb{R}^n$ is an arbitrary vector.

In the case when $u(t) = 0$ from (1.46) we have $x_0 = \bar{E}\bar{E}^D v$. A homogenous equation $E\dot{x}(t) = Ax(t)$ has the solution if and only if $x_0 \in \mathrm{Im}\bar{E}\bar{E}^D$.

Lemma 1.3. [47] The matrices $\bar{E}^D[Ec - A]^{-1}$, $\bar{A}^D[Ec - A]^{-1}$, $\bar{E}^D\bar{E}$, $\bar{E}^D\bar{A}$, $\bar{E}\bar{A}^D$, $\bar{E}^D\bar{B}$, $\bar{A}^D\bar{B}$ and the index q of the matrix $\bar{E}$ do not depend of the choice of c.

Theorem 1.5. [47] The solution (1.45) of the Eq. (1.1a) and the set of consistent initial conditions X_0 do not depend of the choice of c.

Example 1.2. (Continuation of Example 1.1) Consider the descriptor continuous-time linear system (1.1a) with (1.17). According to (1.18) the matrix pencil (E, A) is regular. We choose $c = 0$ and using (1.24b) we obtain

$$\bar{E} = [-A]^{-1}E = \begin{bmatrix} 0.5 & 0 & 0 & 0 \\ 0 & 0 & 0 & 0 \\ 1 & 0 & 1 & 0 \\ 0 & -2 & 0 & 0 \end{bmatrix},$$

$$\bar{A} = [-A]^{-1}A = \begin{bmatrix} -1 & 0 & 0 & 0 \\ 0 & -1 & 0 & 0 \\ 0 & 0 & -1 & 0 \\ 0 & 0 & 0 & -1 \end{bmatrix}, \tag{1.47}$$

$$\bar{B} = [-A]^{-1}B = \begin{bmatrix} 0 & 1 \\ 0 & 1 \\ 2 & 4 \\ 4 & 2 \end{bmatrix}.$$

It is easy to see that rank $\bar{E}^2 = \text{rank}\,\bar{E}^3$, so $q = 2$. The Drazin inverse of $\bar{E}$ can be computed using (1.27d):

$$\begin{aligned} \bar{E} &= T \begin{bmatrix} J & 0 \\ 0 & N \end{bmatrix} T^{-1} \\ &= \begin{bmatrix} 0 & 1 & 0 & 0 \\ 0 & 0 & 1 & 0 \\ 1 & -2 & 0 & 0 \\ 0 & 0 & 0 & 1 \end{bmatrix} \begin{bmatrix} 1 & 0 & 0 & 0 \\ 0 & 0.5 & 0 & 0 \\ 0 & 0 & 0 & 0 \\ 0 & 0 & -2 & 0 \end{bmatrix} \begin{bmatrix} 2 & 0 & 1 & 0 \\ 1 & 0 & 0 & 0 \\ 0 & 1 & 0 & 0 \\ 0 & 0 & 0 & 1 \end{bmatrix} \\ &= \begin{bmatrix} 0.5 & 0 & 0 & 0 \\ 0 & 0 & 0 & 0 \\ 1 & 0 & 1 & 0 \\ 0 & -2 & 0 & 0 \end{bmatrix}, \end{aligned} \tag{1.48a}$$

$$\begin{aligned} \bar{E}^D &= T \begin{bmatrix} J^{-1} & 0 \\ 0 & 0 \end{bmatrix} T^{-1} \\ &= \begin{bmatrix} 0 & 1 & 0 & 0 \\ 0 & 0 & 1 & 0 \\ 1 & -2 & 0 & 0 \\ 0 & 0 & 0 & 1 \end{bmatrix} \begin{bmatrix} 1 & 0 & 0 & 0 \\ 0 & 2 & 0 & 0 \\ 0 & 0 & 0 & 0 \\ 0 & 0 & 0 & 0 \end{bmatrix} \begin{bmatrix} 2 & 0 & 1 & 0 \\ 1 & 0 & 0 & 0 \\ 0 & 1 & 0 & 0 \\ 0 & 0 & 0 & 1 \end{bmatrix} \\ &= \begin{bmatrix} 2 & 0 & 0 & 0 \\ 0 & 0 & 0 & 0 \\ -2 & 0 & 1 & 0 \\ 0 & 0 & 0 & 0 \end{bmatrix}. \end{aligned} \tag{1.48b}$$

Note that $\bar{A}^D = \bar{A}^{-1} = \bar{A}$. Next, we compute the matrices

$$\bar{E}\bar{E}^D = \begin{bmatrix} 1&0&0&0 \\ 0&0&0&0 \\ 0&0&1&0 \\ 0&0&0&0 \end{bmatrix}, \quad \bar{E}^D\bar{A} = \begin{bmatrix} -2&0&0&0 \\ 0&0&0&0 \\ 2&0&-1&0 \\ 0&0&0&0 \end{bmatrix},$$

$$\bar{E}^D\bar{B} = \begin{bmatrix} 0&2 \\ 0&0 \\ 2&2 \\ 0&0 \end{bmatrix}, \quad (\bar{E}\bar{E}^D - \mathbb{I}_4) = \begin{bmatrix} 0&0&0&0 \\ 0&-1&0&0 \\ 0&0&0&0 \\ 0&0&0&-1 \end{bmatrix} \tag{1.49}$$

$$\bar{A}^D\bar{E} = \begin{bmatrix} -0.5&0&0&0 \\ 0&0&0&0 \\ -1&0&-1&0 \\ 0&2&0&0 \end{bmatrix}, \quad \bar{A}^D\bar{B} = \begin{bmatrix} 0&-1 \\ 0&-1 \\ -2&-4 \\ -4&-2 \end{bmatrix}$$

and

$$e^{\bar{E}^D\bar{A}t} = \begin{bmatrix} e^{-2t} & 0&0&0 \\ 0&1&0&0 \\ 2e^{-2t}(e^t-1) & 0 & e^{-t} & 0 \\ 0&0&0&1 \end{bmatrix}. \tag{1.50}$$

Taking into account that $q = 2$ and using (1.38), (1.49) and (1.50) we obtain

$$\begin{aligned} x(t) &= e^{\bar{E}^D\bar{A}t}\bar{E}\bar{E}^D x_0 + \int_0^t e^{\bar{E}^D\bar{A}(t-\tau)}\bar{E}^D\bar{B}u(\tau)d\tau + (\bar{E}\bar{E}^D - \mathbb{I}_4)\bar{A}^D\bar{E}x_0\delta(t) \\ &+ (\bar{E}\bar{E}^D - \mathbb{I}_4)\bar{A}^D\bar{B}u(t) + (\bar{E}\bar{E}^D - \mathbb{I}_4)\bar{A}^D\bar{E}\bar{A}^D\bar{B}u^{(1)}(t) \\ &= \begin{bmatrix} e^{-2t} & 0&0&0 \\ 0&0&0&0 \\ 2e^{-2t}(e^t-1) & 0 & e^{-t} & 0 \\ 0&0&0&0 \end{bmatrix} x_0 \\ &+ \int_0^t \begin{bmatrix} 0 & 2e^{-2(t-\tau)} \\ 0 & 0 \\ 2e^{-(t-\tau)} & 2e^{-2(t-\tau)}(3e^{t-\tau}-2) \\ 0 & 0 \end{bmatrix} u(\tau)d\tau \\ &+ \begin{bmatrix} 0&0&0&0 \\ 0&0&0&0 \\ 0&0&0&0 \\ 0&-2&0&0 \end{bmatrix} x_0\delta(t) + \begin{bmatrix} 0&0 \\ 0&1 \\ 0&0 \\ 4&2 \end{bmatrix} u(t) + \begin{bmatrix} 0&0 \\ 0&0 \\ 0&0 \\ 0&2 \end{bmatrix} u^{(1)}(t), \end{aligned} \tag{1.51}$$

which is equivalent to (1.23). The equation above shows the solution (1.38) of the state equation of the descriptor linear system (1.1a) with (1.17). In further considerations we will use the solution (1.45), which would have been obtained neglecting the third component of (1.51).

1.1.2.3 Weierstrass-Kronecker Decomposition Method

In this section the solution to the Eq. (1.1a) will be given using the Weierstrass-Kronecker decomposition method [45, 47, 96, 176]. Two forms of the solution will be considered: the one that contain Dirac impulses and their derivatives and the one without these components.

It is well-known [47, 48] that if the condition (1.2) holds, then there exists a pair of nonsingular matrices $P, Q \in \mathbb{R}^{n\times n}$ such that

$$P[Es - A]Q = \begin{bmatrix} \mathbb{I}_{n_1}s - A_1 & 0 \\ 0 & Ns - \mathbb{I}_{n_2} \end{bmatrix}, \quad A_1 \in \mathbb{R}^{n_1\times n_1}, \quad N \in \mathbb{R}^{n_2\times n_2}, \tag{1.52}$$

where $n_1 = \deg\{\det[Es - A]\}$, $n_2 = n - n_1$ and N is a nilpotent matrix with the nilpotency index μ, i.e. $N^{\mu-1} \neq 0$ and $N^{\mu} = 0$. The matrices P and Q can be computed using one of procedures given in [47, 48, 168].

Premultiplying (1.1a) by the matrix P and introducing the new state vector

$$\bar{x}(t) = \begin{bmatrix} \bar{x}_1(t) \\ \bar{x}_2(t) \end{bmatrix} = Q^{-1}x(t), \quad \bar{x}_1(t) \in \mathbb{R}^{n_1}, \quad \bar{x}_2(t) \in \mathbb{R}^{n_2}, \tag{1.53}$$

we obtain

$$PEQQ^{-1}\dot{x}(t) = PAQQ^{-1}x(t) + PBu(t), \tag{1.54a}$$

$$y(t) = CQQ^{-1}x(t) + Du(t) \tag{1.54b}$$

and using (1.52) we get

$$\dot{\bar{x}}_1(t) = A_1\bar{x}_1(t) + B_1u(t), \tag{1.55a}$$

$$N\dot{\bar{x}}_2(t) = \bar{x}_2(t) + B_2u(t), \tag{1.55b}$$

$$y(t) = C_1\bar{x}_1(t) + C_2\bar{x}_2(t) + Du(t), \tag{1.55c}$$

where

$$PB = \begin{bmatrix} B_1 \\ B_2 \end{bmatrix}, \quad B_1 \in \mathbb{R}^{n_1\times m}, \quad B_2 \in \mathbb{R}^{n_2\times m}, \tag{1.56a}$$

$$CQ = \begin{bmatrix} C_1 & C_2 \end{bmatrix}, \quad C_1 \in \mathbb{R}^{p\times n_1}, \quad C_2 \in \mathbb{R}^{p\times n_2}. \tag{1.56b}$$

Theorem 1.6. [47] Solution to the Eq. (1.55a) for given initial condition $\bar{x}_{10} = \bar{x}_1(0) \in \mathbb{R}^{n_1}$ and input $u(t) \in U$ has the form

$$\bar{x}_1(t) = e^{A_1t}\bar{x}_{10} + \int_0^t e^{A_1(t-\tau)}B_1u(\tau)d\tau. \tag{1.57}$$

Theorem 1.7. [47] Solution to the Eq. (1.55b) for consistent initial condition $\bar{x}_{20} = \bar{x}_2(0) \in \mathbb{R}^{n_2}$ and input $u(t) \in U$ is given by the formula

$$\bar{x}_2(t) = -\sum_{k=1}^{\mu-1} \delta^{(k-1)}(t) N^k \bar{x}_{20} - \sum_{k=0}^{\mu-1} N^k B_2 u^{(k)}(t), \tag{1.58}$$

where $\delta^{(k)}(t)$ is the k-th distributional derivative of the Dirac impulse and $u^{(k)}(t)$ is the k-th time derivative of the input vector.

Similarly to the previously considered methods, the solution (1.58) contains the component with Dirac impulse and its derivatives. We will show that the formula (1.58) without this component, which will be called the non-impulse solution to the state equation, also satisfies the equality (1.55b).

Theorem 1.8. Non-impulse solution to the Eq. (1.55b) for $u(t) \in U$ has the form

$$\bar{x}_2(t) = -\sum_{k=0}^{\mu-1} N^k B_2 u^{(k)}(t) \tag{1.59}$$

and the consistent initial conditions of the subsystem (1.55b) are given by the formula

$$\bar{x}_{20} = -\sum_{k=0}^{\mu-1} N^k B_2 u^{(k)}(0), \tag{1.60}$$

where $u^{(k)}(t)$ is the k-th time derivative of the input vector.

Proof. From (1.55b) and (1.59) we have

$$\begin{aligned} N\dot{\bar{x}}_2(t) &= -\sum_{k=0}^{\mu-1} N^{k+1} B_2 u^{(k+1)}(t) = -\sum_{k=0}^{\mu-1} N^k B_2 u^{(k)}(t) + B_2 u(t) \\ &= \bar{x}_2(t) + B_2 u(t) \end{aligned} \tag{1.61}$$

since $N^\mu = 0$. Substituting $t = 0$ into (1.59) we obtain (1.60). □

Assume that the matrix N in (1.55b) has the form

$$N = \begin{bmatrix} 0\,1\,0 \ldots 0 \\ 0\,0\,1 \ldots 0 \\ \vdots\;\vdots\;\vdots\;\ddots\;\vdots \\ 0\,0\,0 \ldots 1 \\ 0\,0\,0 \ldots 0 \end{bmatrix} \tag{1.62}$$

and

$$\bar{x}_2(t) = \begin{bmatrix} \bar{x}_{21}(t) \\ \bar{x}_{22}(t) \\ \vdots \\ \bar{x}_{2n_2}(t) \end{bmatrix}, \quad B_2 = \begin{bmatrix} B_{21} \\ \vdots \\ B_{2n_2} \end{bmatrix}. \tag{1.63}$$

From (1.55b), (1.62) and (1.63) we have

$$\begin{bmatrix} 0\,1\,0 \dots 0 \\ 0\,0\,1 \dots 0 \\ \vdots\;\vdots\;\vdots\;\ddots\;\vdots \\ 0\,0\,0 \dots 1 \\ 0\,0\,0 \dots 0 \end{bmatrix} \frac{d}{dt} \begin{bmatrix} \bar{x}_{21}(t) \\ \bar{x}_{22}(t) \\ \vdots \\ \bar{x}_{2n_2}(t) \end{bmatrix} = \begin{bmatrix} \bar{x}_{21}(t) \\ \bar{x}_{22}(t) \\ \vdots \\ \bar{x}_{2n_2}(t) \end{bmatrix} + \begin{bmatrix} B_{21} \\ \vdots \\ B_{2n_2} \end{bmatrix} u(t) \tag{1.64}$$

and

$$\begin{aligned} 0 &= \bar{x}_{2n_2}(t) + B_{2n_2}u(t), \\ \dot{\bar{x}}_{2n_2}(t) &= \bar{x}_{2n_2-1}(t) + B_{2n_2-1}u(t), \\ &\vdots \\ \dot{\bar{x}}_{22}(t) &= \bar{x}_{21}(t) + B_{21}u(t). \end{aligned} \tag{1.65}$$

Solving the Eq. (1.65) with respect to components of the vector $\bar{x}_2(t)$ we obtain

$$\begin{aligned} \bar{x}_{2n_2}(t) &= -B_{2n_2}u(t), \\ \bar{x}_{2n_2-1}(t) &= -B_{2n_2}\dot{u}(t) - B_{2n_2-1}u(t), \\ &\vdots \\ \bar{x}_{21}(t) &= -B_{2n_2}u^{(n_2-1)}(t) - \dots - B_{21}u(t), \end{aligned} \tag{1.66}$$

where $u^{(k)}(t)$ is the k-th time derivative of the input vector.

Consistent initial conditions of the system (1.55) can be computed from (1.66) for $t = 0$. Solution to the Eq. (1.55b) for $u(t) \in U$ and consistent initial condition $\bar{x}_{20} = \bar{x}_2(0) \in \mathbb{R}^{n_2}$ is given by (1.66). Considerations can be easily extended to the case when the matrix N has the form

$$N = \text{blockdiag}\left[N_1 \dots N_j \right], \quad j > 1, \tag{1.67}$$

where N_k, $k = 1, 2, \dots, j$ are given by (1.62).

Example 1.3. (Continuation of Examples 1.1 and 1.2) Consider the descriptor continuous-time linear system (1.1a) with (1.17). According to (1.18) the matrix pencil (E, A) is regular. In this case

$$P = \begin{bmatrix} -1 & 3 & 0 & 1 \\ 0 & -3 & 0 & 2 \\ 1 & 0 & 1 & 0 \\ 0 & 1 & 0 & 1 \end{bmatrix}, \quad Q = \begin{bmatrix} 0 & 1 & 0 & 0 \\ 0 & 0 & 0 & 1 \\ 2 & 0 & 0 & 0 \\ 0 & 0 & 2 & 0 \end{bmatrix} \tag{1.68}$$

and

$$\begin{bmatrix} \mathbb{I}_{n_1} & 0 \\ 0 & N \end{bmatrix} = PEQ = \begin{bmatrix} 1 & 0 & 0 & 0 \\ 0 & 1 & 0 & 0 \\ 0 & 0 & 0 & 1 \\ 0 & 0 & 0 & 0 \end{bmatrix},$$

$$\begin{bmatrix} A_1 & 0 \\ 0 & \mathbb{I}_{n_2} \end{bmatrix} = PAQ = \begin{bmatrix} -1 & 1 & 0 & 0 \\ 0 & -2 & 0 & 0 \\ 0 & 0 & 1 & 0 \\ 0 & 0 & 0 & 1 \end{bmatrix}, \tag{1.69}$$

$$\begin{bmatrix} B_1 \\ B_2 \end{bmatrix} = PB = \begin{bmatrix} 1 & 1 \\ 0 & 2 \\ -2 & -1 \\ 0 & -1 \end{bmatrix}.$$

Next we compute the transition matrix

$$e^{A_1 t} = \begin{bmatrix} e^{-t} & e^{-2t}(e^t - 1) \\ 0 & e^{-2t} \end{bmatrix}. \tag{1.70}$$

Using (1.57), (1.69) and (1.70) we obtain

$$\begin{aligned} \bar{x}_1(t) = & \begin{bmatrix} e^{-t} & e^{-2t}(e^t - 1) \\ 0 & e^{-2t} \end{bmatrix} \bar{x}_{10} \\ & + \int_0^t \begin{bmatrix} e^{-(t-\tau)} & 3e^{-(t-\tau)} - 2e^{-2(t-\tau)} \\ 0 & 2e^{-2(t-\tau)} \end{bmatrix} u(\tau)d\tau. \end{aligned} \tag{1.71}$$

Taking into account that $\mu = 2$ and using (1.58), (1.69) we obtain

$$\begin{aligned} \bar{x}_2(t) &= -N\delta(t)\bar{x}_{20} - B_2 u(t) - N B_2 u^{(1)}(t) \\ &= \begin{bmatrix} 0 & -1 \\ 0 & 0 \end{bmatrix} \delta(t)\bar{x}_{20} + \begin{bmatrix} 2 & 1 \\ 0 & 1 \end{bmatrix} u(t) + \begin{bmatrix} 0 & 1 \\ 0 & 0 \end{bmatrix} u^{(1)}(t). \end{aligned} \tag{1.72}$$

From (1.53), (1.71) and (1.72) we have

$$\bar{x}(t) = \begin{bmatrix} e^{-t} & e^{-2t}(e^{t}-1) & 0 & 0 \\ 0 & e^{-2t} & 0 & 0 \\ 0 & 0 & 0 & 0 \\ 0 & 0 & 0 & 0 \end{bmatrix} \bar{x}_0 + \int_0^t \begin{bmatrix} e^{-(t-\tau)} & 3e^{-(t-\tau)} - 2e^{-2(t-\tau)} \\ 0 & 2e^{-2(t-\tau)} \\ 0 & 0 \\ 0 & 0 \end{bmatrix} u(\tau)d\tau$$
$$+ \begin{bmatrix} 0&0&0&0 \\ 0&0&0&0 \\ 0&0&0&-1 \\ 0&0&0&0 \end{bmatrix} \bar{x}_0\delta(t) + \begin{bmatrix} 0&0 \\ 0&0 \\ 2&1 \\ 0&1 \end{bmatrix} u(t) + \begin{bmatrix} 0&0 \\ 0&0 \\ 0&1 \\ 0&0 \end{bmatrix} u^{(1)}(t). \quad (1.73)$$

Taking into account that $\bar{x}(t) = Q^{-1}x(t)$ we have to premultiply the Eq. (1.73) by Q and substitute $\bar{x}_0 = Q^{-1}x_0$. Thus we obtain the formula (1.23) (or (1.51)), which means that all considered methods of the descriptor systems analysis give the same result. In further considerations the component with Dirac impulse in (1.73) will be neglected.

Remark 1.2. Comparing all methods of the descriptor systems analysis we can notice that the numbers: n_0 in (1.3a) and (1.12), q in (1.38) and (1.45), μ in (1.58) and (1.59) are equivalent.

1.1.3 Positive Descriptor Continuous-Time Linear Systems

In positive systems inputs, outputs and state variables take only nonnegative values [40, 49]. The following definition of the positivity is suitable for all methods of the descriptor systems analysis.

Definition 1.3. The descriptor continuous-time linear system (1.1) is called (internally) positive if $x(t) \in \mathbb{R}_+^n$ and $y(t) \in \mathbb{R}_+^p$, $t \in [0, +\infty)$ for any consistent initial condition $x_0 \in X_0 \subset \mathbb{R}_+^n$ and all admissible inputs $u(t) \in U \subset \mathbb{R}_+^m$, $t \in [0, +\infty)$ such that $u^{(k)}(t) \in \mathbb{R}_+^m$, $k = 1, \ldots, l-1$, $t \in [0, +\infty)$, where $l = n_0 = q = \mu$.

1.1.3.1 Laurent Series Expansion Method

In this section sufficient positivity conditions of the descriptor continuous-time linear system (1.1) will be established using the Laurent series expansion method. It is assumed that the solution to the state Eq. (1.1a) has the form (1.12).

Theorem 1.9. The descriptor continuous-time linear system (1.1) is (internally) positive if

$$
\begin{aligned}
T_kE \in \mathbb{R}_+^{n\times n}, \ k = 0, 1, \dots,\\
T_kB \in \mathbb{R}_+^{n\times m}, \ k = -n_0, -n_0+1, \dots, 0, 1, \dots,\\
C \in \mathbb{R}_+^{p\times n}, \ D \in \mathbb{R}_+^{p\times m},
\end{aligned}
\tag{1.74}
$$

where the matrices $T_k \in \mathbb{R}^{n\times n}$ are given by (1.3b).

Proof. From (1.12) we have $\frac{t^k}{k!} \geq 0$ for $k \in \mathbb{Z}_+$ and $t \in [0, +\infty)$. According to Definition 1.3 we have $x_0 \in \mathbb{R}_+^n$ and $u^{(k)}(t) \in \mathbb{R}_+^m$, $k = 0, 1, \dots, n_0 - 1$, $t \in [0, +\infty)$. Therefore, $x(t) \in \mathbb{R}_+^n$ if $T_kE \in \mathbb{R}_+^{n\times n}$, $k = 0, 1, \dots$ and $T_kB \in \mathbb{R}_+^{n\times m}$, $k = -n_0$, $-n_0+1, \dots, 0, 1, \dots$. Substituting (1.12) into (1.1b) we obtain

$$
\begin{aligned}
y(t) &= C\sum_{k=0}^{\infty}\left(\frac{T_kEt^k}{k!}x_0 + \int_0^t \frac{T_k(t-\tau)^k}{k!}Bu(\tau)d\tau\right)\\
&\quad + C\sum_{k=1}^{n_0} T_{-k}Bu^{(k-1)}(t) + Du(t).
\end{aligned}
\tag{1.75}
$$

From (1.75) it follows that $y(t) \in \mathbb{R}_+^p$, $t \in [0, +\infty)$ if $C \in \mathbb{R}_+^{p\times n}$ and $D \in \mathbb{R}_+^{p\times m}$. □

Example 1.4. Consider the descriptor continuous-time linear system (1.1) with

$$
E = \begin{bmatrix} 1 & 0 & 0\\ 0 & 1 & 0\\ 0 & 0 & 0 \end{bmatrix}, \quad A = \begin{bmatrix} 0 & 2 & 0\\ 2 & 0 & -1\\ 0 & 0 & 1 \end{bmatrix}, \quad B = \begin{bmatrix} 0\\ 2\\ -1 \end{bmatrix},
$$
$$
C = \begin{bmatrix} 1 & 2 & 0 \end{bmatrix}, \quad D = 0. \tag{1.76}
$$

The matrix pencil (E, A) of (1.76) is regular since

$$
\det[Es - A] = \begin{vmatrix} s & -2 & 0\\ -2 & s & 1\\ 0 & 0 & -1 \end{vmatrix} = -s^2 + 4 \neq 0. \tag{1.77}
$$

The matrix $[Es - A]^{-1}$ has the form

$$
[Es - A]^{-1} = \begin{bmatrix} s & -2 & 0\\ -2 & s & 1\\ 0 & 0 & -1 \end{bmatrix}^{-1} = \begin{bmatrix} \frac{s}{s^2-4} & \frac{2}{s^2-4} & \frac{2}{s^2-4}\\ \frac{2}{s^2-4} & \frac{s}{s^2-4} & \frac{s}{s^2-4}\\ 0 & 0 & -1 \end{bmatrix}. \tag{1.78}
$$

Using (1.6) the matrix (1.78) can be written in the form

$$
[Es - A]^{-1} = \sum_{k=-1}^{\infty} T_k s^{-(k+1)}, \tag{1.79a}
$$

where

$$T_k = \begin{cases} \begin{bmatrix} 0 & 0 & 0 \\ 0 & 0 & 0 \\ 0 & 0 & -1 \end{bmatrix} & \text{for } k = -1, \\ \begin{bmatrix} 2^k & 0 & 0 \\ 0 & 2^k & 2^k \\ 0 & 0 & 0 \end{bmatrix} & \text{for } k = 0, 2, 4, 6, \ldots, \\ \begin{bmatrix} 0 & 2^k & 2^k \\ 2^k & 0 & 0 \\ 0 & 0 & 0 \end{bmatrix} & \text{for } k = 1, 3, 5, 7, \ldots. \end{cases} \tag{1.79b}$$

Using (1.76) and (1.79b) we have

$$T_k E = \begin{cases} \begin{bmatrix} 2^k & 0 & 0 \\ 0 & 2^k & 0 \\ 0 & 0 & 0 \end{bmatrix} & \text{for } k = 0, 2, 4, 6, \ldots, \\ \begin{bmatrix} 0 & 2^k & 0 \\ 2^k & 0 & 0 \\ 0 & 0 & 0 \end{bmatrix} & \text{for } k = 1, 3, 5, 7, \ldots, \end{cases} \tag{1.80a}$$

$$T_k B = \begin{cases} \begin{bmatrix} 0 \\ 0 \\ 1 \end{bmatrix} & \text{for } k = -1, \\ \begin{bmatrix} 0 \\ 2^k \\ 0 \end{bmatrix} & \text{for } k = 0, 2, 4, 6, \ldots, \\ \begin{bmatrix} 2^k \\ 0 \\ 0 \end{bmatrix} & \text{for } k = 1, 3, 5, 7, \ldots. \end{cases} \tag{1.80b}$$

It is easy to see that $T_k E \in \mathbb{R}_+^{3\times3}$, $k = 0, 1, \ldots$ and $T_k B \in \mathbb{R}_+^{3\times1}$, $k = -1, 0, 1, \ldots$, $C \in \mathbb{R}_+^{1\times3}$, $D \in \mathbb{R}_+^{1\times1}$. Therefore, by Theorem 1.9 the descriptor continuous-time linear system (1.1) with (1.76) is positive.

1.1.3.2 Drazin Inverse Matrix Method

In this section necessary and sufficient positivity conditions of the descriptor continuous-time linear system (1.24) (or equivalently (1.1)) will be established using the Drazin inverse matrix method. This problem has been also considered in [79, 141, 142, 172]. It is assumed that the solution to the state Eq. (1.24a) has the form (1.45).

Lemma 1.4. [13] The matrices $\bar{E}^D$ and $e^{\bar{E}^D\bar{A}t}$ satisfy the equality

$$\bar{E}^D e^{\bar{E}^D\bar{A}t} = e^{\bar{E}^D\bar{A}t}\bar{E}^D. \tag{1.81}$$

Proof. The matrix $e^{\bar{E}^D\bar{A}t}$ can be expanded into a series

$$e^{\bar{E}^D\bar{A}t} = \mathbb{I}_n + \bar{E}^D\bar{A}t + \frac{(\bar{E}^D\bar{A})^2 t^2}{2!} + \dots. \tag{1.82}$$

Substituting (1.82) into (1.81) we have

$$\begin{aligned}
\bar{E}^D e^{\bar{E}^D\bar{A}t} &= \bar{E}^D\left(\mathbb{I}_n + \bar{E}^D\bar{A}t + \frac{\bar{E}^D\bar{A}\bar{E}^D\bar{A}t^2}{2!} + \dots\right)\\
&= \bar{E}^D + \bar{E}^D\bar{E}^D\bar{A}t + \frac{\bar{E}^D\bar{E}^D\bar{A}\bar{E}^D\bar{A}t^2}{2!} + \dots\\
&= \left(\mathbb{I}_n + \bar{E}^D\bar{A}t + \frac{\bar{E}^D\bar{A}\bar{E}^D\bar{A}t^2}{2!} + \dots\right)\bar{E}^D = e^{\bar{E}^D\bar{A}t}\bar{E}^D,
\end{aligned} \tag{1.83}$$

since $\bar{E}^D\bar{A} = \bar{A}\bar{E}^D$. □

Lemma 1.5. [49] Let $\bar{E}^D\bar{A} \in \mathbb{M}_n$. Then

$$e^{\bar{E}^D\bar{A}t} \in \mathbb{R}_+^{n\times n}, \quad t \in [0, +\infty). \tag{1.84}$$

Lemma 1.6. The matrices $\bar{E}^D\bar{A}$ and $\bar{E}^D\bar{A} + G(\mathbb{I}_n - \bar{E}\bar{E}^D)$, where $G \in \mathbb{R}^{n\times n}$ is arbitrary, satisfy the equality

$$e^{\bar{E}^D\bar{A}t}\bar{E}\bar{E}^D = e^{[\bar{E}^D\bar{A}+G(\mathbb{I}_n-\bar{E}\bar{E}^D)]t}\bar{E}\bar{E}^D. \tag{1.85}$$

Proof. The matrix $e^{[\bar{E}^D\bar{A}+G(\mathbb{I}_n-\bar{E}\bar{E}^D)]t}\bar{E}\bar{E}^D$ can be written in the form

$$\begin{aligned}
&e^{[\bar{E}^D\bar{A}+G(\mathbb{I}_n-\bar{E}\bar{E}^D)]t}\bar{E}\bar{E}^D\\
&= \left(\mathbb{I}_n + [\bar{E}^D\bar{A} + G(\mathbb{I}_n - \bar{E}\bar{E}^D)]t + \frac{[\bar{E}^D\bar{A} + G(\mathbb{I}_n - \bar{E}\bar{E}^D)]^2 t^2}{2!} + \dots\right)\bar{E}\bar{E}^D\\
&= \left(\mathbb{I}_n + \bar{E}^D\bar{A}t + \frac{(\bar{E}^D\bar{A})^2 t^2}{2!} + \dots\right)\bar{E}\bar{E}^D = e^{\bar{E}^D\bar{A}t}\bar{E}\bar{E}^D,
\end{aligned} \tag{1.86}$$

since $(\mathbb{I}_n - \bar{E}\bar{E}^D)\bar{E}\bar{E}^D = 0$. □

Theorem 1.10. Let

$$\bar{F}_k = (\bar{E}\bar{A}^D)^k\bar{A}^D\bar{B}, \quad k = 0, 1, \dots, q-1. \tag{1.87}$$

The descriptor continuous-time linear system (1.24) (or equivalently (1.1)) is positive if and only if there exists a matrix $G \in \mathbb{R}^{n\times n}$ [79, 141] such that

$$\bar{E}^D\bar{A} + G(\mathbb{I}_n - \bar{E}\bar{E}^D) \in \mathbb{M}_n \tag{1.88a}$$

and

$$\mathrm{Im}\bar{E}\bar{E}^D \subset \mathbb{R}^n_+, \quad \bar{E}^D\bar{B} \in \mathbb{R}^{n\times m}_+, \quad C \in \mathbb{R}^{p\times n}_+, \quad D \in \mathbb{R}^{p\times m}_+, \tag{1.88b}$$

$$(\bar{E}\bar{E}^D - \mathbb{I}_n)\bar{F}_k \in \mathbb{R}^{n\times m}_+, \quad k = 0, 1, \ldots, q-1. \tag{1.88c}$$

Proof. Using the superposition principle we will consider components of the solution (1.45) independently:

1. If $\mathrm{Im}\bar{E}\bar{E}^D \subset \mathbb{R}^{n\times n}_+$, then $\bar{E}\bar{E}^D x_0 \subset \mathbb{R}^n_+$. Therefore, for the matrix G such that $\bar{E}^D\bar{A} + G(\mathbb{I}_n - \bar{E}\bar{E}^D) \in \mathbb{M}_n$ according to Lemma 1.5 we have $e^{[\bar{E}^D\bar{A}+G(\mathbb{I}_n-\bar{E}\bar{E}^D)]t} \in \mathbb{R}^{n\times n}_+$ and using Lemma 1.6 we obtain $e^{[\bar{E}^D\bar{A}+G(\mathbb{I}_n-\bar{E}\bar{E}^D)]t}\bar{E}\bar{E}^D x_0 = e^{\bar{E}^D\bar{A}t}\bar{E}\bar{E}^D x_0 \in \mathbb{R}^n_+, t \in [0, +\infty)$.
2. Using Lemma 1.4 we can write the second component in the form $\int_0^t e^{\bar{E}^D\bar{A}(t-\tau)}\bar{E}^D\bar{B}u(\tau)d\tau$. If $\bar{E}^D\bar{A} + G(\mathbb{I}_n - \bar{E}\bar{E}^D) \in \mathbb{M}_n$, $\bar{E}^D\bar{B} \in \mathbb{R}^{n\times m}_+$ and $u(t) \in U \subset \mathbb{R}^m_+$ we have $\bar{E}^D\int_0^t e^{\bar{E}^D\bar{A}(t-\tau)}\bar{B}u(\tau)d\tau \in \mathbb{R}^n_+$.
3. If $(\bar{E}\bar{E}^D - \mathbb{I}_n)\bar{F}_k \in \mathbb{R}^{n\times m}_+, k = 0, 1, \ldots, q-1$, where the matrices $\bar{F}_k$ are given by (1.87) and $u^{(k)}(t) \in \mathbb{R}^m_+, k = 0, 1, \ldots, q-1, t \in [0, +\infty)$, then the last component of the solution (1.45) is also nonnegative.

Therefore, we have $x(t) \in \mathbb{R}^n_+$, $t \in [0, +\infty)$ if $\mathrm{Im}\bar{E}\bar{E}^D \subset \mathbb{R}^{n\times n}_+$, $\bar{E}^D\bar{A} + G(\mathbb{I}_n - \bar{E}\bar{E}^D) \in \mathbb{M}_n$, $\bar{E}^D\bar{B} \in \mathbb{R}^{n\times m}_+$ and $(\bar{E}\bar{E}^D - \mathbb{I}_n)\bar{F}_k \in \mathbb{R}^{n\times m}_+, k = 0, 1, \ldots, q-1$. Substituting (1.45) into (1.1b) it is easy to see that $y(t) \in \mathbb{R}^p_+, t \in [0, +\infty)$ if $C \in \mathbb{R}^{p\times n}_+$ and $D \in \mathbb{R}^{p\times m}_+$. □

Remark 1.3. [79] Note that the descriptor continuous-time linear system (1.24) can be positive even though the matrix $\bar{E}^D\bar{A}$ is not a Metzler matrix. In the special case for $G = 0$ from (1.88a) we obtain the positivity condition $\bar{E}^D\bar{A} \in \mathbb{M}_n$.

Example 1.5. (Continuation of Example 1.4) Consider the descriptor continuous-time linear system with (1.76). According to (1.77) the matrix pencil (E, A) is regular. We choose $c = 0$ and using (1.24b) we obtain

$$\bar{E} = [-A]^{-1}E = \begin{bmatrix} 0 & -0.5 & 0 \\ -0.5 & 0 & 0 \\ 0 & 0 & 0 \end{bmatrix}, \quad \bar{A} = [-A]^{-1}A = \begin{bmatrix} -1 & 0 & 0 \\ 0 & -1 & 0 \\ 0 & 0 & -1 \end{bmatrix},$$
$$\bar{B} = [-A]^{-1}B = \begin{bmatrix} -0.5 \\ 0 \\ 1 \end{bmatrix}. \tag{1.89}$$

To compute the Drazin inverse of $\bar{E}$ we can use Method 1 from Appendix B. Note that $\bar{E} = \bar{E}_1$, where $\bar{E}_{11} = \begin{bmatrix} 0 & -0.5 \\ -0.5 & 0 \end{bmatrix}$, $\bar{E}_{12} = \begin{bmatrix} 0 \\ 0 \end{bmatrix}$ and the index of $\bar{E}$ is $q = 1$ since $\operatorname{rank}\bar{E} = \operatorname{rank}\bar{E}^2 = 2$.

Using (B.2) we compute

$$\bar{E}^D = \begin{bmatrix} \bar{E}_{11}^{-1} & \bar{E}_{11}^{-2}\bar{E}_{12} \\ 0 & 0 \end{bmatrix} = \begin{bmatrix} 0 & -2 & 0 \\ -2 & 0 & 0 \\ 0 & 0 & 0 \end{bmatrix},$$
$$\bar{A}^D = \bar{A}^{-1} = \bar{A} = \begin{bmatrix} -1 & 0 & 0 \\ 0 & -1 & 0 \\ 0 & 0 & -1 \end{bmatrix} \tag{1.90}$$

and

$$\operatorname{Im}\bar{E}\bar{E}^D = \operatorname{Im}\begin{bmatrix} 1 & 0 & 0 \\ 0 & 1 & 0 \\ 0 & 0 & 0 \end{bmatrix} \subset \mathbb{R}_+^{3\times 3} \text{ for } v = \begin{bmatrix} v_1 \\ v_2 \\ 0 \end{bmatrix} \subset \mathbb{R}_+^3, \tag{1.91a}$$

$$\bar{E}^D\bar{A} = \begin{bmatrix} 0 & 2 & 0 \\ 2 & 0 & 0 \\ 0 & 0 & 0 \end{bmatrix} \in \mathbb{M}_3, \quad \bar{E}^D\bar{B} = \begin{bmatrix} 0 \\ 1 \\ 0 \end{bmatrix} \in \mathbb{R}_+^{3\times 1}. \tag{1.91b}$$

According to Lemma 1.3 the matrices (1.91) and the index of $\bar{E}$ do not depend of the choice of c. From (1.87) we have

$$\bar{F}_0 = (\bar{E}\bar{A}^D)^0\bar{A}^D\bar{B} = \begin{bmatrix} 0.5 \\ 0 \\ -1 \end{bmatrix}. \tag{1.92}$$

From (1.91a) and (1.92) we have

$$(\bar{E}\bar{E}^D - \mathbb{I}_3)\bar{F}_0 = \begin{bmatrix} 0 \\ 0 \\ 1 \end{bmatrix} \in \mathbb{R}_+^{3\times 1}. \tag{1.93}$$

Therefore, the descriptor continuous-time linear system with (1.76) is positive.

1.1.3.3 Weierstrass-Kronecker Decomposition Method

In this section necessary and sufficient positivity conditions of the descriptor continuous-time linear system (1.55) (or equivalently (1.1)) will be established using the Weierstrass-Kronecker decomposition method with the assumption that the matrix Q determined by (1.52) is a monomial matrix [59, 84]. This problem has been also considered in [19]. It is assumed that the solution to the Eq. (1.55b) has the form (1.59).

Definition 1.4. [49] A square matrix $A \in \mathbb{R}_+^{n\times n}$ is called monomial if each its row and each its column contains only one positive entry and the remaining entries are zero.

Remark 1.4. [49] An inverse of a monomial matrix is also monomial.

Let the decomposition (1.52) of the positive descriptor continuous-time linear system (1.1) be possible for a monomial matrix $Q \in \mathbb{R}_+^{n\times n}$. In this case $x(t) = Q\bar{x}(t) \in \mathbb{R}_+^n$ if and only if $\bar{x}(t) \in \mathbb{R}_+^n$. It is also well-known that premultiplying the Eq. (1.1a) by a nonsingular matrix P does not change the solution $x(t)$.

Theorem 1.11. [19, 59, 84] Let the decomposition (1.52) of the system (1.1) be possible for a monomial matrix $Q \in \mathbb{R}_+^{n\times n}$. The descriptor continuous-time linear system (1.55) (or equivalently (1.1)) is positive if and only if

$$\begin{gathered} A_1 \in \mathbb{M}_{n_1},\quad B_1 \in \mathbb{R}_+^{n_1\times m},\quad C \in \mathbb{R}_+^{p\times n},\quad D \in \mathbb{R}_+^{p\times m}, \\ -N^k B_2 \in \mathbb{R}_+^{n_2\times m},\quad k = 0, 1, \ldots, \mu - 1. \end{gathered} \tag{1.94}$$

Proof. For a monomial matrix $Q \in \mathbb{R}_+^{n\times n}$ we have $x(t) \in \mathbb{R}_+^n$ if $\bar{x}(t) \in \mathbb{R}_+^n$. According to Lemma 1.5 for $A_1 \in \mathbb{M}_{n_1}$ we have $e^{A_1 t} \in \mathbb{R}_+^{n\times n}$. If $A_1 \in \mathbb{M}_{n_1}$ and $B_1 \in \mathbb{R}_+^{n_1\times m}$, then from (1.57) we obtain $\bar{x}_1 \in \mathbb{R}_+^{n_1}$ since from Definition 1.3 $\bar{x}_{10} = Q^{-1}x_{10} \in \mathbb{R}_+^{n_1}$ and $u(t) \in \mathbb{R}_+^m$. If $-N^k B_2 \in \mathbb{R}_+^{n_2\times m}$, $k = 0, 1, \ldots, \mu - 1$, then from (1.59) we obtain $\bar{x}_2 \in \mathbb{R}_+^{n_2}$ since from Definition 1.3 $u^{(k)}(t) \in \mathbb{R}_+^m$, $k = 0, 1, \ldots, \mu - 1$. From (1.54b) for $C \in \mathbb{R}_+^{p\times n}$ and $D \in \mathbb{R}_+^{p\times m}$ we have $y(t) \in \mathbb{R}_+^p$ since $x(t) \in \mathbb{R}_+^n$ and $u(t) \in \mathbb{R}_+^m$. □

Example 1.6. (Continuation of Example 1.4) Consider the descriptor continuous-time linear system with (1.76). According to (1.77) the matrix pencil (E, A) is regular. In this case

$$P = \begin{bmatrix} 1 & 0 & 0 \\ 0 & 1 & 1 \\ 0 & 0 & 1 \end{bmatrix},\quad Q = \begin{bmatrix} 1 & 0 & 0 \\ 0 & 1 & 0 \\ 0 & 0 & 1 \end{bmatrix} \tag{1.95}$$

and

$$\begin{gathered} \begin{bmatrix} \mathbb{I}_{n_1} & 0 \\ 0 & N \end{bmatrix} = PEQ = \begin{bmatrix} 1 & 0 & 0 \\ 0 & 1 & 0 \\ 0 & 0 & 0 \end{bmatrix}, \\ \begin{bmatrix} A_1 & 0 \\ 0 & \mathbb{I}_{n_2} \end{bmatrix} = PAQ = \begin{bmatrix} 0 & 2 & 0 \\ 2 & 0 & 0 \\ 0 & 0 & 1 \end{bmatrix}, \\ \begin{bmatrix} B_1 \\ B_2 \end{bmatrix} = PB = \begin{bmatrix} 0 \\ 1 \\ -1 \end{bmatrix}. \end{gathered} \tag{1.96}$$

Hence, $n_1 = 2$ and $n_2 = 1$. The matrix $Q \in \mathbb{R}_+^{3\times 3}$ in (1.95) is monomial. From (1.76) and (1.96) we have

$$A_1 = \begin{bmatrix} 0 & 2 \\ 2 & 0 \end{bmatrix} \in \mathbb{M}_2, \quad B_1 = \begin{bmatrix} 0 \\ 1 \end{bmatrix} \in \mathbb{R}_+^{2\times 1}, \quad -B_2 = 1 \in \mathbb{R}_+^{1\times 1}, \\ C = \begin{bmatrix} 1 & 2 & 0 \end{bmatrix} \in \mathbb{R}_+^{1\times 3}, \quad D = 0 \in \mathbb{R}_+^{1\times 1}. \tag{1.97}$$

By Theorem 1.11 the descriptor continuous-time linear system with (1.76) is positive.

1.1.4 Stability of Descriptor Continuous-Time Linear Systems

In the following considerations it is assumed that $\det A \neq 0$, i.e. the descriptor continuous-time linear system (1.1) has exactly one equilibrium point [47].

1.1.4.1 Laurent Series Expansion Method

In this section necessary and sufficient stability conditions of the descriptor continuous-time linear system (1.1) will be established using the Laurent series expansion method. It is assumed that the solution to the state Eq. (1.1a) has the form (1.12).

Definition 1.5. The descriptor continuous-time linear system (1.1) is called asymptotically stable if

$$\lim_{t\to\infty} x(t) = 0 \tag{1.98}$$

for all consistent initial conditions $x_0 \in X_0$ and $u(t) = 0$.

Let $\Psi(t)$ be the matrix given by

$$\Psi(t) = \sum_{k=0}^{\infty} \frac{T_k E t^k}{k!}. \tag{1.99}$$

Definition 1.6. The characteristic equation of the matrix pair (E, A) has the form

$$p_{(E,A)}(s) = \det[Es - A] = a_r s^r + a_{r-1} s^{r-1} + \ldots + a_1 s + a_0 = 0, \tag{1.100}$$

where $r = \operatorname{rank}\Psi(t)$ and $p_{(E,A)}(s)$ is the characteristic polynomial of the matrix pair (E, A).

Theorem 1.12. [47] The matrix pair (E, A) (or equivalently the system (1.1)) is asymptotically stable if and only if its eigenvalues s_k, $k = 1, \ldots, r$ (roots of the characteristic equation) satisfy the condition

$$\mathrm{Re}s_k < 0, \quad k = 1, \ldots, r. \tag{1.101}$$

Theorem 1.13. The descriptor continuous-time linear system (1.1) is asymptotically stable if and only if

$$\lim_{t\to\infty} \Psi(t) = 0. \tag{1.102}$$

Proof. Substituting $u(t) = 0$ into (1.12) we obtain

$$x(t) = \sum_{k=0}^{\infty} \frac{T_k E t^k}{k!} x_0. \tag{1.103}$$

Next, by Definition 1.5 we have

$$\lim_{t\to\infty} x(t) = \lim_{t\to\infty} \sum_{k=0}^{\infty} \frac{T_k E t^k}{k!} x_0 = \lim_{t\to\infty} \Psi(t) x_0. \tag{1.104}$$

Therefore, the descriptor system (1.1) is asymptotically stable for $x_0 \in X_0$ if and only if the condition (1.102) holds. □

From the above considerations we have the following theorem.

Theorem 1.14. The descriptor continuous-time linear system (1.1) is asymptotically stable if and only if one of the following equivalent conditions is satisfied:

1. Roots of the characteristic Eq. (1.100) satisfy the condition (1.101).
2. The matrix $\Psi(t)$ defined by (1.99) satisfies the condition (1.102).

Example 1.7. (Continuation of Example 1.4) Consider the descriptor continuous-time linear system (1.1) with (1.76). Using (1.80a) and (1.99) we have

$$\begin{aligned}\Psi(t) = \sum_{k=0}^{\infty} \frac{T_k E t^k}{k!} &= \sum_{k=0}^{\infty} \begin{bmatrix} \frac{2^{2k} t^{2k}}{(2k)!} & \frac{2^{2k+1} t^{2k+1}}{(2k+1)!} & 0 \\ \frac{2^{2k+1} t^{2k+1}}{(2k+1)!} & \frac{2^{2k} t^{2k}}{(2k)!} & 0 \\ 0 & 0 & 0 \end{bmatrix} \\ &= \begin{bmatrix} \cosh(2t) & \sinh(2t) & 0 \\ \sinh(2t) & \cosh(2t) & 0 \\ 0 & 0 & 0 \end{bmatrix}.\end{aligned} \tag{1.105}$$

According to the second condition of Theorem 1.14 the descriptor system (1.1) with (1.76) is unstable since $\lim_{t\to\infty} \cosh(2t) = \infty$ and $\lim_{t\to\infty} \sinh(2t) = \infty$.

The characteristic equation of the system (1.76) has the form

$$p_{(E,A)}(s) = \det[Es - A] = \begin{vmatrix} s & -2 & 0 \\ -2 & s & 1 \\ 0 & 0 & -1 \end{vmatrix} = -s^2 + 4 = 0. \tag{1.106}$$

Eigenvalues of the pair (E, A) are $s_1 = -2, s_2 = 2$. Therefore, by the first condition of Theorem 1.14 the descriptor system (1.1) with (1.76) is unstable since $\mathrm{Re}s_2 = 2 > 0$.

Example 1.8. Consider the descriptor continuous-time linear system (1.1) with

$$E = \begin{bmatrix} 1 & 0 \\ 0 & 0 \end{bmatrix}, \quad A = \begin{bmatrix} -2 & 1 \\ 1 & -2 \end{bmatrix}. \tag{1.107}$$

The matrix pencil of (1.107) is regular since

$$\det[Es - A] = \begin{vmatrix} s+2 & -1 \\ -1 & 2 \end{vmatrix} = 2s + 3 \neq 0. \tag{1.108}$$

The matrix $[Es - A]^{-1}$ has the form

$$[Es - A]^{-1} = \begin{bmatrix} s+2 & -1 \\ -1 & 2 \end{bmatrix}^{-1} = \begin{bmatrix} \frac{2}{2s+3} & \frac{1}{2s+3} \\ \frac{1}{2s+3} & \frac{s+2}{2s+3} \end{bmatrix}. \tag{1.109}$$

Using (1.6) for $n_0 = 1$ the matrix (1.109) can be written in the form

$$[Es - A]^{-1} = \sum_{k=-1}^{\infty} T_k s^{-(k+1)}, \tag{1.110a}$$

where

$$T_k = \begin{cases} \begin{bmatrix} 0 & 0 \\ 0 & 0.5 \end{bmatrix} & \text{for } k = -1, \\ \begin{bmatrix} (-1.5)^k & 0.5(-1.5)^k \\ 0.5(-1.5)^k & 0.25(-1.5)^k \end{bmatrix} & \text{for } k = 0, 1, 2, \ldots. \end{cases} \tag{1.110b}$$

Using (1.107) and (1.110b) we have

$$T_k E = \begin{bmatrix} (-1.5)^k & 0 \\ 0.5(-1.5)^k & 0 \end{bmatrix} \quad \text{for } k = 0, 1, 2, \ldots \tag{1.111}$$

and

$$\Psi(t) = \sum_{k=0}^{\infty} \frac{T_k E t^k}{k!} = \sum_{k=0}^{\infty} \begin{bmatrix} \frac{(-1.5)^k t^k}{k!} & 0 \\ 0.5\frac{(-1.5)^k t^k}{k!} & 0 \end{bmatrix} = \begin{bmatrix} e^{-1.5t} & 0 \\ 0.5e^{-1.5t} & 0 \end{bmatrix}. \tag{1.112}$$

By the second condition of Theorem 1.14 the descriptor system (1.1) with (1.107) is asymptotically stable since $\lim_{t\to\infty} e^{-1.5t} = 0$.

The characteristic equation of the system (1.107) is given by

$$p_{(E,A)}(s) = \det[Es - A] = \begin{vmatrix} s+2 & -1 \\ -1 & 2 \end{vmatrix} = 2s + 3 = 0. \quad (1.113)$$

From (1.113) we can see that the pair (E, A) has one eigenvalue $s = -1.5$ and by Theorem 1.14 the descriptor system (1.1) with (1.107) is asymptotically stable since $\mathrm{Re}\, s = -1.5 < 0$.

1.1.4.2 Drazin Inverse Matrix Method

In this section necessary and sufficient stability conditions of the descriptor continuous-time linear system (1.24) (or equivalently (1.1)) will be established using the Drazin inverse matrix method. It is assumed that the solution to the state Eq. (1.24a) has the form (1.45).

Definition 1.5 of the asymptotic stability of the descriptor system is suitable for this method since the solutions of (1.1) and (1.24) are equivalent.

Definition 1.7. The characteristic equation of the matrix pair $(\bar{E}, \bar{A})$ is given by

$$p_{(\bar{E},\bar{A})}(s) = \det[\bar{E}s - \bar{A}] = \bar{a}_r s^r + \bar{a}_{r-1} s^{r-1} + \ldots + \bar{a}_1 s + \bar{a}_0 = 0, \quad (1.114)$$

where $r = \mathrm{rank}\, \bar{E}^D \bar{A}$ and $p_{(\bar{E},\bar{A})}(s)$ is the characteristic polynomial of the matrix pair $(\bar{E}, \bar{A})$.

Lemma 1.7. The characteristic polynomials of the matrix pairs (E, A) and $(\bar{E}, \bar{A})$ are related by

$$p_{(\bar{E},\bar{A})}(s) = \frac{p_{(E,A)}(s)}{\det[Ec - A]}. \quad (1.115)$$

Proof. The characteristic polynomial of the pair $(\bar{E}, \bar{A})$ has the form

$$\begin{aligned} p_{(\bar{E},\bar{A})}(s) &= \det[\bar{E}s - \bar{A}] = \det\{[Ec - A]^{-1} Es - [Ec - A]^{-1} A\} \\ &= \det\{[Ec - A]^{-1}[Es - A]\} = \det[Ec - A]^{-1} \det[Es - A] \quad (1.116) \\ &= \{\det[Ec - A]\}^{-1} p_{(E,A)}(s), \end{aligned}$$

which is equivalent to (1.115). □

Remark 1.5. The characteristic equations of the matrix pairs (E, A) and $(\bar{E}, \bar{A})$ have the same form. Therefore, both pairs have the same set of eigenvalues and Theorem 1.12 can be used for the roots of the Eq. (1.114).

Lemma 1.8. The matrix $\bar{E}^D \bar{A}$ has r eigenvalues of the pair $(\bar{E}, \bar{A})$ (or (E, A)) and additionally $n - r$ zero eigenvalues, i.e. its characteristic equation has the form

$$p_{\bar{E}^D \bar{A}}(s) = \det[\mathbb{I}_n s - \bar{E}^D \bar{A}] = s^{n-r} p_{(\bar{E},\bar{A})}(s) = 0, \quad (1.117)$$

where $p_{\bar{E}^D\bar{A}}(s)$ is the characteristic polynomial of the matrix $\bar{E}^D\bar{A}$.

Proof. Using (1.27d) we have

$$\begin{aligned} p_{\bar{E}^D\bar{A}}(s) &= \det[\mathbb{I}_n s - \bar{E}^D\bar{A}] = \det\left\{T\begin{bmatrix}\mathbb{I}_{n_1}s - J^{-1}A_1 & 0\\ 0 & \mathbb{I}_{n_2}s\end{bmatrix}T^{-1}\right\}\\ &= \det T\det[\mathbb{I}_{n_1}s - J^{-1}A_1]\det[\mathbb{I}_{n_2}s]\det T^{-1}\\ &= s^{n_2}\det[\mathbb{I}_{n_1}s - J^{-1}A_1], \end{aligned} \tag{1.118}$$

since $\det T\det T^{-1} = \mathbb{I}_n$. It is easy to see that $n_1 = r$ and $n_2 = n - r$.

Using again (1.27d) we can write

$$\begin{aligned} p_{(\bar{E},\bar{A})}(s) &= \det[\bar{E}s - \bar{A}] = \det\left\{T\begin{bmatrix}Js - A_1 & 0\\ 0 & Ns - A_2\end{bmatrix}T^{-1}\right\}\\ &= \det T\det Js - A_1]\det[Ns - A_2]\det T^{-1}\\ &= \det[Js - A_1]\det[Ns - A_2], \end{aligned} \tag{1.119}$$

since $\det T\det T^{-1} = \mathbb{I}_n$.

Note that $\bar{E}\bar{A} = \bar{A}\bar{E}$. Then from (1.27d) we have $NA_2 = A_2N$ if and only if $A_2 = \alpha\mathbb{I}_{n_2}$, where $\alpha \in \mathbb{R}$, i.e. A_2 is a scalar matrix (a nilpotent matrix N can not be scalar). Hence, the Eq. (1.119) can be written in the form

$$\begin{aligned} p_{(\bar{E},\bar{A})}(s) &= \det[Js - A_1]\det[Ns - \alpha\mathbb{I}_{n_2}]\\ &= (-\alpha)^{n_2}\det[Js - A_1] = (-\alpha)^{n_2}\det[\mathbb{I}_{n_1}s - J^{-1}A_1], \end{aligned} \tag{1.120}$$

since $\det[Ns - \alpha\mathbb{I}_{n_2}] = (-\alpha)^{n_2}$.

Equating (1.118) and (1.120) to zero we have

$$p_{\bar{E}^D\bar{A}}(s) = s^{n_2}\det[\mathbb{I}_{n_1}s - J^{-1}A_1] = 0, \tag{1.121a}$$

$$p_{(\bar{E},\bar{A})}(s) = \det[\mathbb{I}_{n_1}s - J^{-1}A_1] = 0, \tag{1.121b}$$

which is equivalent to (1.117). □

Theorem 1.15. The descriptor continuous-time linear system (1.1) is asymptotically stable if and only if the matrix $\bar{E}^D\bar{A}$ has r stable eigenvalues (satisfying the condition (1.101)) and $n - r$ zero eigenvalues.

Proof. The proof follows directly from Lemma 1.8. □

Based on the above considerations we have the following Theorem.

Theorem 1.16. The descriptor continuous-time linear system (1.1) is asymptotically stable if and only if one of the following equivalent conditions is satisfied:

1. Roots of the characteristic Eq. (1.114) satisfy the condition (1.101).

2. The matrix $\bar{E}^D\bar{A}$ has r stable eigenvalues and $n-r$ zero eigenvalues.

Example 1.9. (Continuation of Examples 1.5 and 1.7) Consider the descriptor continuous-time linear system (1.1) with (1.76). Using (1.89) and (1.114) we have

$$p_{(\bar{E},\bar{A})}(s)=\det[\bar{E}s-\bar{A}]=\begin{vmatrix}1 & -0.5s & 0\\ -0.5s & 1 & 0\\ 0 & 0 & 1\end{vmatrix}=-0.25s^2+1=0. \tag{1.122}$$

Eigenvalues of the pair $(\bar{E},\bar{A})$ are $s_1=-2, s_2=2$ and are equal to the eigenvalues of the pair (E,A). By Theorem 1.16 the descriptor system (1.1) with (1.76) is unstable since $\text{Re}s_2=2>0$.

Using (1.91) and (1.117) we have

$$p_{\bar{E}^D\bar{A}}(s)=\det[\mathbb{I}_3 s-\bar{E}^D\bar{A}]=\begin{vmatrix}s & -2 & 0\\ -2 & s & 0\\ 0 & 0 & s\end{vmatrix}=s(s^2-4)=0. \tag{1.123}$$

It is easy to see that $r=\text{rank}\bar{E}^D\bar{A}=2$. According to (1.123) the matrix $\bar{E}^D\bar{A}$ has $r=2$ eigenvalues of the pair $(\bar{E},\bar{A})$ equal to $s_1=-2$, $s_2=2$ and $n-r=1$ zero eigenvalue $s_3=0$. Therefore, by Theorem 1.16 the considered system is unstable.

1.1.4.3 Weierstrass-Kronecker Decomposition Method

In this section necessary and sufficient stability conditions of the descriptor continuous-time linear system (1.55) (or equivalently (1.1)) using the Weierstrass-Kronecker decomposition method. It is assumed that the solution to the Eq. (1.55b) has the form (1.59).

Note that the state vector of the subsystem (1.55b) $\bar{x}_2(t)=0$ for $u(t)=0$. Therefore, the stability of the system (1.55) (or equivalently (1.1)) depends entirely on the subsystem (1.55a).

Definition 1.8. The descriptor continuous-time linear system (1.55) (or equivalently (1.1)) is called asymptotically stable if

$$\lim_{t\to\infty}\bar{x}_1(t)=0 \tag{1.124}$$

for all initial conditions $\bar{x}_{10}\in\mathbb{R}^{n_1}$ and $u(t)=0$.

Definition 1.9. [47] The characteristic equation of the matrix A_1 is given by

$$p_{A_1}(s)=\det[\mathbb{I}_{n_1}s-A_1]=s^{n_1}+\tilde{a}_{n_1-1}s^{n_1-1}+\ldots+\tilde{a}_1 s+\tilde{a}_0=0, \tag{1.125}$$

where $n_1=\deg\{\det[Es-A]\}$ and $p_{A_1}(s)$ is the characteristic polynomial of the matrix A_1.

Lemma 1.9. [11] The characteristic polynomials of the matrix pair (E, A) and of the matrix A_1 are related by

$$p_{(E,A)}(s) = \frac{(-1)^{n_2}}{\det(PQ)} p_{A_1}(s). \tag{1.126}$$

Proof. From (1.52) we have

$$\det\{P[Es - A]Q\} = \det[\mathbb{I}_{n_1}s - A_1]\det[Ns - \mathbb{I}_{n_2}] \tag{1.127}$$

and

$$\det P \det[Es - A]\det Q = (-1)^{n_2}\det[\mathbb{I}_{n_1}s - A_1], \tag{1.128}$$

since $\det[Ns - \mathbb{I}_{n_2}] = (-1)^{n_2}$. The matrices P and Q are nonsingular, and therefore $\det P \neq 0$ and $\det Q \neq 0$. Dividing both sides of the Eq. (1.128) by $\det P \det Q = \det(PQ)$ we obtain (1.126). □

Remark 1.6. The characteristic equations of the matrix pair (E, A) and of the matrix A_1 have the same form. Therefore, the matrix pair (E, A) and the matrix A_1 have the same set of eigenvalues and Theorem 1.12 can be used for the roots of the Eq. (1.125).

Theorem 1.17. [47] The matrix A_1 (or equivalently the system (1.55)) is asymptotically stable if and only if its eigenvalues satisfy the condition (1.101).

Example 1.10. (Continuation of Examples 1.6 and 1.7) Consider the descriptor continuous-time linear system (1.1) with (1.76). Using (1.97) and (1.125) we have

$$p_{A_1}(s) = \det[\mathbb{I}_2 s - A_1] = \begin{vmatrix} s & -2 \\ -2 & s \end{vmatrix} = s^2 - 4. \tag{1.129}$$

Eigenvalues of the matrix A_1 are $s_1 = -2$, $s_2 = 2$ and are equal to the eigenvalues of the pair (E, A). By Theorem 1.17 the considered descriptor system is unstable.

1.1.5 Superstability of Descriptor Continuous-Time Linear Systems

The asymptotic stability of a dynamical system ensures that its free response decreases to zero for $t \to \infty$, however its value may increase significantly in the initial part of the state vector trajectory (it is so-called "peak effect"). In superstable systems the norm of the state vector decreases monotonically to zero for $t \to \infty$, which prevents such undesirable effects [70, 137, 138].

1.1.5.1 Properties of Superstable Continuous-Time Linear Systems

The following norms will be used:

1) ∞-norm of a vector $x = [x_i] \in \mathbb{R}^n$

$$\|x\| = \max_{1 \le i \le n} |x_i|, \tag{1.130}$$

2) 1-norm of a matrix $A = [a_{ij}] \in \mathbb{R}^{n \times n}$

$$\|A\| = \max_{1 \le i \le n} \left(\sum_{j=1}^{n} |a_{ij}| \right). \tag{1.131}$$

Consider the standard continuous-time linear system

$$\dot{x}(t) = Ax(t), \tag{1.132}$$

where $x(t) \in \mathbb{R}^n$ is the state vector and $A \in \mathbb{R}^{n \times n}$.

Definition 1.10. [137] The matrix $A \in \mathbb{R}^{n \times n}$ of the continuous-time linear system (1.132) is called superstable if

$$\sigma(A) = \sigma = \min_{1 \le i \le n} \left(-a_{ii} - \sum_{\substack{j=1 \\ j \ne i}}^{n} |a_{ij}| \right) > 0. \tag{1.133}$$

Quantity $\sigma(A)$ is called the superstability degree of the matrix A. If the matrix is superstable, then it is also stable, however the reverse implication does not hold.

Lemma 1.10. [137] For a superstable matrix A we have

$$\|e^{At}\| \le e^{-\sigma t}. \tag{1.134}$$

Theorem 1.18. [137] If the system (1.132) is superstable, then

$$\|x(t)\| \le \|x_0\| e^{-\sigma t}, \quad t \in [0, +\infty). \tag{1.135}$$

Note that the superstability ensure monotonic decrease of the norm of the state vector, however particular state variables can oscillate. The main difference is that for asymptotic stable systems the Eq. (1.135) is replaced by

$$\|x(t)\| \le c(A, \nu) \|x_0\| e^{-\nu t}, \quad 0 < \nu < \min_{1 \le i \le n} \{-\mathrm{Re} s_i\}, \tag{1.136}$$

where the constant $c(A, \nu)$ can take significant values in the initial part of the state vector trajectory. In superstable systems there are no such undesirable "peaks" [137].

Example 1.11. Consider the continuous-time linear system (1.132) with

$$A = \begin{bmatrix} -3 & 2 \\ 0 & -2 \end{bmatrix}. \tag{1.137}$$

The system is asymptotically stable since the eigenvalues of (1.137) are $s_1 = -3$, $s_2 = -2$. By the condition (1.133) we have $\sigma = 1 > 0$ and the system is also superstable. Assuming the initial condition $x_0 = [\ 1 \quad 2\ ]^T$ we present time plots of the state variables in Fig. 1.1, from which we can see that the norm of the state vector decreases monotonically for $t \to \infty$. Now let us consider the continuous-time linear system (1.132) with

$$A = \begin{bmatrix} -3 & 6 \\ 0 & -2 \end{bmatrix}. \tag{1.138}$$

Note that the eigenvalues of (1.137) and (1.138) are the same (i.e. $s_1 = -3, s_2 = -2$), but in the second case the system is not superstable since by the condition (1.133) we have $\sigma = -3 < 0$. Thus, this property of dynamical systems does not depend of their eigenvalues. From Fig. 1.2 we can see that the value of the state variable $x_1(t)$ increases highly above the initial condition and the norm of the state vector does not decrease monotonically for $t \to \infty$.

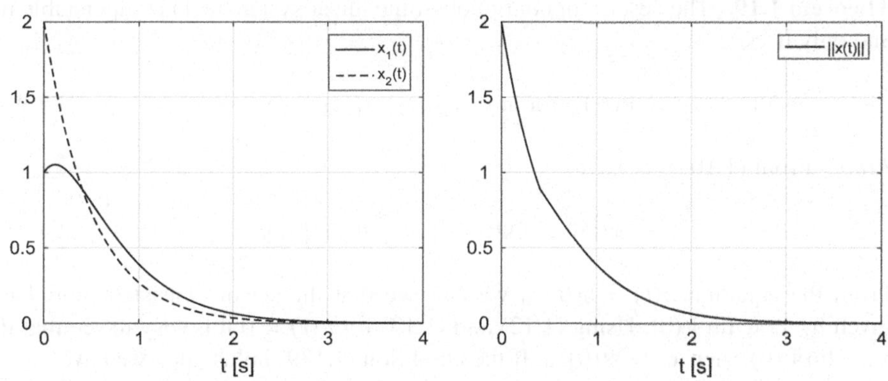

Fig. 1.1 State variables (on the left) and norm of the state vector (on the right) of the system (1.132) with (1.137) for $x_0 = [\ 1 \quad 2\ ]^T$

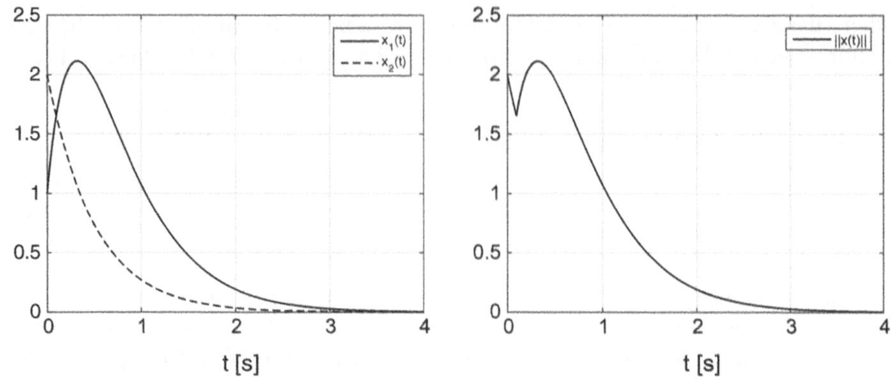

Fig. 1.2 State variables (on the left) and norm of the state vector (on the right) of the system (1.132) with (1.138) for $x_0 = [\,1 \quad 2\,]^T$

1.1.5.2 Laurent Series Expansion Method

In this section necessary and sufficient superstability conditions of the descriptor continuous-time linear system (1.1) will be established using the Laurent series expansion method. It is assumed that the solution to the state Eq. (1.1a) has the form (1.12).

Consider the descriptor continuous-time linear system (1.1). Solution to the Eq. (1.1a) for $u(t) = 0$ is given by (1.103). Let $\Psi(t)$ be the matrix defined by (1.99).

Theorem 1.19. The descriptor continuous-time linear system (1.1) is superstable if and only if

$$\|\Psi(t_1)\| > \|\Psi(t_2)\| \text{ for any } t_1 < t_2,\ \ t_1, t_2 \in [0, +\infty). \tag{1.139}$$

Proof. From (1.103) we have

$$\|x(t)\| = \|\Psi(t)x_0\| \le \|\Psi(t)\|\|x_0\|. \tag{1.140}$$

From the equation $x(t) = \Psi(t)x_0$ we can see that the set of consistent initial is given by $x_0 \in \operatorname{Im}\Psi(0)$. Using (1.12) and (1.13) for $u(t) = 0$ it is easy to see that if $x_0 \in \operatorname{Im}\Psi(0)$, then $x_0 = \Psi(0)x_0$. If the condition (1.139) holds, then we have

$$\max_t \|\Psi(t)x_0\| = \|\Psi(0)x_0\| = \|x_0\| \le \|\Psi(0)\|\|x_0\| \tag{1.141}$$

and

$$\|x(t_1)\| > \|x(t_2)\| \text{ for any } t_1 < t_2,\ \ t_1, t_2 \in [0, +\infty). \tag{1.142}$$

From the condition (1.139) it also follows that $\lim_{t\to\infty} \|\Psi(t)\| = 0$. Therefore, the norm of the state vector decreases monotonically to zero for $t \to \infty$. □

Example 1.12. Consider the descriptor continuous-time linear system (1.1) with

$$E = \begin{bmatrix} 0 & -1 & 0 \\ 0.5 & 0.5 & 0 \\ 0 & 0 & 0 \end{bmatrix}, \quad A = \begin{bmatrix} -1 & 3 & 0 \\ 0 & -1.5 & 0 \\ 0 & 0 & 1 \end{bmatrix}. \tag{1.143}$$

The matrix pencil (E, A) of (1.143) is regular since

$$\det[Es - A] = \begin{vmatrix} 1 & -(s+3) & 0 \\ 0.5s & 0.5(s+3) & 0 \\ 0 & 0 & -1 \end{vmatrix} = -0.5s^2 - 2s - 1.5 \neq 0. \tag{1.144}$$

The matrix $[Es - A]^{-1}$ has the form

$$[Es - A]^{-1} = \begin{bmatrix} 1 & -(s+3) & 0 \\ 0.5s & 0.5(s+3) & 0 \\ 0 & 0 & -1 \end{bmatrix}^{-1} = \begin{bmatrix} \frac{1}{s+1} & \frac{2}{s+1} & 0 \\ -\frac{s}{(s+1)(s+3)} & \frac{2}{(s+1)(s+3)} & 0 \\ 0 & 0 & -1 \end{bmatrix}. \tag{1.145}$$

Using (1.6) the matrix (1.145) can be written in the form

$$[Es - A]^{-1} = \sum_{k=-1}^{\infty} T_k s^{-(k+1)}, \tag{1.146a}$$

where

$$T_k = \begin{cases} \begin{bmatrix} 0 & 0 & 0 \\ 0 & 0 & 0 \\ 0 & 0 & -1 \end{bmatrix} & \text{for } k = -1, \\ \begin{bmatrix} (-1)^k & 2(-1)^k & 0 \\ 0.5(-1)^{k+1}(3^{k+1} - 1) & (-1)^{k+1}(3^k - 1) & 0 \\ 0 & 0 & 0 \end{bmatrix} & \text{for } k = 0, 1, 2, \ldots. \end{cases} \tag{1.146b}$$

Using (1.143) and (1.146b) we have

$$T_k E = \begin{bmatrix} (-1)^k & 0 & 0 \\ 0.5(-1)^{k+1}(3^k - 1) & (-3)^k & 0 \\ 0 & 0 & 0 \end{bmatrix} \text{ for } k = 0, 1, 2, \ldots \tag{1.147}$$

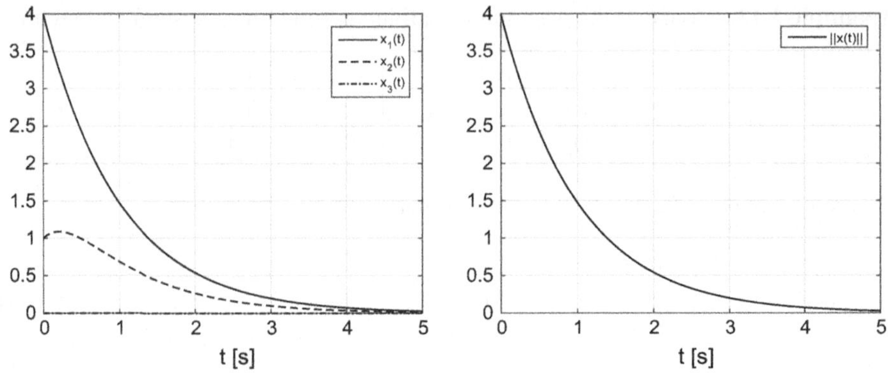

Fig. 1.3 State variables (on the left) and norm of the state vector (on the right) of the system (1.1a) with (1.143) for $x_0 = [\,4 \ \ 1 \ \ 0\,]^T$

and

$$\Psi(t) = \sum_{k=0}^{\infty} \frac{T_k E t^k}{k!} = \sum_{k=0}^{\infty} \begin{bmatrix} \frac{(-1)^k t^k}{k!} & 0 & 0 \\ 0.5\frac{(-1)^{k+1}(3^k-1)t^k}{k!} & \frac{(-3)^k t^k}{k!} & 0 \\ 0 & 0 & 0 \end{bmatrix}$$
$$= \begin{bmatrix} e^{-t} & 0 & 0 \\ 0.5(e^{-t} - e^{-3t}) & e^{-3t} & 0 \\ 0 & 0 & 0 \end{bmatrix}. \tag{1.148}$$

Analyzing (1.148) we can see that

$$\max_t \|\Psi(t)\| = \|\Psi(0)\| = \left\| \begin{bmatrix} 1 & 0 & 0 \\ 0 & 1 & 0 \\ 0 & 0 & 0 \end{bmatrix} \right\| = 1 \tag{1.149}$$

and $\|\Psi(t_1)\| > \|\Psi(t_2)\|$ for any $t_1 < t_2$, $t_1, t_2 \in [0, +\infty)$. By Theorem 1.19 the descriptor continuous-time linear system (1.1) with (1.143) is superstable. Using (1.13) we can determine consistent initial conditions of the system

$$x_0 = T_0 E v = \begin{bmatrix} 1 & 0 & 0 \\ 0 & 1 & 0 \\ 0 & 0 & 0 \end{bmatrix} \begin{bmatrix} v_1 \\ v_2 \\ v_3 \end{bmatrix} = \begin{bmatrix} v_1 \\ v_2 \\ 0 \end{bmatrix}, \tag{1.150}$$

where $v_1, v_2 \in \mathbb{R}$ are any numbers. From the time plots presented in Fig. 1.3 for $x_0 = [\,4 \ \ 1 \ \ 0\,]^T$ we can see that the norm of the state vector decreases monotonically to zero for $t \to \infty$.

1.1.5.3 Drazin Inverse Matrix Method

In this section necessary and sufficient superstability conditions of the descriptor continuous-time linear system (1.24) (or equivalently (1.1)) will be established using the Drazin inverse matrix method. It is assumed that the solution to the state Eq. (1.24a) has the form (1.45).

Consider the descriptor continuous-time linear system (1.24). Solution to the Eq. (1.24a) for $u(t) = 0$ has the form

$$x(t) = e^{\bar{E}^D \bar{A} t} \bar{E} \bar{E}^D x_0. \tag{1.151}$$

Using Lemma 1.6 we can write

$$x(t) = e^{\bar{E}^D \bar{A} t} \bar{E} \bar{E}^D x_0 = e^{[\bar{E}^D \bar{A} + G(\mathbb{I}_n - \bar{E}\bar{E}^D)]t} \bar{E} \bar{E}^D x_0, \tag{1.152}$$

where $G \in \mathbb{R}^{n \times n}$ is an arbitrary matrix.

Taking into account (1.45) and (1.46) for $u(t) = 0$ it is easy to see that if $x_0 \in \operatorname{Im} \bar{E}\bar{E}^D$, then $x_0 = \bar{E}\bar{E}^D x_0$. Therefore, from (1.152) we have

$$\|x(t)\| \leq \|e^{[\bar{E}^D \bar{A} + G(\mathbb{I}_n - \bar{E}\bar{E}^D)]t}\| \|x_0\|. \tag{1.153}$$

If the matrix $\bar{E}^D \bar{A} + G(\mathbb{I}_n - \bar{E}\bar{E}^D)$ is superstable (i.e. it satisfies the condition (1.133)), then by Lemma 1.10 we have $\|e^{[\bar{E}^D \bar{A} + G(\mathbb{I}_n - \bar{E}\bar{E}^D)]t}\| \leq e^{-\sigma t}$. From the inequality (1.153) we obtain

$$\|x(t)\| \leq \|x_0\| e^{-\sigma t}, \tag{1.154}$$

and the norm of the state vector decreases monotonically to zero. Therefore, the following theorem has been proved.

Theorem 1.20. The descriptor continuous-time linear system (1.24) (or equivalently (1.1)) is superstable if and only if there exists a matrix $G \in \mathbb{R}^{n \times n}$ such that the matrix $\bar{E}^D \bar{A} + G(\mathbb{I}_n - \bar{E}\bar{E}^D)$ satisfies the condition (1.133).

Example 1.13. (Continuation of Example 1.12) Consider the descriptor continuous-time linear system (1.1) with (1.143). Using (1.24) for $c = 0$ we obtain

$$\bar{E} = [-A]^{-1} E = \begin{bmatrix} 1 & 0 & 0 \\ \frac{1}{3} & \frac{1}{3} & 0 \\ 0 & 0 & 0 \end{bmatrix}, \quad \bar{A} = [-A]^{-1} A = \begin{bmatrix} -1 & 0 & 0 \\ 0 & -1 & 0 \\ 0 & 0 & -1 \end{bmatrix}. \tag{1.155}$$

Next, from (B.1) and (B.2) we have

$$\bar{E}^D = \begin{bmatrix} 1 & 0 & 0 \\ -1 & 3 & 0 \\ 0 & 0 & 0 \end{bmatrix} \tag{1.156}$$

and

$$\bar{E}^D\bar{A} = \begin{bmatrix} -1 & 0 & 0 \\ 1 & -3 & 0 \\ 0 & 0 & 0 \end{bmatrix}. \tag{1.157}$$

For the matrix $G = [g_{ij}] \in \mathbb{R}^{3\times 3}$ we obtain

$$\bar{E}^D\bar{A} + G(\mathbb{I}_3 - \bar{E}\bar{E}^D) = \begin{bmatrix} -1 & 0 & g_{13} \\ 1 & -3 & g_{23} \\ 0 & 0 & g_{33} \end{bmatrix}. \tag{1.158}$$

Assuming $g_{13} = 0$, $g_{23} = 0$ and $g_{33} < 0$ it is easy to see that the matrix (1.158) satisfies the condition (1.135) since $\sigma = 1 > 0$ for $g_{33} \le -1$ and $\sigma = g_{33} > 0$ for $g_{33} \in (-1; 0)$. Therefore, the considered descriptor system is superstable. Consistent initial conditions of the system are determined by (1.150). The time plots of the state variables and the norm of the state vector for $x_0 = [\,4 \quad 1 \quad 0\,]$ are presented in Fig. 1.3.

1.1.5.4 Weierstrass-Kronecker Decomposition Method

In this section necessary and sufficient superstability conditions of the descriptor continuous-time linear system (1.55) (or equivalently (1.1)) using the Weierstrass-Kronecker decomposition method. It is assumed that the solution to the Eq. (1.55b) has the form (1.59).

Note that every superstable system is also asymptotically stable. State vector of the subsystem (1.55b) $\bar{x}_2 = 0$ for $u(t) = 0$. Using (1.53) and (1.57) we obtain

$$\begin{aligned} x(t) &= Q\bar{x}(t) = Q\begin{bmatrix} \bar{x}_1(t) \\ \bar{x}_2(t) \end{bmatrix} = Q\begin{bmatrix} e^{A_1 t}\bar{x}_{10} \\ 0 \end{bmatrix} \\ &= Q\begin{bmatrix} e^{A_1 t} & 0 \\ 0 & e^{Mt} \end{bmatrix}\begin{bmatrix} \bar{x}_{10} \\ 0 \end{bmatrix} = Q\begin{bmatrix} e^{A_1 t} & 0 \\ 0 & e^{Mt} \end{bmatrix} Q^{-1}x_0 \\ &= \exp\left(Q\begin{bmatrix} A_1 & 0 \\ 0 & M \end{bmatrix} Q^{-1}\right)x_0 = e^{A_M t}x_0, \end{aligned} \tag{1.159}$$

where the matrix $M \in \mathbb{R}^{n_2\times n_2}$ is arbitrary and

$$A_M = Q\begin{bmatrix} A_1 & 0 \\ 0 & M \end{bmatrix} Q^{-1}. \tag{1.160}$$

From the Eq. (1.159) we have

$$\|x(t)\| = \|e^{A_M t}x_0\| \le \|e^{A_M t}\|\|x_0\|. \tag{1.161}$$

If the matrix A_M is superstable (i.e. it satisfies the condition (1.133)), then by Lemma 1.10 we have $\|e^{A_M t}\| \leq e^{-\sigma t}$. From the inequality (1.161) we obtain

$$\|x(t)\| \leq \|x_0\| e^{-\sigma t}, \tag{1.162}$$

and the norm of the state vector decreases monotonically to zero. Therefore, the following theorem has been proved.

Theorem 1.21. The descriptor continuous-time linear system (1.55) (or equivalently (1.1)) is superstable if and only if there exists a matrix $M \in \mathbb{R}^{n_2 \times n_2}$ such that the matrix A_M defined by (1.160) satisfies the condition (1.135).

Example 1.14. (Continuation of Example 1.12) Consider the descriptor continuous-time linear system (1.1) with (1.155). In this case

$$P = \begin{bmatrix} -1 & 0 & 0 \\ 1 & 2 & 0 \\ 0 & 0 & 1 \end{bmatrix}, \quad Q = \begin{bmatrix} 0 & 1 & 0 \\ 1 & 0 & 0 \\ 0 & 0 & 1 \end{bmatrix} \tag{1.163}$$

and

$$\begin{aligned} \begin{bmatrix} \mathbb{I}_{n_1} & 0 \\ 0 & N \end{bmatrix} &= PEQ = \begin{bmatrix} 1 & 0 & 0 \\ 0 & 1 & 0 \\ 0 & 0 & 0 \end{bmatrix}, \\ \begin{bmatrix} A_1 & 0 \\ 0 & \mathbb{I}_{n_2} \end{bmatrix} &= PAQ = \begin{bmatrix} -3 & 1 & 0 \\ 0 & -1 & 0 \\ 0 & 0 & 1 \end{bmatrix}, \end{aligned} \tag{1.164}$$

where $n_1 = 2$ and $n_2 = 1$. Using (1.160) for the matrices (1.163), (1.164) and $M = [m] \in \mathbb{R}^{1 \times 1}$ we obtain

$$A_M = \begin{bmatrix} 0 & 1 & 0 \\ 1 & 0 & 0 \\ 0 & 0 & 1 \end{bmatrix} \begin{bmatrix} -3 & 1 & 0 \\ 0 & -1 & 0 \\ 0 & 0 & m \end{bmatrix} \begin{bmatrix} 0 & 1 & 0 \\ 1 & 0 & 0 \\ 0 & 0 & 1 \end{bmatrix} = \begin{bmatrix} -1 & 0 & 0 \\ 1 & -3 & 0 \\ 0 & 0 & m \end{bmatrix}. \tag{1.165}$$

Assuming $m < 0$ it is easy to see that the matrix (1.165) satisfies the condition (1.135) since $\sigma = 1 > 0$ for $m \leq 1$ and $\sigma = m > 0$ for $m \in (-1; 0)$. Therefore, the considered descriptor system is superstable.

1.2 Discrete-Time Systems

1.2.1 State Equations of Descriptor Discrete-Time Linear Systems

A discrete approximation of the derivative of the state vector in (1.1a) can be obtained using the Euler method:

$$\dot{x}(t) \approx \frac{x_{i+1} - x_i}{h}, \tag{1.166}$$

where $h > 0$ is the discretization period and x_i, $i \in \mathbb{Z}_+$ are the values of $x(t)$ for $t = 0, h, 2h, \ldots$. Other methods of approximation are given in [100].

Discretizing the Eq. (1.1) and using the approximation (1.166) we obtain

$$Ex_{i+1} = A_d x_i + B_d u_i, \tag{1.167a}$$

$$y_i = Cx_i + Du_i, \tag{1.167b}$$

where

$$A_d = Ah + E \in \mathbb{R}^{n\times n}, \quad B_d = Bh \in \mathbb{R}^{n\times m}, \tag{1.167c}$$

$x_i \in \mathbb{R}^n$, $u_i \in \mathbb{R}^m$, $y_i \in \mathbb{R}^p$ are state, input and output vectors and $E \in \mathbb{R}^{n\times n}$, $C \in \mathbb{R}^{p\times n}$, $D \in \mathbb{R}^{p\times m}$. It is assumed that $\det E = 0$ and

$$\det[Ez - A_d] \neq 0 \text{ for some } z \in \mathbb{C}. \tag{1.168}$$

Consider the matrix A_d defined by (1.167c). The condition (1.168) takes the form

$$\det[Ez - A_d] = \det[E(z-1) - Ah] = h^n \det[E(z-1)h^{-1} - A] \neq 0. \tag{1.169}$$

If the condition (1.2) holds, then the condition (1.169) is also met for $h > 0$. If the matrix E is nonsingular ($\det E \neq 0$), then we can transform (1.167a) to the state equation of a standard discrete-time linear system through premultiplication of (1.167a) by E^{-1}. The system (1.167) with the singular matrix E ($\det E = 0$) will be called a descriptor discrete-time linear system.

1.2.2 Solution to the State Equation of Descriptor Continuous-Time Linear Systems

Let U be the set of admissible input vectors $u_i \in U \subset \mathbb{R}^m$ and $X_0 \subset \mathbb{R}^n$ the set of consistent initial conditions $x(0) = x_0 \in X_0$ for which the Eq. (1.167a) has the solution x_i with $u_i \in U$.

1.2.2.1 Laurent Series Expansion Method

In this section the solution to the Eq. (1.167a) will be given using the Laurent series expansion method [26, 47, 49, 111, 147]. In addition, formulae linking together the matrices of the continuous-time system (1.1) and its discrete-time equivalent (1.167) will be given.

Theorem 1.22. [47] Solution to the Eq. (1.167a) for $x_0 \in X_0$ and $u_i \in U, i \in \mathbb{Z}_+$ is given by the formula

$$x_i = T_{d,i}Ex_0 + \sum_{k=0}^{n_0+i-1} T_{d,i-k-1}B_d u_k, \tag{1.170a}$$

where the matrices $T_{d,k} \in \mathbb{R}^{n\times n}$ are determined by

$$\begin{cases} ET_{d,k} - A_d T_{d,k-1} = \mathbb{I}_n \text{ for } k = 0, \\ ET_{d,k} - A_d T_{d,k-1} = 0 \text{ for } k \neq 0, \\ ET_{d,k} = 0 \qquad\qquad \text{for } k \leq -n_0 \end{cases} \tag{1.170b}$$

and n_0 is called the index of the system. From (1.170a) for $i = 0$ we have

$$x_0 = T_{d,0}Ev + \sum_{k=0}^{n_0-1} T_{d,-k-1}B_d u_k, \tag{1.171}$$

where $v \in \mathbb{R}^n$ is an arbitrary vector.

Remark 1.7. If the matrix $E = \mathbb{I}_n$, then $n_0 = 0$ and $T_{d,i} = A_d^i$. Therefore, from the Eq. (1.170) we have

$$x_i = A_d^i x_0 + \sum_{k=0}^{i-1} A_d^{i-k-1} B_d u_k. \tag{1.172}$$

The formula (1.172) is the solution to the state equation of a standard discrete-time linear system.

Similarly as for continuous-time systems, the matrices $T_{d,k}$ can be computed from the equality

$$[Ez - A_d]^{-1} = \sum_{k=-n_0}^{\infty} T_{d,k} z^{-(k+1)}. \tag{1.173}$$

Using (1.167c) we can write the above equation in the form

$$
\begin{aligned}
[Ez - A_d]^{-1} &= [Ez - (Ah + E)]^{-1} = h^{-1}\left[E\frac{z-1}{h} - A\right]^{-1} \\
&= h^{-1}\sum_{k=-n_0}^{\infty} T_k(z-1)^{-(k+1)}h^{k+1} \\
&= \sum_{k=-n_0}^{\infty} h^k T_k(z-1)^{-(k+1)},
\end{aligned}
\tag{1.174}
$$

where T_k are determined by (1.3b) (they are matrices of the continuous-time system). Expanding the expressions $(z-1)^{-(k+1)}$ in the series with respect to z^{-k}, $k = -n_0, \ldots, 0, 1, \ldots$ we can write (1.174) in the form (1.173), where

$$
T_{d,k} = \sum_{j=0}^{k} \binom{k}{j} h^j T_j, \tag{1.175a}
$$

$$
T_{d,-k} = \sum_{j=k}^{n_0} (-1)^{j-k} \binom{j-1}{k-1} h^{-j} T_{-j} \tag{1.175b}
$$

and

$$
\binom{k}{j} = \frac{k!}{(k-j)!j!}. \tag{1.175c}
$$

The formulae (1.175) link together the matrices T_k of the continuous-time system and the matrices $T_{d,k}$ of the discrete-time system for the assumed approximation of the derivative (1.166).

Example 1.15. (Continuation of Example 1.1) Consider the descriptor discrete-time linear system (1.1) with (1.17). The discrete-time equivalent of the continuous-time system is described by (1.167a), where the matrix E is determined by (1.17) and

$$
A_d = Ah + E = \begin{bmatrix} -0.2h - 0.4 & 1.8h & 0.5h - 0.5 & 0 \\ 0.4h - 0.2 & 0.4h & 0 & 0 \\ 0.2h + 0.4 & -1.8h + 1 & -0.5h + 0.5 & 0.5h \\ -0.4h + 0.2 & 0.6h & 0 & 0 \end{bmatrix},
$$

$$
B_d = Bh = \begin{bmatrix} -h & -3.6h \\ 0 & -0.8h \\ -h & 2.6h \\ 0 & -0.2h \end{bmatrix}. \tag{1.176}
$$

The matrix pencil (E, A_d) is regular since

$$\begin{aligned}\det[Ez - A_d] = &- 0.05h^2z^2 + (-0.15h^3 + 0.1h^2)z \\ &- 0.1h^4 + 0.15h^3 - 0.05h^2 \neq 0.\end{aligned} \tag{1.177}$$

The matrix $[Ez - A_d]^{-1}$ has the form

$$[Ez - A_d]^{-1} = \begin{bmatrix} 0 & -\frac{3}{z+2h-1} & 0 & \frac{2}{z+2h-1} \\ 0 & -\frac{1}{h} & 0 & -\frac{1}{h} \\ -\frac{2}{z+h-1} & \frac{6}{z+2h-1} & 0 & \frac{2(z+4h-1)}{(z+h-1)(z+2h-1)} \\ -\frac{2}{h} & -\frac{2(z-1)}{h^2} & -\frac{2}{h} & -\frac{2(z-1)}{h^2} \end{bmatrix}. \tag{1.178}$$

Using (1.173) the matrix (1.178) can be written in the form

$$[Ez - A_d]^{-1} = \sum_{k=-2}^{\infty} T_{d,k} z^{k+1}, \tag{1.179a}$$

where

$$T_{d,k} = \begin{cases} \begin{bmatrix} 0 & 0 & 0 & 0 \\ 0 & 0 & 0 & 0 \\ 0 & 0 & 0 & 0 \\ 0 & -\frac{2}{h^2} & 0 & -\frac{2}{h^2} \end{bmatrix} & \text{for } k = -2, \\ \begin{bmatrix} 0 & 0 & 0 & 0 \\ 0 & -\frac{1}{h} & 0 & -\frac{1}{h} \\ 0 & 0 & 0 & 0 \\ -\frac{2}{h} & \frac{2}{h^2} & -\frac{2}{h} & \frac{2}{h^2} \end{bmatrix} & \text{for } k = -1, \\ \begin{bmatrix} 0 & -3(1-2h)^k & 0 & 2(1-2h)^k \\ 0 & 0 & 0 & 0 \\ -2(1-h)^k & 6(1-2h)^k & 0 & 6(1-h)^k - 4(1-2h)^k \\ 0 & 0 & 0 & 0 \end{bmatrix} & \text{for } k = 0, 1, \ldots. \end{cases} \tag{1.179b}$$

Using (1.17) and (1.179b) we have

$$T_{d,k}E = \begin{bmatrix} (1-2h)^k & 0 & 0 & 0 \\ 0 & 0 & 0 & 0 \\ 2(1-h)^k - 2(1-2h)^k & 0 & (1-h)^k & 0 \\ 0 & 0 & 0 & 0 \end{bmatrix} \quad \text{for } k = 0, 1, 2, \ldots, \tag{1.180a}$$

$$T_{d,k}B_d = \begin{cases} \begin{bmatrix} 0 & 0 \\ 0 & 0 \\ 0 & 0 \\ 0 & \frac{2}{h} \end{bmatrix} & \text{for } k = -2, \\ \begin{bmatrix} 0 & 0 \\ 0 & 1 \\ 0 & 0 \\ 4 & 2 - \frac{2}{h} \end{bmatrix} & \text{for } k = -1, \\ \begin{bmatrix} 0 & 2h(1-2h)^k \\ 0 & 0 \\ 2h(1-h)^k & 6h(1-h)^k - 4h(1-2h)^k \\ 0 & 0 \end{bmatrix} & \text{for } k = 0, 1, 2, \ldots. \end{cases} \tag{1.180b}$$

Using (1.170a), (1.17), (1.176), (1.180) and taking into account that $n_0 = 2$ we obtain

$$\begin{aligned} x_i &= T_{d,i}Ex_0 + \sum_{k=0}^{i+1} T_{d,i-k-1}B_d u_k = T_{d,i}Ex_0 + \sum_{k=0}^{i-1} T_{d,i-k-1}B_d u_k + T_{d,-1}B_d u_i \\ &+ T_{d,-2}B_d u_{i+1} = \begin{bmatrix} (1-2h)^i & 0 & 0 & 0 \\ 0 & 0 & 0 & 0 \\ 2(1-h)^i - 2(1-2h)^i & 0 & (1-h)^i & 0 \\ 0 & 0 & 0 & 0 \end{bmatrix} x_0 \\ &+ \sum_{k=0}^{i-1} \begin{bmatrix} 0 & 2h(1-2h)^{i-k-1} \\ 0 & 0 \\ 2h(1-h)^{i-k-1} & 6h(1-h)^{i-k-1} - 4h(1-2h)^{i-k-1} \\ 0 & 0 \end{bmatrix} u_k \\ &+ \begin{bmatrix} 0 & 0 \\ 0 & 1 \\ 0 & 0 \\ 4 & 2 - \frac{2}{h} \end{bmatrix} u_i + \begin{bmatrix} 0 & 0 \\ 0 & 0 \\ 0 & 0 \\ 0 & \frac{2}{h} \end{bmatrix} u_{i+1}. \end{aligned} \tag{1.181}$$

The equation above shows the solution (1.170a) of the state equation of the discrete-time equivalent (1.167a) of the system (1.1a) with (1.17).

1.2.2.2 Drazin Inverse Matrix Method

In this section the solution to the Eq. (1.167a) will be given using the Drazin inverse matrix method [26, 27, 34, 47]. In addition, formulae linking together the matrices of the continuous-time system (1.24) and its discrete-time equivalent (1.182) will be given.

Consider the Eq. (1.167a) and assume that the condition (1.168) is met. There exists a number $c \in \mathbb{C}$ such that $\det[Ec - A_d] \neq 0$. Premultiplying the Eq. (1.167a) by the matrix $[Ec - A_d]^{-1}$ we obtain

$$\bar{E}_d x_{i+1} = \bar{A}_d x_i + \bar{B}_d u_i, \tag{1.182a}$$

where

$$\bar{E}_d = [Ec - A_d]^{-1} E, \ \bar{A}_d = [Ec - A_d]^{-1} A_d, \ \bar{B}_d = [Ec - A_d]^{-1} B_d. \tag{1.182b}$$

Note that the Eqs. (1.167a) and (1.182a) have the same solution x_i. Basic definitions and properties concerning Drazin inverses are given in Sect. 1.1.2.2.

Theorem 1.23. [47] Solution to the Eq. (1.167a) for $x_0 \in X_0$ and $u_i \in U, i \in \mathbb{Z}_+$ has the form

$$\begin{aligned} x_i &= (\bar{E}_d^D \bar{A}_d)^i \bar{E}_d^D \bar{E}_d x_0 + \sum_{k=0}^{i-1} \bar{E}_d^D (\bar{E}_d^D \bar{A}_d)^{i-k-1} \bar{B}_d u_k \\ &+ (\bar{E}_d \bar{E}_d^D - \mathbb{I}_n) \sum_{k=0}^{q-1} (\bar{E}_d \bar{A}_d^D)^k \bar{A}_d^D \bar{B}_d u_{i+k}, \end{aligned} \tag{1.183}$$

where q is the index of $\bar{E}_d$.

From (1.183) for $i = 0$ we have

$$x_0 = \bar{E}_d \bar{E}_d^D v + (\bar{E}_d \bar{E}_d^D - \mathbb{I}_n) \sum_{k=0}^{q-1} (\bar{E}_d \bar{A}_d^D)^k \bar{A}_d^D \bar{B}_d u_k, \tag{1.184}$$

where $v \in \mathbb{R}^n$ is an arbitrary vector.

For discrete-time systems we can formulate a similar theorem to Theorem 1.5, i.e. the solution (1.183) of the Eq. (1.167a) and the set of consistent initial conditions X_0 do not depend of the choice of c [47].

Using (1.167c) the matrices (1.182b) can be written in the form

$$\bar{E}_d = [Ec - (Ah + E)]^{-1} E = h^{-1} \left[E\frac{c-1}{h} - A \right]^{-1} E = h^{-1} [E\bar{c} - A]^{-1} E, \tag{1.185a}$$

$$\begin{aligned} \bar{A}_d &= [Ec - (Ah + E)]^{-1} (Ah + E) = h^{-1} \left[E\frac{c-1}{h} - A \right]^{-1} (Ah + E) \\ &= [E\bar{c} - A]^{-1} A + h^{-1} [E\bar{c} - A]^{-1} E, \end{aligned} \tag{1.185b}$$

$$\bar{B}_d = [Ec - (Ah + E)]^{-1} Bh = h^{-1} \left[E\frac{c-1}{h} - A \right]^{-1} Bh = [E\bar{c} - A]^{-1} B, \tag{1.185c}$$

where $\bar{c} = \frac{c-1}{h}$.

Taking into account that we can always find $c, \bar{c} \in \mathbb{C}$ such that $[Ec - A] = [E\bar{c} - A]$ and $\det[Ec - A] = \det[E\bar{c} - A] \neq 0$ from (1.185) we obtain

$$\bar{E}_d = h^{-1}\bar{E},\quad \bar{A}_d = \bar{A} + h^{-1}\bar{E},\quad \bar{B}_d = \bar{B}, \tag{1.186}$$

where $\bar{E}$, $\bar{A}$, $\bar{B}$ are the matrices of the continuous-time system defined by (1.24b).

The formulae (1.186) link together the matrices $\bar{E}$, $\bar{A}$, $\bar{B}$ of the continuous-time system and the matrices $\bar{E}_d$, $\bar{A}_d$, $\bar{B}_d$ of the discrete-time system for the assumed approximation of the derivative (1.166).

Example 1.16. (Continuation of Examples 1.2 and 1.15) Consider the descriptor continuous-time linear system (1.24a) with (1.47). The discrete-time equivalent of the continuous-time system is described by (1.182a). Using (1.186) we obtain

$$\bar{E}_d = h^{-1}\bar{E} = \begin{bmatrix} \frac{1}{2h} & 0 & 0 & 0 \\ 0 & 0 & 0 & 0 \\ \frac{1}{h} & 0 & \frac{1}{h} & 0 \\ 0 & -\frac{2}{h} & 0 & 0 \end{bmatrix},$$

$$\bar{A}_d = \bar{A} + h^{-1}\bar{E} = \begin{bmatrix} \frac{1}{2h} - 1 & 0 & 0 & 0 \\ 0 & -1 & 0 & 0 \\ \frac{1}{h} & 0 & \frac{1}{h} - 1 & 0 \\ 0 & \frac{-2}{h} & 0 & -1 \end{bmatrix}, \tag{1.187}$$

$$\bar{B}_d = \bar{B} = \begin{bmatrix} 0 & 1 \\ 0 & 1 \\ 2 & 4 \\ 4 & 2 \end{bmatrix}.$$

The matrix $\bar{E}^D$ of the continuous-time system has the form (1.48b). Taking into account that $\bar{E}_d^D = (h^{-1}\bar{E})^D = h\bar{E}^D$ we have

$$\bar{E}_d^D = h\bar{E}^D = \begin{bmatrix} 2h & 0 & 0 & 0 \\ 0 & 0 & 0 & 0 \\ -2h & 0 & h & 0 \\ 0 & 0 & 0 & 0 \end{bmatrix}. \tag{1.188}$$

The matrix $\bar{A}_d$ determined by (1.187) is nonsingular and then

$$\bar{A}_d^D = \bar{A}_d^{-1} = \begin{bmatrix} -\frac{2h}{2h-1} & 0 & 0 & 0 \\ 0 & -1 & 0 & 0 \\ -\frac{2h}{(2h-1)(h-1)} & 0 & -\frac{h}{h-1} & 0 \\ 0 & \frac{2}{h} & 0 & -1 \end{bmatrix}. \tag{1.189}$$

Next, we compute the matrices

$$
\bar{E}_d\bar{E}_d^D = \begin{bmatrix} 1 & 0 & 0 & 0 \\ 0 & 0 & 0 & 0 \\ 0 & 0 & 1 & 0 \\ 0 & 0 & 0 & 0 \end{bmatrix}, \quad \bar{E}_d^D\bar{A}_d = \begin{bmatrix} 1-2h & 0 & 0 & 0 \\ 0 & 0 & 0 & 0 \\ 2h & 0 & 1-h & 0 \\ 0 & 0 & 0 & 0 \end{bmatrix},
$$

$$
\bar{E}_d^D\bar{B}_d = \begin{bmatrix} 0 & 2h \\ 0 & 0 \\ 2h & 2h \\ 0 & 0 \end{bmatrix}, \quad (\bar{E}_d\bar{E}_d^D - \mathbb{I}_4) = \begin{bmatrix} 0 & 0 & 0 & 0 \\ 0 & -1 & 0 & 0 \\ 0 & 0 & 0 & 0 \\ 0 & 0 & 0 & -1 \end{bmatrix},
$$

$$
\bar{A}_d^D\bar{E}_d = \begin{bmatrix} -\frac{1}{2h-1} & 0 & 0 & 0 \\ 0 & 0 & 0 & 0 \\ -\frac{2h}{(h-1)(2h-1)} & 0 & -\frac{1}{h-1} & 0 \\ 0 & \frac{2}{h} & 0 & 0 \end{bmatrix},
\tag{1.190}
$$

$$
\bar{A}_d^D\bar{B}_d = \begin{bmatrix} 0 & -\frac{2h}{2h-1} \\ 0 & -1 \\ -\frac{2h}{h-1} & -\frac{2h(4h-1)}{(2h-1)(h-1)} \\ -4 & \frac{2}{h-2} \end{bmatrix}.
$$

Taking into account that $q = 2$ and using (1.183), (1.190) we obtain

$$
\begin{aligned}
x_i &= (\bar{E}_d^D\bar{A}_d)^i\bar{E}_d^D\bar{E}_d x_0 + \sum_{k=0}^{i-1}\bar{E}_d^D(\bar{E}_d^D\bar{A}_d)^{i-k-1}\bar{B}_d u_k + (\bar{E}_d\bar{E}_d^D - \mathbb{I}_4)\bar{A}_d^D u_i \\
&\quad + (\bar{E}_d\bar{E}_d^D - \mathbb{I}_4)\bar{E}_d\bar{A}_d^D\bar{A}_d^D\bar{B}_d u_{i+1} \\
&= \begin{bmatrix} (1-2h)^i & 0 & 0 & 0 \\ 0 & 0 & 0 & 0 \\ 2(1-h)^i - 2(1-2h)^i & 0 & (1-h)^i & 0 \\ 0 & 0 & 0 & 0 \end{bmatrix} x_0 \\
&\quad + \sum_{k=0}^{i-1}\begin{bmatrix} 0 & 2h(1-2h)^{i-k-1} \\ 0 & 0 \\ 2h(1-h)^{i-k-1} & 6h(1-h)^{i-k-1} - 4h(1-2h)^{i-k-1} \\ 0 & 0 \end{bmatrix} u_k \\
&\quad + \begin{bmatrix} 0 & 0 \\ 0 & 1 \\ 0 & 0 \\ 4 & 2-\frac{2}{h} \end{bmatrix} u_i + \begin{bmatrix} 0 & 0 \\ 0 & 0 \\ 0 & 0 \\ 0 & \frac{2}{h} \end{bmatrix} u_{i+1}.
\end{aligned}
\tag{1.191}
$$

The equation above shows the solution (1.183) of the state equation of the discrete-time equivalent (1.182a) of the system (1.24a) with (1.47). Note that the Eqs. (1.181) and (1.191) are equivalent.

1.2.2.3 Weierstrass-Kronecker Decomposition Method

In this section the solution to the Eq. (1.167a) will be given using the Weierstrass-Kronecker decomposition method [45, 47, 96, 176]. In addition, formulae linking together the matrices of the continuous-time system (1.55) and its discrete-time equivalent (1.208) will be given.

If the condition (1.168) holds, then there exists a pair of nonsingular matrices $P_d, Q_d \in \mathbb{R}^{n \times n}$ such that [47, 48]

$$P_d[Ez - A_d]Q_d = \begin{bmatrix} \mathbb{I}_{n_1} z - A_{1d} & 0 \\ 0 & Nz - \mathbb{I}_{n_2} \end{bmatrix}, \quad A_{1d} \in \mathbb{R}^{n_1 \times n_1}, \quad N \in \mathbb{R}^{n_2 \times n_2}, \tag{1.192}$$

where $n_1 = \deg\{\det[Ez - A_d]\}$, $n_2 = n - n_1$ and N is a nilpotent matrix with the nilpotency index μ, i.e. $N^{\mu-1} \neq 0$ and $N^{\mu} = 0$. The matrices P_d and Q_d can be computed using one of procedures given in [47, 48, 168].

Premultiplying (1.167a) by the matrix P_d and introducing the new state vector

$$\bar{x}_i = \begin{bmatrix} \bar{x}_{1,i} \\ \bar{x}_{2,i} \end{bmatrix} = Q_d^{-1} x_i, \quad \bar{x}_{1,i} \in \mathbb{R}^{n_1}, \quad \bar{x}_{2,i} \in \mathbb{R}^{n_2}, \tag{1.193}$$

we obtain

$$P_d E Q_d Q_d^{-1} x_{i+1} = P_d A_d Q_d Q_d^{-1} x_i + P_d B_d u_i, \tag{1.194a}$$

$$y_i = C Q_d Q_d^{-1} x_i + D u_i \tag{1.194b}$$

and using (1.192) we get

$$\bar{x}_{1,i+1} = A_{1d} \bar{x}_{1,i} + B_{1d} u_i, \tag{1.195a}$$

$$N \bar{x}_{2,i+1} = \bar{x}_{2,i} + B_{2d} u_i, \tag{1.195b}$$

$$y_i = C_{1d} \bar{x}_{1,i} + C_{2d} \bar{x}_{2,i} + D u_i, \tag{1.195c}$$

where

$$P_d B_d = \begin{bmatrix} B_{1d} \\ B_{2d} \end{bmatrix}, \quad B_{1d} \in \mathbb{R}^{n_1 \times m}, \quad B_{2d} \in \mathbb{R}^{n_2 \times m}, \tag{1.196a}$$

$$C Q_d = \begin{bmatrix} C_{1d} & C_{2d} \end{bmatrix}, \quad C_{1d} \in \mathbb{R}^{p \times n_1}, \quad C_{2d} \in \mathbb{R}^{p \times n_2}. \tag{1.196b}$$

Theorem 1.24. [47] Solution to the Eq. (1.195a) for given initial condition $\bar{x}_{10} \in \mathbb{R}^{n_1}$ and input $u_i \in U$, $i \in \mathbb{Z}_+$ has the form

$$\bar{x}_{1,i} = A_{1d}^i \bar{x}_{10} + \sum_{k=0}^{i-1} A_{1d}^{i-k-1} B_{1d} u_k. \tag{1.197}$$

Theorem 1.25. Solution to the Eq. (1.195b) for given input $u_i \in U, i \in \mathbb{Z}_+$ has the form

$$\bar{x}_{2,i} = -\sum_{k=0}^{\mu-1} N^k B_{2d} u_{i+k}, \tag{1.198}$$

and the consistent initial conditions of the subsystem (1.195b) are given by the formula

$$\bar{x}_{20} = -\sum_{k=0}^{\mu-1} N^k B_{2d} u_k. \tag{1.199}$$

Proof. From (1.195b) and (1.198) we have

$$\begin{aligned} N\bar{x}_{2,i+1} &= -\sum_{k=0}^{\mu-1} N^{k+1} B_{2d} u_{i+k+1} = -\sum_{k=0}^{\mu-1} N^k B_{2d} u_{i+k} + B_{2d} u_i \\ &= \bar{x}_{2,i} + B_{2d} u_i, \end{aligned} \tag{1.200}$$

since $N^\mu = 0$. Substituting $i = 0$ into (1.198) we obtain (1.199). □

Assume that the matrix N in (1.195b) has the form (1.62) and

$$\bar{x}_{2,i} = \begin{bmatrix} \bar{x}_{21,i} \\ \bar{x}_{22,i} \\ \vdots \\ \bar{x}_{2n_2,i} \end{bmatrix}, \quad B_{2d} = \begin{bmatrix} B_{21d} \\ \vdots \\ B_{2n_2d} \end{bmatrix}. \tag{1.201}$$

From (1.195b), (1.201) and (1.62) we have

$$\begin{bmatrix} 0\ 1\ 0 \dots 0 \\ 0\ 0\ 1 \dots 0 \\ \vdots\ \vdots\ \vdots\ \ddots\ \vdots \\ 0\ 0\ 0 \dots 1 \\ 0\ 0\ 0 \dots 0 \end{bmatrix} \begin{bmatrix} \bar{x}_{21,i+1} \\ \bar{x}_{22,i+1} \\ \vdots \\ \bar{x}_{2n_2,i+1} \end{bmatrix} = \begin{bmatrix} \bar{x}_{21,i} \\ \bar{x}_{22,i} \\ \vdots \\ \bar{x}_{2n_2,i} \end{bmatrix} + \begin{bmatrix} B_{21d} \\ \vdots \\ B_{2n_2d} \end{bmatrix} u_i \tag{1.202}$$

and

$$\begin{aligned} 0 &= \bar{x}_{2n_2,i} + B_{2n_2d} u_i, \\ \bar{x}_{2n_2,i+1} &= \bar{x}_{2n_2-1,i} + B_{2n_2-1d} u_i, \\ &\vdots \\ \bar{x}_{22,i+1} &= \bar{x}_{21,i} + B_{21d} u_i. \end{aligned} \tag{1.203}$$

Solving the Eq. (1.203) with respect to components of the vector $\bar{x}_{2,i}$ we obtain

$$\begin{aligned}\bar{x}_{2n_2,i} &= -B_{2n_2d}u_i,\\ \bar{x}_{2n_2-1,i} &= -B_{2n_2d}u_{i+1} - B_{2n_2-1d}u_i,\\ &\vdots\\ \bar{x}_{21,i} &= -B_{2n_2d}u_{i+n_2-1} - \ldots - B_{21d}u_i.\end{aligned} \tag{1.204}$$

Consistent initial conditions of the system (1.195) can be computed from (1.204) for $i = 0$. Solution to the Eq. (1.195b) for $u_i \in U$ and consistent initial condition $\bar{x}_{20} \in \mathbb{R}^{n_2}$ is given by (1.204). Considerations can be easily extended to the case when the matrix N has the form (1.67).

Assuming $P_d = P$ and $Q_d = Q$ determined by (1.52) and using (1.167c) the equality (1.192) takes the form

$$P[Ez - (Ah + E)]Q = \begin{bmatrix} \mathbb{I}_{n_1}z - (A_1h + \mathbb{I}_{n_1}) & 0 \\ 0 & Nz - (\mathbb{I}_{n_2}h + N) \end{bmatrix}, \tag{1.205}$$

where n_1 and n_2 are defined by (1.192).

Premultiplying (1.167a) by the matrix P and introducing the new state vector

$$\tilde{x}_i = \begin{bmatrix} \tilde{x}_{1,i} \\ \tilde{x}_{2,i} \end{bmatrix} = Q^{-1}x_i, \ \tilde{x}_{1,i} \in \mathbb{R}^{n_1}, \ \tilde{x}_{2,i} \in \mathbb{R}^{n_2}, \tag{1.206}$$

we obtain

$$PEQQ^{-1}x_{i+1} = (PAQh + PEQ)Q^{-1}x_i + PBhu_i, \tag{1.207a}$$

$$y_i = CQQ^{-1}x_i + Du_i \tag{1.207b}$$

and using (1.52) we get

$$\tilde{x}_{1,i+1} = (A_1h + \mathbb{I}_{n_1})\tilde{x}_{1,i} + B_1hu_i, \tag{1.208a}$$

$$N\tilde{x}_{2,i+1} = (\mathbb{I}_{n_2}h + N)\tilde{x}_{2,i} + B_2hu_i, \tag{1.208b}$$

$$y_i = C_1\tilde{x}_{1,i} + C_2\tilde{x}_{2,i} + Du_i, \tag{1.208c}$$

where $A_1 \in \mathbb{R}^{n_1\times n_1}$, $B_1 \in \mathbb{R}^{n_1\times m}$, $B_2 \in \mathbb{R}^{n_2\times m}$, $C_1 \in \mathbb{R}^{p\times n_1}$, $C_2 \in \mathbb{R}^{n_2\times p}$ are the matrices of the continuous-time system determined by (1.52) and (1.56).

Theorem 1.26. Solution to the Eq. (1.208a) for given initial condition $\tilde{x}_{10} \in \mathbb{R}^{n_1}$ and input $u_i \in U, i \in \mathbb{Z}_+$ has the form

$$\tilde{x}_{1,i} = (A_1 h + \mathbb{I}_{n_1})^i \tilde{x}_{10} + \sum_{k=0}^{i-1} (A_1 h + \mathbb{I}_{n_1})^{i-k-1} B_1 h u_k. \tag{1.209}$$

Proof. The proof can be accomplished in a similar way to the proof of Theorem 1.24 [47]. □

Theorem 1.27. Solution to the Eq. (1.208b) for given input $u_i \in U, i \in \mathbb{Z}_+$ has the form

$$\tilde{x}_{2,i} = -\sum_{k=0}^{\mu-1} \bar{N}_k B_2 u_{i+k}, \tag{1.210a}$$

where

$$\bar{N}_k = \sum_{j=k}^{\mu-1} \binom{j}{k} h^{-j} N^j (-1)^{j-k}. \tag{1.210b}$$

The coefficients $\binom{j}{k}$ are given by (1.175c).

Proof. The proof can be accomplished in a similar way to the proof of Theorem 1.25. Without loss of generality we will show that the solution (1.210) satisfies the equality (1.208b) for $\mu = 3$. From (1.208b) and (1.210) we have

$$\begin{aligned} N\tilde{x}_{2,i+1} &= -\sum_{k=0}^{2} \left(\sum_{j=k}^{2} \binom{j}{k} h^{-j} N^{j+1} (-1)^{j-k} B_2 u_{i+k+1} \right) \\ &= -\sum_{j=0}^{2} \binom{j}{0} h^{-j} N^{j+1} (-1)^j B_2 u_{i+1} - \sum_{j=1}^{2} \binom{j}{1} h^{-j} N^{j+1} (-1)^{j-1} B_2 u_{i+2} \\ &= (-NB_2 + h^{-1} N^2 B_2) u_{i+1} - h^{-1} N^2 B_2 u_{i+2}, \end{aligned} \tag{1.211a}$$

$$\begin{aligned} \mathbb{I}_{n_2} h \tilde{x}_{2,i} &= -\sum_{k=0}^{2} \left(\sum_{j=k}^{2} \binom{j}{k} h^{-j+1} N^j (-1)^{j-k} B_2 u_{i+k} \right) \\ &= -\sum_{j=0}^{2} \binom{j}{0} h^{-j+1} N^j (-1)^j B_2 u_i - \sum_{j=1}^{2} \binom{j}{1} h^{-j+1} N^j (-1)^{j-1} B_2 u_{i+1} \\ &\quad - \sum_{j=2}^{2} \binom{j}{2} h^{-j+1} N^j (-1)^{j-2} B_2 u_{i+2} \\ &= (-hB_2 + NB_2 - h^{-1} N^2 B_2) u_i + (-NB_2 + 2h^{-1} N^2 B_2) u_{i+1} \\ &\quad - h^{-1} N^2 B_2 u_{i+2}, \end{aligned} \tag{1.211b}$$

$$
\begin{aligned}
N\tilde{x}_{2,i} &= -\sum_{k=0}^{2}\left(\sum_{j=k}^{2}\binom{j}{k}h^{-j}N^{j+1}(-1)^{j-k}B_2u_{i+k}\right)\\
&= -\sum_{j=0}^{2}\binom{j}{0}h^{-j}N^{j+1}(-1)^{j}B_2u_i - \sum_{j=1}^{2}\binom{j}{1}h^{-j}N^{j+1}(-1)^{j-1}B_2u_{i+1}\\
&= (-NB_2 + h^{-1}N^2B_2)u_i - h^{-1}N^2B_2u_{i+1}
\end{aligned}
\tag{1.211c}
$$

and

$$
(\mathbb{I}_{n_2}h + N)\tilde{x}_{2,i} + B_2hu_i = (-NB_2 + h^{-1}N^2B_2)u_{i+1} - h^{-1}N^2B_2u_{i+2} = N\tilde{x}_{2,i+1}. \tag{1.212}
$$

In a similar way we can show that the solution (1.210) satisfies the equality (1.208b) for any μ. □

Example 1.17. (Continuation of Examples 1.3, 1.15 and 1.16) Consider the descriptor continuous-time linear system (1.55) with (1.69). The discrete-time equivalent of the continuous-time system is described by (1.208), where

$$
A_1h + \mathbb{I}_2 = \begin{bmatrix} 1-h & h \\ 0 & 1-2h \end{bmatrix}, \quad B_1h = \begin{bmatrix} h & h \\ 0 & 2h \end{bmatrix}. \tag{1.213}
$$

Using (1.209) and (1.213) we obtain

$$
\begin{aligned}
\tilde{x}_{1,i} &= \begin{bmatrix} (1-h)^i & (1-h)^i - (1-2h)^i \\ 0 & (1-2h)^i \end{bmatrix}\tilde{x}_{10}\\
&+ \sum_{k=0}^{i-1}\begin{bmatrix} h(1-h)^{i-k-1} & 3h(1-h)^{i-k-1} - 2h(1-h)^{i-k-1} \\ 0 & 2h(1-2h)^{i-k-1} \end{bmatrix}u_k.
\end{aligned}
\tag{1.214}
$$

Taking into account that $\mu = 2$ and using (1.210), (1.69) we get

$$
\begin{aligned}
\tilde{x}_{2,i} &= -\bar{N}_0B_2u_i - \bar{N}_1B_2u_{i+1} = (-B_2 + h^{-1}NB_2)u_i - h^{-1}NB_2u_{i+1}\\
&= \begin{bmatrix} 2 & 1-\frac{1}{h} \\ 0 & 1 \end{bmatrix}u_i + \begin{bmatrix} 0 & \frac{1}{h} \\ 0 & 0 \end{bmatrix}u_{i+1}.
\end{aligned}
\tag{1.215}
$$

From (1.206), (1.214) and (1.215) we have

$$\tilde{x}_i = \begin{bmatrix} (1-h)^i & (1-h)^i - (1-2h)^i & 0 & 0 \\ 0 & (1-2h)^i & 0 & 0 \\ 0 & 0 & 0 & 0 \\ 0 & 0 & 0 & 0 \end{bmatrix} \tilde{x}_0$$

$$+ \sum_{k=0}^{i-1} \begin{bmatrix} h(1-h)^{i-k-1} & 3h(1-h)^{i-k-1} - 2h(1-h)^{i-k-1} \\ 0 & 2h(1-2h)^{i-k-1} \\ 0 & 0 \\ 0 & 0 \end{bmatrix} u_k \tag{1.216}$$

$$+ \begin{bmatrix} 0 & 0 \\ 0 & 0 \\ 2 & 1-\frac{1}{h} \\ 0 & 1 \end{bmatrix} u_i + \begin{bmatrix} 0 & 0 \\ 0 & 0 \\ 0 & \frac{1}{h} \\ 0 & 0 \end{bmatrix} u_{i+1}.$$

The equation above shows the solutions (1.209) and (1.210) of the state equations of the discrete-time equivalent (1.208) of the system (1.55) with (1.69). Taking into account that $\tilde{x}_i = Q^{-1}x_i$ we have to premultiply the Eq. (1.216) by Q and substitute $\tilde{x}_0 = Q^{-1}x_0$. Thus we obtain the formula, which is equivalent to (1.181) and (1.191).

1.2.3 Positive Descriptor Discrete-Time Linear Systems

In positive systems inputs, outputs and state variables take only nonnegative values [40, 49]. The following definition of the positivity is suitable for all methods of the descriptor systems analysis.

Definition 1.11. The descriptor discrete-time linear system (1.167a) is called (internally) positive if $x_i \in \mathbb{R}_+^n$ and $y_i \in \mathbb{R}_+^p$, $i \in \mathbb{Z}_+$ for any consistent initial condition $x_0 \in X_0 \subset \mathbb{R}_+^n$ and all admissible inputs $u_i \in U \subset \mathbb{R}_+^m$, $i \in \mathbb{Z}_+$.

1.2.3.1 Laurent Series Expansion Method

In this section necessary and sufficient positivity conditions of the descriptor discrete-time linear system (1.167) will be established using the Laurent series expansion method. Considerations on the influence of the discretization period h on the positivity of the discrete-time equivalent (1.167) of the continuous-time model (1.1) will be presented.

Theorem 1.28. The descriptor discrete-time linear system (1.167a) is (internally) positive if and only if

$$\begin{gathered} T_{d,k}E \in \mathbb{R}_+^{n\times n}, \quad k = 0, 1, \ldots, \\ T_{d,k}B \in \mathbb{R}_+^{n\times m}, \quad k = -n_0, -n_0+1, \ldots, 0, 1, \ldots \\ C \in \mathbb{R}_+^{p\times n}, \quad D \in \mathbb{R}_+^{p\times m}, \end{gathered} \tag{1.217}$$

where the matrices $T_{d,k} \in \mathbb{R}^{n\times n}$ are given by (1.170b) (or (1.175)).

Proof. By Definition 1.11 $x_0 \in X_0 \subset \mathbb{R}_+^n$ and $u_i \in U \subset \mathbb{R}_+^m$, $i \in \mathbb{Z}_+$. From (1.170) we have $x_i \in \mathbb{R}_+^n$, $i \in \mathbb{Z}_+$ for $u_i = 0$, $i \in \mathbb{Z}_+$ if and only if $T_{d,i}E \in \mathbb{R}_+^{n\times n}$, $i \in \mathbb{Z}_+$ and for $x_0 = 0$ if and only if $T_{d,i}B \in \mathbb{R}_+^{n\times m}$. Similarly, from the Eq. (1.167b) it follows that $y_i \in \mathbb{R}_+^p$ if and only if $C \in \mathbb{R}_+^{p\times n}$ i $D \in \mathbb{R}_+^{p\times m}$. □

Lemma 1.11. If the matrices T_kE and T_kB of the continuous-time system (1.1) satisfy the conditions (1.74) for $k \geq 0$, then the matrices $T_{d,k}E$ and $T_{d,k}B$ of its discrete-time equivalent (1.167) satisfy the conditions (1.217) for $k \geq 0$.

Proof. The proof follows immediately from (1.175a). The matrices $T_{d,k}$ are linear combinations (with nonnegative coefficients) of the matrices T_k for $k \geq 0$. □

From the above considerations it follows that if the positive continuous-time system (1.1) satisfies the conditions (1.74), then analysis of the positivity of its discrete-time equivalent (1.167) should be done only for the matrices $T_{d,-k}$, $k = 1, \ldots, n_0$ since by (1.175b) they may contain negative elements.

Example 1.18. (Continuation of Example 1.4) Consider the descriptor continuous-time linear system (1.1) with (1.76). The discrete-time equivalent of the continuous-time system is described by (1.167), where the matrices E, C, D are determined by (1.76) and

$$A_d = Ah + E = \begin{bmatrix} 1 & 2h & 0 \\ 2h & 1 & -h \\ 0 & 0 & h \end{bmatrix}, \quad B_d = Bh = \begin{bmatrix} 0 \\ 2h \\ -h \end{bmatrix}. \tag{1.218}$$

The matrix pencil (E, A_d) is regular since

$$\det[Ez - A_d] = -hz^2 + 2hz + 4h^3 - h \neq 0. \tag{1.219}$$

The matrices $T_{d,k}$ of the discrete-time system can be computed using (1.173) or (1.175) basing on the matrices T_k of the continuous-time system given by (1.79b). From (1.80) it follows that $T_kE \in \mathbb{R}_+^{n\times n}$ and $T_kB \in \mathbb{R}_+^{n\times m}$ for $k \geq 0$ and by Lemma 1.11 the matrices $T_{d,k}E \in \mathbb{R}_+^{n\times n}$ and $T_{d,k}B \in \mathbb{R}_+^{n\times m}$ for $k \geq 0$. Therefore, to determine the positivity of the discrete-time system we should calculate the matrix $T_{d,-1}$ using (1.79b) and (1.175b):

$$T_{d,-1} = h^{-1}T_{-1} = \begin{bmatrix} 0 & 0 & 0 \\ 0 & 0 & 0 \\ 0 & 0 & -h^{-1} \end{bmatrix}. \tag{1.220}$$

Hence, we have

$$T_{d,-1}B = \begin{bmatrix} 0 \\ 0 \\ h^{-1} \end{bmatrix} \in \mathbb{R}_+^{3\times 1}. \tag{1.221}$$

Therefore, the discrete-time equivalent (1.167) of the continuous-time system (1.1) with (1.76) and (1.218) is positive for any discretization period $h > 0$.

1.2.3.2 Drazin Inverse Matrix Method

In this section necessary and sufficient positivity conditions of the descriptor discrete-time linear system (1.182) (or equivalently (1.167)) will be established using the Drazin inverse matrix method. This problem has been also investigated in [19, 142, 172]. Considerations on the influence of the discretization period h on the positivity of the discrete-time equivalent (1.182) of the continuous-time model (1.24) will be presented.

Theorem 1.29. Let

$$\bar{F}_{d,k} = (\bar{E}_d \bar{A}_d^D)^k \bar{A}_d^D \bar{B}_d, \quad k = 0, 1, \ldots, q-1. \tag{1.222}$$

The descriptor discrete-time linear system (1.182) (or equivalently (1.167)) is positive if and only if there exists a matrix $G_d \in \mathbb{R}^{n\times n}$ [142] such that

$$\bar{E}_d^D \bar{A}_d + G_d(\mathbb{I}_n - \bar{E}_d \bar{E}_d^D) \in \mathbb{R}_+^{n\times n}, \tag{1.223a}$$

and

$$\operatorname{Im} \bar{E}_d \bar{E}_d^D \subset \mathbb{R}_+^n, \quad \bar{E}_d^D \bar{B}_d \in \mathbb{R}_+^{n\times m}, \quad C \in \mathbb{R}_+^{p\times n}, \quad D \in \mathbb{R}_+^{p\times m} \tag{1.223b}$$

$$(\bar{E}_d \bar{E}_d^D - \mathbb{I}_n)\bar{F}_{d,k} \in \mathbb{R}_+^{n\times m}, \quad k = 0, 1, \ldots, q-1. \tag{1.223c}$$

Proof. The proof can be accomplished in a similar way to the proof of Theorem 1.10. Using the superposition principle we will consider components of the solution (1.183) independently:

1. If $\operatorname{Im} \bar{E}_d \bar{E}_d^D \subset \mathbb{R}_+^{n\times n}$, then $\bar{E}_d \bar{E}_d^D x_0 \subset \mathbb{R}_+^n$. Therefore, for the matrix G_d such that $\bar{E}_d^D \bar{A}_d + G_d(\mathbb{I}_n - \bar{E}_d \bar{E}_d^D) \in \mathbb{R}_+^{n\times n}$ we have $[\bar{E}_d^D \bar{A}_d + G_d(\mathbb{I}_n - \bar{E}_d \bar{E}_d^D)]^i \in \mathbb{R}_+^{n\times n}$. Taking into account that $[\bar{E}_d^D \bar{A}_d + G_d(\mathbb{I}_n - \bar{E}_d \bar{E}_d^D)]^i \bar{E}_d \bar{E}_d^D = (\bar{E}_d^D \bar{A}_d)^i \bar{E}_d \bar{E}_d^D$ we obtain $[\bar{E}_d^D \bar{A}_d + G_d(\mathbb{I}_n - \bar{E}_d \bar{E}_d^D)]^i \bar{E}_d \bar{E}_d^D x_0 = (\bar{E}_d^D \bar{A}_d)^i \bar{E}_d \bar{E}_d^D x_0 \in \mathbb{R}_+^n, i \in \mathbb{Z}_+$.
2. Using (1.27a) we can write the second component in the form $\sum_{k=0}^{i-1} (\bar{E}_d^D \bar{A}_d)^{i-k-1} \bar{E}_d^D \bar{B}_d u_k$. If $\bar{E}_d^D \bar{A}_d + G_d(\mathbb{I}_n - \bar{E}_d \bar{E}_d^D) \in \mathbb{R}_+^{n\times n}$ and $\bar{E}_d^D \bar{B}_d \in \mathbb{R}_+^{n\times m}$, then for $u_i \in U \subset \mathbb{R}_+^m$, $i \in \mathbb{Z}_+$ we have $\sum_{k=0}^{i-1} \bar{E}_d^D (\bar{E}_d^D \bar{A}_d)^{i-k-1} \bar{B}_d u_k \in \mathbb{R}_+^n$, $i \in \mathbb{Z}_+$.
3. If $(\bar{E}_d \bar{E}_d^D - \mathbb{I}_n)\bar{F}_{d,k} \in \mathbb{R}_+^{n\times m}$, $k = 0, 1, \ldots, q-1$, where the matrices $\bar{F}_{d,k}$ are given by (1.222) and $u_i \in U \subset \mathbb{R}_+^m$, $i \in \mathbb{Z}_+$, then the last component of the solution (1.183) is also nonnegative.

Therefore, we have $x_i \in \mathbb{R}_+^n$, $i \in \mathbb{Z}_+$ if $\operatorname{Im} \bar{E}_d \bar{E}_d^D \subset \mathbb{R}_+^{n\times n}$, $\bar{E}_d^D \bar{A}_d + G_d(\mathbb{I}_n - \bar{E}_d \bar{E}_d^D) \in \mathbb{R}_+^{n\times n}$, $\bar{E}_d^D \bar{B}_d \in \mathbb{R}_+^{n\times m}$ and $(\bar{E}_d \bar{E}_d^D - \mathbb{I}_n)\bar{F}_{d,k} \in \mathbb{R}_+^{n\times m}$, $k = 0, 1, \ldots, q-1$.

Substituting (1.183) into (1.167b) it is easy to see that $y_i \in \mathbb{R}_+^p$, $i \in \mathbb{Z}_+$ if $C \in \mathbb{R}_+^{p\times n}$ and $D \in \mathbb{R}_+^{p\times m}$. □

Remark 1.8. Note that the descriptor discrete-time linear system (1.182) can be positive even though the matrix $\bar{E}_d^D \bar{A}_d$ contains negative entries. In the special case for $G_d = 0$ from (1.223a) we obtain the positivity condition $\bar{E}_d^D \bar{A}_d \in \mathbb{R}_+^{n\times n}$.

Now let us consider the continuous-time system (1.24). It is assumed that the system is positive, i.e. it satisfies the conditions (1.88). Using (1.186) we can write

$$\bar{E}_d \bar{E}_d^D = h^{-1}\bar{E}h\bar{E}^D = \bar{E}\bar{E}^D, \tag{1.224a}$$

since $\bar{E}_d^D = (h^{-1}\bar{E})^D = h\bar{E}^D$, which follows from (1.27d). Moreover, we have

$$\bar{E}_d^D \bar{A}_d = h\bar{E}^D(\bar{A} + h^{-1}\bar{E}) = h\bar{E}^D\bar{A} + \bar{E}\bar{E}^D \tag{1.224b}$$

and

$$\bar{E}_d^D \bar{B}_d = h\bar{E}^D\bar{B}. \tag{1.224c}$$

Using (1.224) we can easily obtain the formulae linking together the positivity conditions of the continuous-time system (1.88) and its discrete-time equivalent (1.223).

Analysis of the conditions (1.88c) and (1.223c) is much more difficult. Using (1.186) and (1.222) we can write

$$\begin{aligned}(\bar{E}_d\bar{E}_d^D - \mathbb{I}_n)F_{d,k} &= (\bar{E}_d\bar{E}_d^D - \mathbb{I}_n)(\bar{E}_d\bar{A}_d^D)^k\bar{A}_d^D\bar{B}_d \\ &= (\bar{E}\bar{E}^D - \mathbb{I}_n)[h^{-1}\bar{E}h(\bar{A}h + \bar{E})^D]^k h(\bar{A}h + \bar{E})^D\bar{B} \\ &= h(\bar{E}\bar{E}^D - \mathbb{I}_n)[\bar{E}(\bar{A}h + \bar{E})^D]^k(\bar{A}h + \bar{E})^D\bar{B},\end{aligned} \tag{1.225}$$

since $[h^{-1}(\bar{A}h + \bar{E})]^D = h(\bar{A}h + \bar{E})^D$.

Theorem 1.30. If the descriptor continuous-time linear system (1.24) (or equivalently (1.1)) is positive, i.e. it satisfies the conditions (1.88), then its discrete-time counterpart (1.182) (or equivalently (1.167)) is also positive if and only if

$$0 < h \le \min_{1\le i\le n} \frac{-\bar{e}_{ii}^d}{\bar{a}_{ii}^d} \text{ for } \bar{a}_{ii}^d < 0 \tag{1.226a}$$

or

$$h > 0 \text{ for } \bar{a}_{ii}^d > 0, \tag{1.226b}$$

where $\bar{e}_{ij}^d$ i $\bar{a}_{ij}^d$, $i, j = 1, \ldots, n$ are the elements of the matrices $\bar{E}\bar{E}^D$ and $\bar{E}^D\bar{A}$, respectively and

$$(\bar{E}\bar{E}^D - \mathbb{I}_n)[\bar{E}(\bar{A}h + \bar{E})^D]^k(\bar{A}h + \bar{E})^D\bar{B} \in \mathbb{R}_+^{n\times m}, \quad k = 0, 1, \ldots, q-1 \tag{1.227}$$

for h satisfying (1.226).

Proof. If the continuous-time system (1.24) is positive, then $\mathrm{Im}\bar{E}\bar{E}^D \subset \mathbb{R}_+^{n\times n}$, $\bar{E}^D\bar{A} + G(\mathbb{I}_n - \bar{E}\bar{E}^D) \in \mathbb{M}_n$ and $\bar{E}^D\bar{B} \in \mathbb{R}_+^{n\times m}$. Therefore, from (1.224a) and (1.224c) we have $\mathrm{Im}\bar{E}_d\bar{E}_d^D \subset \mathbb{R}_+^{n\times n}$ and $\bar{E}_d^D\bar{B}_d \in \mathbb{R}_+^{n\times m}$ for any $h > 0$. In the case of the matrix (1.224b) we have $\bar{E}_d^D\bar{A}_d + G_d(\mathbb{I}_n - \bar{E}_d\bar{E}_d^D) \in \mathbb{R}_+^{n\times n}$ for $G_d = G$ if and only if

$$h\bar{a}_{ii}^d + \bar{e}_{ii}^d \geq 0, \ i = 1, \ldots, n. \tag{1.228}$$

Solving (1.228) with respect to h we obtain (1.226). If there exists such a value of h in the range (1.226), for which the condition (1.227) is met, then the discrete-time equivalent (1.182) of the positive continuous-time system (1.24) is also positive. □

Example 1.19. (Continuation of Example 1.5) Consider the descriptor continuous-time linear system (1.24) with (1.89). According to (1.91) and (1.93) the system is positive. Its discrete-time equivalent is described by (1.182). Using (1.186) we have

$$\mathrm{Im}\bar{E}_d\bar{E}_d^D = \mathrm{Im}\bar{E}\bar{E}^D = \mathrm{Im}\begin{bmatrix} 1 & 0 & 0 \\ 0 & 1 & 0 \\ 0 & 0 & 0 \end{bmatrix} \subset \mathbb{R}_+^{3\times 3} \text{ for } v = \begin{bmatrix} v_1 \\ v_2 \\ 0 \end{bmatrix} \subset \mathbb{R}_+^3, \tag{1.229a}$$

$$\bar{E}_d^D\bar{A}_d = h\bar{E}^D\bar{A} + \bar{E}\bar{E}^D = \begin{bmatrix} 1 & 2h & 0 \\ 2h & 1 & 0 \\ 0 & 0 & 0 \end{bmatrix} \in \mathbb{R}_+^{3\times 3}, \tag{1.229b}$$

$$\bar{E}_d^D\bar{B}_d = h\bar{E}^D\bar{B} = \begin{bmatrix} 0 \\ h \\ 0 \end{bmatrix} \in \mathbb{R}_+^{3\times 1} \tag{1.229c}$$

for any $h > 0$. By (1.186) the indices of the matrices $\bar{E}_d$ and $\bar{E}$ are equivalent and so $q = 1$. According to (1.227) we shall check if all entries of the matrix $(\bar{E}\bar{E}^D - \mathbb{I}_3)(\bar{A}h + \bar{E})^D\bar{B}$ are nonnegative. It is easy to show that

$$(\bar{E}\bar{E}^D - \mathbb{I}_3)(\bar{A}h + \bar{E})^D\bar{B} = \begin{bmatrix} 0 \\ 0 \\ h^{-1} \end{bmatrix} \in \mathbb{R}_+^{3\times 1}. \tag{1.230}$$

Therefore, the discrete-time equivalent (1.229) of the continuous-time system (1.89) is positive for any discretization period $h > 0$.

1.2.3.3 Weierstrass-Kronecker Decomposition Method

In this section necessary and sufficient positivity conditions of the descriptor discrete-time linear system (1.195) (or equivalently (1.167)) will be established using the Weierstrass-Kronecker decomposition method with the assumption that the matrix Q_d determined by (1.192) is a monomial matrix [59, 66]. This problem has been also investigated in [19]. Considerations on the influence of the discretization period h on the positivity of the discrete-time equivalent (1.208) of the continuous-time model (1.55) will be presented.

Let the decomposition (1.192) of the positive descriptor discrete-time linear system (1.167) be possible for a monomial matrix $Q_d \in \mathbb{R}_+^{n\times n}$. In this case $x_i = Q_d\bar{x}_i \in \mathbb{R}_+^n$ if and only if $\bar{x}_i \in \mathbb{R}_+^n$. It is also well-known that premultiplying the Eq. (1.167a) by a nonsingular matrix P does not change the solution x_i.

Theorem 1.31. [19, 59, 66] Let the decomposition (1.192) of the system (1.167) be possible for a monomial matrix $Q_d \in \mathbb{R}_+^{n\times n}$. The descriptor discrete-time linear system (1.195) (or equivalently (1.167)) is positive if and only if

$$\begin{gathered} A_{1d} \in \mathbb{R}_+^{n_1\times n_1},\quad B_{1d} \in \mathbb{R}_+^{n_1\times m},\quad C \in \mathbb{R}_+^{p\times n},\quad D \in \mathbb{R}_+^{p\times m},\\ -N^k B_{2d} \in \mathbb{R}_+^{n_2\times m},\quad k = 0, 1, \ldots, \mu - 1. \end{gathered} \tag{1.231}$$

Proof. For a monomial matrix $Q_d \in \mathbb{R}_+^{n\times n}$ we have $x_i \in \mathbb{R}_+^n$, $i \in \mathbb{Z}_+$ if $\bar{x}_i \in \mathbb{R}_+^n$, $i \in \mathbb{Z}_+$. If $A_{1d} \in \mathbb{R}_+^{n_1\times n_1}$ and $B_{1d} \in \mathbb{R}_+^{n_1\times m}$, then from (1.197) we obtain $\bar{x}_{1,i} \in \mathbb{R}_+^{n_1}$, $i \in \mathbb{Z}_+$ since by Definition 1.11 $\bar{x}_{10} = Q^{-1}x_{10} \in \mathbb{R}_+^{n_1}$ and $u_i \in \mathbb{R}_+^m$, $i \in \mathbb{Z}_+$. If $-N^k B_{2d} \in \mathbb{R}_+^{n_2\times m}$, $k = 0, 1, \ldots, \mu - 1$, then from (1.198) we obtain $\bar{x}_{2,i} \in \mathbb{R}_+^{n_2}$ since $u_i \in \mathbb{R}_+^m$, $i \in \mathbb{Z}_+$. From (1.194b) for $C \in \mathbb{R}_+^{p\times n}$ and $D \in \mathbb{R}_+^{p\times m}$ we have $y_i \in \mathbb{R}_+^p$, $i \in \mathbb{Z}_+$ since $x_i \in \mathbb{R}_+^n$ and $u_i \in \mathbb{R}_+^m$, $i \in \mathbb{Z}_+$. □

Now let us consider the continuous-time system (1.1), for which the decomposition (1.52) is possible with a monomial matrix $Q \in \mathbb{R}_+^{n\times n}$. The system (1.55) is positive if it satisfies the conditions (1.94). The discrete-time equivalent of the system is described by (1.208). In this case $x_i = Q\tilde{x}_i \in \mathbb{R}_+^n$ if and only if $\tilde{x}_i \in \mathbb{R}_+^n$. Also, premultiplying the Eq. (1.167a) by a nonsingular matrix P does not change the solution x_i.

Theorem 1.32. If the descriptor continuous-time linear system (1.55) (or equivalently (1.1)) is positive, i.e. it satisfies the conditions (1.94), then its discrete-time counterpart (1.208) (or equivalently (1.167)) is also positive if and only if

$$0 < h \le \min_{1\le i\le n} \frac{-1}{a_{ii}^1} \quad \text{for } a_{ii}^1 < 0 \tag{1.232a}$$

or

$$h > 0 \ \text{for } a_{ii}^1 > 0, \tag{1.232b}$$

where a_{ij}^1, $i, j = 1, \ldots, n$ are the entries of the matrix A_1 and

$$-\bar{N}_k B_2 \in \mathbb{R}_+^{n\times m},\quad k = 0, 1, \ldots, \mu - 1 \tag{1.233}$$

for h satisfying (1.232).

Proof. If the continuous-time system (1.55) is positive, then $A_1 \in \mathbb{M}_n$ and $B_1 \in \mathbb{R}_+^{n_1\times m}$. From (1.205) we have $B_1 h \in \mathbb{R}_+^{n\times m}$ for any $h > 0$ and $A_1 h + \mathbb{I}_{n_1} \in \mathbb{R}_+^{n_1\times n_1}$

if and only if

$$1 + ha_{ii}^1 \geq 0, \;\; i = 1, \ldots, n. \tag{1.234}$$

Solving (1.234) with respect to h we obtain (1.232). If there exists such a value of h in the range (1.232), for which the condition (1.233) is met, then the discrete-time equivalent (1.208) of the positive continuous-time system (1.55) is also positive. □

Example 1.20. (Continuation of Example 1.6) Consider the positive descriptor continuous-time linear system (1.55) with (1.97). The discrete-time equivalent of the system is described by (1.208), where

$$A_1 h + \mathbb{I}_2 = \begin{bmatrix} 1 & 2h \\ 2h & 1 \end{bmatrix} \in \mathbb{R}_+^{2\times 2}, \;\; B_1 h = \begin{bmatrix} 0 \\ h \end{bmatrix} \in \mathbb{R}_+^{2\times 1} \tag{1.235}$$

for any $h > 0$. From (1.210b) and (1.233) we have

$$-\bar{N}_0 B_2 = -h^0 N^0 B_2 = -B_2 = 1 \in \mathbb{R}_+^{1\times 1}. \tag{1.236}$$

Therefore, the discrete-time equivalent (1.235) of the continuous-time system (1.97) is positive for any discretization period $h > 0$.

1.2.4 *Stability of Descriptor Discrete-Time Linear Systems*

In the following considerations it is assumed that $\det[E - A_d] \neq 0$, i.e. the descriptor discrete-time linear system (1.167) has exactly one equilibrium point [47]. For the assumed approximation of the derivative (1.166) this condition takes the form $\det Ah \neq 0$. If the continuous-time system (1.1) has one equilibrium point, i.e. $\det A \neq 0$, then its discrete-time equivalent (1.167) also has one equilibrium point since $\det Ah = h^n \det A \neq 0$ for $h > 0$.

1.2.4.1 Laurent Series Expansion Method

In this section necessary and sufficient stability conditions of the descriptor discrete-time linear system (1.167) will be established using the Laurent series expansion method. Considerations on the influence of the discretization period h on the stability of the discrete-time equivalent (1.167) of the continuous-time model (1.1) will be presented.

Definition 1.12. The descriptor discrete-time linear system (1.167) is called asymptotically stable if

$$\lim_{i\to\infty} x_i = 0 \tag{1.237}$$

for all consistent initial conditions $x_0 \in X_0$ and $u_i = 0$, $i \in \mathbb{Z}_+$.

Definition 1.13. The characteristic equation of the matrix pair (E, A_d) has the form

$$p_{(E,A_d)}(z) = \det[Ez - A_d] = a_{d,r}z^r + a_{d,r-1}z^{r-1} + \ldots + a_{d,1}z + a_{d,0} = 0, \tag{1.238}$$

where $r = \text{rank}T_{d,i}E$ and $p_{(E,A_d)}(z)$ is the characteristic polynomial of the matrix pair (E, A_d).

Theorem 1.33. [47] The matrix pair (E, A_d) (or equivalently the system (1.167)) is asymptotically stable if and only if its eigenvalues z_k, $k = 1, \ldots, r$ (roots of the characteristic equation) satisfy the condition

$$|z_k| < 1, \;\; k = 1, \ldots, r. \tag{1.239}$$

Note that for $A_d = Ah + E$ we have $\det[Ez - A_d] = h^n \det[E(z-1)h^{-1} - A]$. Comparing this polynomial with (1.100) we obtain

$$s = (z - 1)h^{-1} \tag{1.240}$$

and therefore

$$z = sh + 1. \tag{1.241}$$

From the above considerations we have the following theorem.

Theorem 1.34. [82] If the descriptor continuous-time linear system (1.1) is asymptotically stable, then its discrete-time equivalent (1.167) is also asymptotically stable if the discretization period h satisfies the condition

$$0 < h < \min_{1 \le k \le r} \frac{2\alpha_k}{\alpha_k^2 + \beta_k^2}, \tag{1.242}$$

where $s_k = -\alpha_k + j\beta_k, k = 1, 2, \ldots, r$ are the eigenvalues of the matrix pair (E, A).

Proof. The eigenvalues of the matrix pairs (E, A) and (E, A_d) are related by (1.241). The discrete-time equivalent (1.167) of the continuous-time system (1.1) is asymptotically stable if and only if the condition (1.239) is satisfied, i.e.

$$|z_k| = |s_k h + 1| = |1 - h\alpha_k + jh\beta_k| < 1, \;\; k = 1, 2, \ldots, r. \tag{1.243}$$

From the inequality (1.243) we have

$$(1 - h\alpha_k)^2 + (h\beta_k)^2 < 1. \tag{1.244}$$

Solving (1.244) with respect to h we obtain (1.242). □

The following theorem can also be used to test the stability of descriptor discrete-time systems.

Theorem 1.35. The descriptor discrete-time linear system (1.167) is asymptotically stable if and only if

$$\lim_{i\to\infty} T_{d,i}E = 0. \tag{1.245}$$

Proof. Substituting $u_i = 0, i \in \mathbb{Z}_+$ into (1.170a) we obtain

$$x_i = T_{d,i}Ex_0. \tag{1.246}$$

By Definition 1.12 we have

$$\lim_{i\to\infty} x_i = \lim_{i\to\infty} [T_{d,i}Ex_0]. \tag{1.247}$$

Therefore, the descriptor system (1.167) is asymptotically stable for $x_0 \in X_0$ if and only if the condition (1.245) is met. □

Example 1.21. (Continuation of Example 1.7) Consider the descriptor continuous-time linear system (1.1) with (1.76). According to (1.106) the eigenvalues of the pair (E, A) are $s_1 = -2$ and $s_2 = 2$ and therefore the system is unstable. The characteristic equation of the pair (E, A_d) of the discrete-time equivalent (1.167) with (1.218) has the form

$$\det[Ez - A_d] = \begin{vmatrix} z-1 & -2h & 0 \\ -2h & z-1 & h \\ 0 & 0 & -h \end{vmatrix} = -hz^2 + 2hz + 4h^3 - h = 0, \tag{1.248}$$

and by (1.241) its roots are

$$z_1 = 2h + 1, \ \ z_2 = -2h + 1, \tag{1.249}$$

From the above considerations it follows that the discrete-time equivalent of the system (1.76) is unstable for any $h > 0$.

Example 1.22. (Continuation of Example 1.8) Consider the descriptor continuous-time linear system (1.1) with (1.107). The system is stable since it has one pole $s = -1.5$. Its discrete-time equivalent (1.167) has one pole $z = -1.5h + 1$. Using (1.242) we have

$$0 < h < \frac{4}{3}. \tag{1.250}$$

For $h = \frac{4}{3}$ we have $z = -1$. Therefore, the discrete-time equivalent (1.167) of the continuous-time system (1.1) with (1.107) is asymptotically stable for the discretization period h satisfying (1.250).

1.2.4.2 Drazin Inverse Matrix Method

In this section necessary and sufficient stability conditions of the descriptor discrete-time linear system (1.182) (or equivalently (1.167)) will be established using the Drazin inverse matrix method. Considerations on the influence of the discretization period h on the stability of the discrete-time equivalent (1.182) of the continuous-time model (1.24) will be presented.

Definition 1.12 of the asymptotic stability of the descriptor system is suitable for this method since the solutions of (1.167) and (1.182) are equivalent.

Definition 1.14. The characteristic equation of the matrix pair $(\bar{E}_d, \bar{A}_d)$ is given by

$$p_{(\bar{E}_d,\bar{A}_d)}(z) = \det[\bar{E}_d z - \bar{A}_d] = \bar{a}_{d,r} z^r + \bar{a}_{d,r-1} z^{r-1} + \ldots + \bar{a}_{d,1} z + \bar{a}_{d,0} = 0, \tag{1.251}$$

where $r = \operatorname{rank} \bar{E}_d^D \bar{A}_d$ and $p_{(\bar{E}_d,\bar{A}_d)}(z)$ is the characteristic polynomial of the matrix pair $(\bar{E}_d, \bar{A}_d)$.

Lemma 1.12. The characteristic polynomials of the matrix pairs (E, A_d) and $(\bar{E}_d, \bar{A}_d)$ are related by

$$p_{(\bar{E}_d,\bar{A}_d)}(z) = \frac{p_{(E,A_d)}(z)}{\det[Ec - A_d]}. \tag{1.252}$$

Proof. The proof is similar to the proof of Lemma 1.7. □

Remark 1.9. The characteristic equations of the matrix pairs (E, A_d) and $(\bar{E}_d, \bar{A}_d)$ have the same form. Therefore, both pairs have the same set of eigenvalues and Theorem 1.33 can be used for the roots of the Eq. (1.251).

Lemma 1.13. The matrix $\bar{E}_d^D \bar{A}_d$ has r eigenvalues of the pair $(\bar{E}_d, \bar{A}_d)$ (or (E, A_d)) and additionally $n - r$ zero eigenvalues, i.e. its characteristic equation has the form

$$p_{(\bar{E}_d^D,\bar{A}_d)}(z) = \det[\mathbb{I}_n z - \bar{E}_d^D \bar{A}_d] = z^{n-r} p_{(\bar{E}_d,\bar{A}_d)}(z) = 0, \tag{1.253}$$

where $p_{(\bar{E}_d^D,\bar{A}_d)}(z)$ is the characteristic polynomial of the matrix $\bar{E}_d^D \bar{A}_d$.

Proof. The proof is similar to the proof of Lemma 1.8. □

Theorem 1.36. The descriptor discrete-time linear system (1.167) is asymptotically stable if and only if the matrix $\bar{E}_d^D \bar{A}_d$ has r stable eigenvalues (satisfying the condition (1.239)) and $n - r$ zero eigenvalues.

Proof. The proof follows immediately from Lemma 1.13. □

To determine such a value of the discretization period h, for which the discrete-time equivalent (1.182) of the stable continuous-time system (1.24) is also stable we can use (1.242) for r stable eigenvalues of the matrix $\bar{E}_d^D \bar{A}_d$.

Example 1.23. (Continuation of Examples 1.9 and 1.19) Consider the descriptor continuous-time linear system (1.1) with (1.76). According to (1.123) the matrix $\bar{E}^D\bar{A}$ has $r = 2$ eigenvalues $s_1 = -2, s_2 = 2$ and $n - r = 1$ eigenvalue $s_3 = 0$. Thus, the considered system is unstable. The matrix $\bar{E}_d^D\bar{A}_d$ of the discrete-time equivalent (1.182) is given by (1.229b) and its characteristic equation has the form

$$p_{\bar{E}_d^D\bar{A}_d}(z) = \det[\mathbb{I}_3 z - \bar{E}_d^D\bar{A}_d] = \begin{vmatrix} z-1 & -2h & 0 \\ -2h & z-1 & 0 \\ 0 & 0 & z \end{vmatrix} \\ = z^3 - 2z^2 + (1 - 4h^2)z = 0. \tag{1.254}$$

According to (1.254) the matrix $\bar{E}_d^D\bar{A}_d$ has $r = 2$ eigenvalues $z_1 = -2h + 1$, $z_2 = 2h + 1$ and $n - r = 1$ eigenvalue $z_3 = 0$. Therefore, the discrete-time equivalent of the continuous-time system is unstable for any $h > 0$.

1.2.4.3 Weierstrass-Kronecker Decomposition Method

In this section necessary and sufficient stability conditions of the descriptor discrete-time linear system (1.195) (or equivalently (1.167)) will be established using the Weierstrass-Kronecker decomposition method. Considerations on the influence of the discretization period h on the stability of the discrete-time equivalent (1.208) of the continuous-time model (1.55) will be presented.

Note that the state vector of the subsystem (1.195b) $\bar{x}_{2,i} = 0$ for $u_i = 0$, $i \in \mathbb{Z}_+$. Therefore, the stability of the system (1.195) (or equivalently (1.167)) depends entirely on the subsystem (1.195a).

Definition 1.15. The descriptor discrete-time linear system (1.195) (or equivalently (1.167)) is called asymptotically stable if

$$\lim_{i\to\infty} \bar{x}_{1,i} = 0 \tag{1.255}$$

for all initial conditions $\bar{x}_{10} \in \mathbb{R}^{n_1}$ and $u_i = 0$, $i \in \mathbb{Z}_+$.

Definition 1.16. [47] The characteristic equation of the matrix A_{1d} is given by

$$p_{A_{1d}}(z) = \det[\mathbb{I}_{n_1} z - A_{1d}] = z^{n_1} + \tilde{a}_{d,n_1-1}z^{n_1-1} + \ldots + \tilde{a}_{d,1}z + \tilde{a}_{d,0} = 0, \tag{1.256}$$

where $n_1 = \deg\{\det[Ez - A_d]\}$ and $p_{A_{d1}}(z)$ is the characteristic polynomial of the matrix A_{1d}.

Lemma 1.14. The characteristic polynomials of the matrix pair (E, A_d) and of the matrix A_{1d} are related by

$$p_{(E,A_d)}(z) = \frac{(-1)^{n_2}}{\det(P_d Q_d)} p_{A_{1d}}(z). \tag{1.257}$$

Proof. The proof is similar to the proof of Lemma 1.9. □

Remark 1.10. The characteristic equations of the matrix pair (E, A_d) and of the matrix A_{1d} have the same form. Therefore, the matrix pair (E, A_d) and the matrix A_{1d} have the same set of eigenvalues and Theorem 1.33 can be used for the roots of the Eq. (1.256).

Theorem 1.37. [47] The matrix A_{1d} (or equivalently the system (1.195)) is asymptotically stable if and only if its eigenvalues satisfy the condition (1.239).

In a similar way we can analyze the system (1.208). For $u_i = 0$, $i \in \mathbb{Z}_+$ we have $\tilde{x}_{2,i} = 0$, $i \in \mathbb{Z}_+$ and the stability of the system (1.208) (or equivalently (1.167)) depends entirely on the vector $\tilde{x}_{1,i}$, $i \in \mathbb{Z}_+$. From (1.205) we have

$$\det P[Ez - (Ah + E)]Q = \det\begin{bmatrix} \mathbb{I}_{n_1}z - (A_1h + \mathbb{I}_{n_1}) & 0 \\ 0 & Nz - (\mathbb{I}_{n_2}h + N) \end{bmatrix} \tag{1.258}$$

and

$$\det(PQ)h^n \det\left[E\frac{z-1}{h} - A\right] = (-1)^{n_2}h^n \det\left[\mathbb{I}_{n_1}\frac{z-1}{h} - A_1\right], \tag{1.259}$$

since

$$\det[Nz - (\mathbb{I}_{n_2}h + N)] = (-h)^{n_2} = (-1)^{n_2}h^{n_2}, \tag{1.260a}$$

$$\det[\mathbb{I}_{n_1}z - (A_1h + \mathbb{I}_{n_1})] = h^{n_1}\det\left[\mathbb{I}_{n_1}\frac{z-1}{h} - A_1\right], \tag{1.260b}$$

$$h^{n_1}h^{n_2} = h^{n_1+n_2} = h^n. \tag{1.260c}$$

From (1.259) we obtain

$$\det\left[E\frac{z-1}{h} - A\right] = \frac{(-1)^{n_2}}{\det(PQ)}\det\left[\mathbb{I}_{n_1}\frac{z-1}{h} - A_1\right]. \tag{1.261}$$

Comparing (1.126) and (1.261) we obtain the formula (1.241). To test the stability of the discrete-time equivalent (1.208) of the stable continuous-time system (1.55) we can use Theorem 1.34 for the eigenvalues of the matrix A_1.

Example 1.24. (Continuation of Examples 1.6 and 1.10) Consider the descriptor continuous-time linear system (1.55) with (1.97). According to (1.129) the eigenvalues of the matrix A_1 are $s_1 = -2$ i $s_2 = 2$. The matrix $A_1h + \mathbb{I}_2$ of the discrete-time equivalent (1.208) has the form (1.235) and its characteristic equation is given by

$$p_{(A_1h+\mathbb{I}_2)}(z) = \det[\mathbb{I}_2 z - (A_1h + \mathbb{I}_2)] = \begin{vmatrix} z-1 & -2h \\ -2h & z-1 \end{vmatrix} = z^2 - 2z + 1 - 4h^2. \tag{1.262}$$

From (1.262) we have $z_1 = -2h + 1$ and $z_2 = 2h + 1$. Therefore, the discrete-time equivalent of the continuous-time system is unstable for any $h > 0$.

1.2.5 Superstability of Descriptor Discrete-Time Linear Systems

The asymptotic stability of a dynamical system ensures that its free response decreases to zero for $i \to \infty$, however its value may increase significantly in the initial part of the state vector trajectory (it is so-called "peak effect"). In superstable systems the norm of the state vector decreases monotonically to zero for $i \to \infty$, which prevents such undesirable effects [70, 137, 138].

1.2.5.1 Properties of Superstable Discrete-Time Linear Systems

In further considerations we will use the norms (1.130) and (1.131) defined in Sect. 1.1.5.1. Consider the standard discrete-time linear system

$$x_{i+1} = A_d x_i, \quad i \in \mathbb{Z}_+, \tag{1.263}$$

where $x_i \in \mathbb{R}^n$ is the state vector and $A_d = [a_{d,jk}] \in \mathbb{R}^{n\times n}$.

Definition 1.17. [137] The matrix $A_d \in \mathbb{R}^{n\times n}$ of the discrete-time linear system (1.263) is called superstable if

$$\sigma_d(A_d) = \sigma_d = 1 - \|A_d\| > 0. \tag{1.264a}$$

or equivalently

$$\|A_d\| < 1. \tag{1.264b}$$

Quantity σ_d is called the superstability degree of the matrix A_d. If the matrix is superstable, then it is also stable, however the reverse implication does not hold.

Theorem 1.38. [137] If the system (1.263) is superstable, then

$$\|x_i\| \leq \sigma_d^i \|x_0\|, \quad i \in \mathbb{Z}_+. \tag{1.265}$$

Note that the superstability ensure monotonic decrease of the norm of the state vector, however particular state variables can oscillate. The main difference is that for asymptotic stable systems the Eq. (1.265) is replaced by

$$\|x_i\| \leq c(\varepsilon)(\rho + \varepsilon)^i \|x_0\|, \quad \varepsilon > 0, \quad \rho + \varepsilon < 1, \tag{1.266}$$

where the constant $c(\varepsilon)$ can take significant values in the initial part of the state vector trajectory. In superstable systems such a situation does not occur [137].

Now let us consider the discrete-time equivalent of the continuous-time system (1.132). Discretizing the Eq. (1.132) and using the approximation (1.166) we obtain (1.263), where

$$A_d = Ah + \mathbb{I}_n. \tag{1.267}$$

Lemma 1.15. For the matrix norm $\|A\|$ defined by (1.131) and the coefficient σ determined by (1.133) the following relationships are true:

$$\sigma_h = \sigma h - 1, \tag{1.268a}$$

$$|\sigma| \leq \|A\| \text{ and } |\sigma_h| \leq \|Ah + \mathbb{I}_n\|, \tag{1.268b}$$

$$\|A\| < 2|a_{ii}^{\|A\|}| \text{ for } a_{ii}^{\|A\|} < 0, \ i = 1, \ldots, n \text{ and } \sigma > 0, \tag{1.268c}$$

where $\sigma_h = \sigma(Ah + \mathbb{I}_n)$ is the quantity (1.133) computed for the matrix (1.267) and $a_{ii}^{\|A\|}$ is the element on the main diagonal of the matrix A in its i-th row corresponding to its norm.

Proof. Using (1.133) for the matrix (1.267) we obtain

$$\begin{aligned} \sigma_h &= \min_{1\leq i\leq n}\left(-(a_{ii}h+1) - \sum_{\substack{j=1\\ j\neq i}}^{n} |a_{ij}h|\right) \\ &= h \min_{1\leq i\leq n}\left(-a_{ii} - \sum_{\substack{j=1\\ j\neq i}}^{n} |a_{ij}|\right) - 1 = \sigma h - 1. \end{aligned} \tag{1.269}$$

From (1.133) and (1.131) we have

$$|\sigma| = \left|\min_{1\leq i\leq n}\left(-a_{ii} - \sum_{\substack{j=1\\ j\neq i}}^{n} |a_{ij}|\right)\right| = \left|-\max_{1\leq i\leq n}\left(a_{ii} + \sum_{\substack{j=1\\ j\neq i}}^{n} |a_{ij}|\right)\right| \tag{1.270}$$

and

$$\|A\| = \max_{1\leq i\leq n}\left(\sum_{j=1}^{n} |a_{ij}|\right) = \max_{1\leq i\leq n}\left(|a_{ii}| + \sum_{\substack{j=1\\ j\neq i}}^{n} |a_{ij}|\right). \tag{1.271}$$

Basing on (1.270) and (1.271) we have $\|A\| = -\sigma$ for $a_{ii} \geq 0$, $i = 1, \ldots, n$ and $\|A\| > \sigma$ for $a_{ii} < 0$, $i = 1, \ldots, n$. Therefore, $|\sigma| \leq \|A\|$. In a similar way it can be shown that $|\sigma_h| \leq \|Ah + \mathbb{I}_n\|$.

Using again (1.271) we have

$$\|A\| = \max_{1 \leq i \leq n} \left(|a_{ii}| + \sum_{\substack{j=1 \\ j \neq i}}^{n} |a_{ij}| \right) = |a_{ii}^{\|A\|}| + \sum_{\substack{j=1 \\ j \neq i}}^{n} |a_{ij}^{\|A\|}|, \tag{1.272}$$

where $a_{ij}^{\|A\|}, i, j = 1, \ldots, n$ are components of the norm of the matrix A. If the system (1.132) is superstable, then

$$|a_{ii}^{\|A\|}| > \sum_{\substack{j=1 \\ j \neq i}}^{n} |a_{ij}^{\|A\|}|. \tag{1.273}$$

Taking into account (1.272) and (1.273) we obtain the inequality (1.268c). □

Theorem 1.39. If the continuous-time linear system (1.132) is not superstable, then its discrete-time equivalent (1.263) with (1.267) is also not superstable.

Proof. If the continuous-time system (1.132) is not superstable, then $\sigma < 0$ and $|\sigma_h| = |\sigma h - 1| > 1$ and therefore from (1.268b) we have $\|Ah + \mathbb{I}_n\| \geq |\sigma_h| > 1$. According to (1.264b) the discrete-time equivalent (1.263) of the system (1.132) is also not superstable. □

Theorem 1.40. If the continuous-time linear system (1.132) is superstable, then its discrete-time equivalent (1.263) with (1.267) is also superstable if and only if

$$0 < h < \frac{2}{\|A\|}. \tag{1.274}$$

Proof. We shall show that if the continuous-time system (1.132) is superstable, then

$$\|Ah + \mathbb{I}_n\| < 1 \tag{1.275}$$

for h satisfying (1.274). Consider the discretization period h equal to

$$h = \frac{p}{\|A\|}, \tag{1.276}$$

where $p \in \mathbb{R}_+$. Substituting (1.276) into (1.275) we obtain

$$\left\| \frac{pA}{\|A\|} + \mathbb{I}_n \right\| = \max_{1 \le i \le n} \left(\left| \frac{p}{\|A\|} a_{ii} + 1 \right| + \sum_{\substack{j=1 \\ j \neq i}}^{n} \frac{p}{\|A\|} |a_{ij}| \right)$$
$$= \max_{1 \le i \le n} \left(\left| \frac{p}{\|A\|} a_{ii} + 1 \right| - \frac{p}{\|A\|} |a_{ii}| + p \right) \tag{1.277}$$

If the system (1.132) is superstable, then the condition (1.268c) holds. In this case the expression (1.277):

1) is equal to 1 for $p = 2$,
2) is greater than 1 for $p > 2$,
3) is less than 1 for $p \in (0, 2)$.

Therefore, the discrete-time equivalent of the superstable continuous-time system (1.132) is also superstable for h satisfying (1.274). □

Example 1.25. (Continuation of Example 1.11) Consider the superstable continuous-time linear system (1.132) with (1.137), which norm (1.131) is $\|A\| = 5$. According to (1.274) the discrete-time equivalent (1.263) of the system with

$$A_d = Ah + \mathbb{I}_2 = \begin{bmatrix} -3h + 1 & 2h \\ 0 & -2h + 1 \end{bmatrix} \tag{1.278}$$

will also be superstable if and only if

$$0 < h < 0.4. \tag{1.279}$$

It is easy to see that for $h = 0.4$ we have

$$A_d = 0.4A + \mathbb{I}_2 = \begin{bmatrix} -0.2 & 0.8 \\ 0 & 0.2 \end{bmatrix}, \quad \|A_d\| = 1. \tag{1.280}$$

We will consider two cases. First we assume $x_0 = [\,1 \quad 2\,]^T$ and $h = 0.3$. By (1.279) the system is superstable. From Fig. 1.4 we can see that the norm of the state vector decreases monotonically.

In second case we assume $x_0 = [\,1 \quad 2\,]^T$ and $h = 0.6$. By (1.279) the system is not superstable. From Fig. 1.5 we can see that the norm of the state vector does not decrease monotonically and the state variable $x_{1,i}$ highly oscillate.

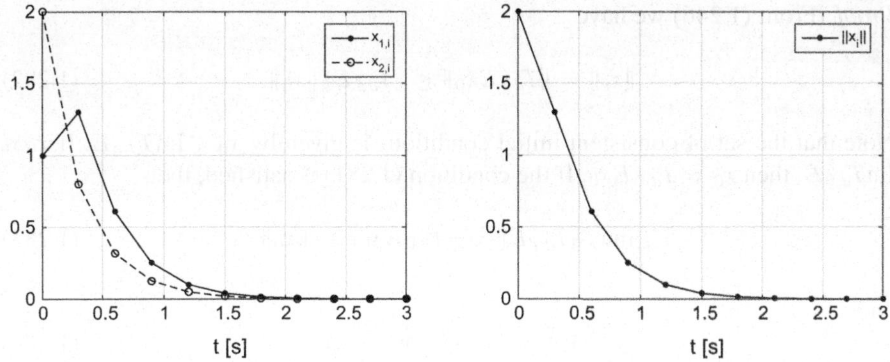

Fig. 1.4 State variables (on the left) and norm of the state vector (on the right) of the system (1.263) with (1.278) for $x_0 = [\,1 \quad 2\,]^T$ and $h = 0.3$

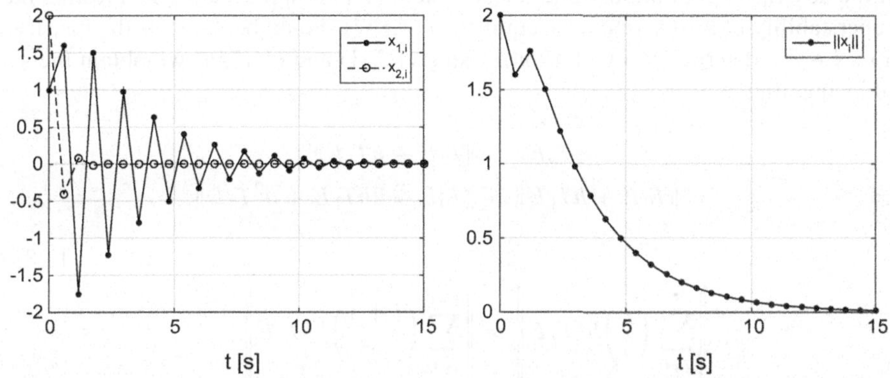

Fig. 1.5 State variables (on the left) and norm of the state vector (on the right) of the system (1.263) with (1.278) for $x_0 = [\,1 \quad 2\,]^T$ and $h = 0.6$

1.2.5.2 Laurent Series Expansion Method

In this section necessary and sufficient superstability conditions of the descriptor discrete-time linear system (1.167) will be established using the Laurent series expansion method. Considerations on the influence of the discretization period h on the superstability of the discrete-time equivalent (1.167) of the continuous-time model (1.1) will be presented.

Consider the descriptor discrete-time linear system (1.167). Solution to the Eq. (1.167a) for $u_i = 0, i \in \mathbb{Z}_+$ is given by (1.246).

Theorem 1.41. The descriptor discrete-time linear system (1.167) is superstable if and only if

$$\|T_{d,i}E\| > \|T_{d,i+1}E\| \text{ for } i \in \mathbb{Z}_+. \tag{1.281}$$

Proof. From (1.246) we have

$$\|x_i\| = \|T_{d,i}Ex_0\| \leq \|T_{d,i}E\|\|x_0\|. \tag{1.282}$$

Note that the set of consistent initial conditions is given by $x_0 \in \mathrm{Im}T_{d,0}E$. If $x_0 \in \mathrm{Im}T_{d,0}E$, then $x_0 = T_{d,0}Ex_0$. If the condition (1.281) is satisfied, then

$$\max_i \|T_{d,i}E\| = \|T_{d,0}Ex_0\| = \|x_0\| \tag{1.283}$$

and

$$\|x_i\| > \|x_{i+1}\| \text{ for } i \in \mathbb{Z}_+. \tag{1.284}$$

Therefore, the norm of the state vector decreases monotonically to zero. □

If the descriptor continuous-time linear system (1.1) is superstable, then testing the superstability of its discrete-time equivalent (1.167) should be done for the matrices $T_{d,i}$, $i \in \mathbb{Z}_+$ determined by (1.175a). Using (1.281) and (1.175a) we obtain the set of inequalities

$$\begin{aligned}
\|T_0E\| &> \|T_0E + hT_1E\|,\\
\|T_0E + hT_1E\| &> \|T_0E + 2hT_1E + h^2T_2E\|,\\
&\vdots\\
\left\|\sum_{j=0}^{i}\binom{i}{j}h^jT_jE\right\| &> \left\|\sum_{j=0}^{i+1}\binom{i+1}{j}h^jT_jE\right\|,
\end{aligned} \tag{1.285}$$

where T_k are the matrices of the continuous-time system. The discrete-time equivalent (1.167) of the continuous-time system (1.1) is superstable for such a value of discretization period h, for which the inequalities (1.285) are satisfied.

Example 1.26. (Continuation of Example 1.12) Consider the superstable descriptor continuous-time linear system (1.143). The matrices T_kE are given by (1.147). Analyzing the form of these matrices and using (1.285) we can check that for $h = 0.5$ we have

$$\left\|\sum_{j=0}^{i}\binom{i}{j}h^jT_jE\right\| = \left\|\sum_{j=0}^{i+1}\binom{i+1}{j}h^jT_jE\right\|, \quad i = 0, 2, 4, \ldots \tag{1.286}$$

and for

$$0 < h < 0.5 \tag{1.287}$$

we obtain

$$\left\| \sum_{j=0}^{i} \binom{i}{j} h^j T_j E \right\| > \left\| \sum_{j=0}^{i+1} \binom{i+1}{j} h^j T_j E \right\|, \quad i \in \mathbb{Z}_+. \tag{1.288}$$

Therefore, the discrete-time equivalent of the superstable continuous-time linear system (1.143) is also superstable for h satisfying (1.287). Taking into account (1.171) and the fact that $T_{d,0} = T_0$ consistent initial conditions of the system can be computed from (1.150).

In Figs. 1.6 and 1.7 we present time plots of the state variables and the norm of the state vector for $x_0 = [\,2 \;\; 3 \;\; 0\,]^T$ and two values of the discretization period: $h = 0.3$ satisfying (1.287) and $h = 0.6$ not satisfying (1.287). In the first case the norm of

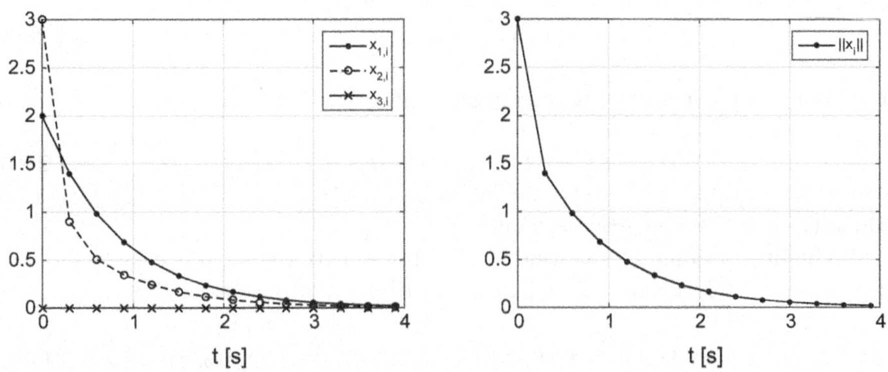

Fig. 1.6 State variables (on the left) and norm of the state vector (on the right) of the discrete-time equivalent (1.167a) of the continuous-time system (1.1a) with (1.143) for $x_0 = [\,2 \;\; 3 \;\; 0\,]^T$ and $h = 0.3$

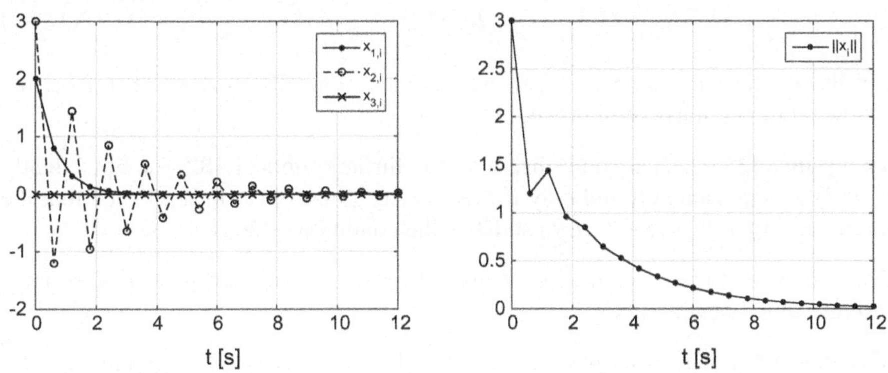

Fig. 1.7 State variables (on the left) and norm of the state vector (on the right) of the discrete-time equivalent (1.167a) of the continuous-time system (1.1a) with (1.143) for $x_0 = [\,2 \;\; 3 \;\; 0\,]^T$ and $h = 0.6$

the state vector decreases monotonically to zero and the discrete-time equivalent preserves superstability (Fig. 1.6), while in the second case too high value of the discretization period causes the system to lose this property (Fig. 1.7).

1.2.5.3 Drazin Inverse Matrix Method

In this section necessary and sufficient superstability conditions of the descriptor discrete-time linear system (1.182) (or equivalently (1.167)) will be established using the Drazin inverse matrix method. Considerations on the influence of the discretization period h on the superstability of the discrete-time equivalent (1.182) of the continuous-time model (1.24) will be presented.

Consider the descriptor discrete-time linear system (1.182). Solution to the Eq. (1.182a) for $u_i = 0$, $i \in \mathbb{Z}_+$ has the form

$$x_i = (\bar{E}_d^D \bar{A}_d)^i \bar{E}_d^D \bar{E}_d x_0. \tag{1.289}$$

The Eq. (1.289) can also be written in the form

$$x_i = [\bar{E}_d^D \bar{A}_d + G_d(\mathbb{I}_n - \bar{E}_d \bar{E}_d^D)]^i \bar{E}_d^D \bar{E}_d x_0, \tag{1.290}$$

where $G_d \in \mathbb{R}^{n\times n}$ is an arbitrary matrix.

Taking into account (1.183) and (1.184) for $u_i = 0$, $i \in \mathbb{Z}_+$ it is easy to see that if $x_0 \in \mathrm{Im}\bar{E}_d\bar{E}_d^D$, then $x_0 = \bar{E}_d\bar{E}_d^D x_0$. Thus, from (1.290) we have

$$\|x_i\| \le \|[\bar{E}_d^D \bar{A}_d + G_d(\mathbb{I}_n - \bar{E}_d \bar{E}_d^D)]^i\| \|x_0\| \le \|\bar{E}_d^D \bar{A}_d + G_d(\mathbb{I}_n - \bar{E}_d \bar{E}_d^D)\|^i \|x_0\|. \tag{1.291}$$

If the matrix $\bar{E}_d^D \bar{A}_d + G_d(\mathbb{I}_n - \bar{E}_d \bar{E}_d^D)$ is superstable (i.e. it satisfies the condition (1.264)), then from the inequality (1.291) we obtain

$$\|x_i\| \le \|\bar{E}_d^D \bar{A}_d + G_d(\mathbb{I}_n - \bar{E}_d \bar{E}_d^D)\|^i \|x_0\| \le \sigma_d^i \|x_0\|, \quad i \in \mathbb{Z}_+ \tag{1.292}$$

and the norm of the state vector decreases monotonically to zero. Therefore, the following theorem has been proved.

Theorem 1.42. The descriptor discrete-time linear system (1.182) (or equivalently (1.167)) is superstable if and only if there exists a matrix $G_d \in \mathbb{R}^{n\times n}$ such that the matrix $\bar{E}_d^D \bar{A}_d + G_d(\mathbb{I}_n - \bar{E}_d \bar{E}_d^D)$ satisfies the condition (1.264).

The following theorem can also be used to test the superstability of descriptor discrete-time linear systems.

Theorem 1.43. The descriptor discrete-time linear system (1.182) (or equivalently (1.167)) is superstable if and only if

$$\|\bar{E}_d^D \bar{A}_d\| < \|\bar{E}_d \bar{E}_d^D\|. \tag{1.293}$$

Proof. Using (1.27d) and (1.293) we can write

$$\left\| T \begin{bmatrix} \hat{E}_d & 0 \\ 0 & 0 \end{bmatrix} T^{-1} \right\| < \left\| T \begin{bmatrix} \mathbb{I}_{n_1} & 0 \\ 0 & 0 \end{bmatrix} T^{-1} \right\|, \tag{1.294}$$

where $\hat{E}_d = J_d^{-1} A_{1d}$. Given that both matrices in the inequality (1.294) are transformed by the same similarity transformation, this condition does not depend on the matrix T. Consider $T = \mathbb{I}_n$. Therefore, from (1.294) we have $\|\hat{E}_d\| < \|\mathbb{I}_{n_1}\| = 1$, which is similar to (1.264b). In the general case for any matrix T we obtain the formula (1.293). □

Now let us consider the descriptor continuous-time linear system (1.24) and assume that it is superstable, i.e. there exists a matrix $G \in \mathbb{R}^{n \times n}$ such that the matrix $\bar{E}^D \bar{A} + G(\mathbb{I}_n - \bar{E}\bar{E}^D)$ satisfies the condition (1.133). Using (1.224) and (1.293) we obtain

$$\|h\bar{E}^D \bar{A} + \bar{E}\bar{E}^D\| < \|\bar{E}\bar{E}^D\|. \tag{1.295}$$

Theorem 1.44. If the descriptor continuous-time linear system (1.24) (or equivalently (1.1)) is superstable, i.e. there exists a matrix $G \in \mathbb{R}^{n \times n}$ such that the matrix $\bar{E}^D \bar{A} + G(\mathbb{I}_n - \bar{E}\bar{E}^D)$ satisfies the condition (1.133), then its discrete-time counterpart (1.182) (or equivalently (1.167)) is also superstable if and only if

$$0 < h < \frac{2\|\bar{E}\bar{E}^D\|}{\|\bar{E}^D \bar{A}\|}. \tag{1.296}$$

Proof. We shall show that if the descriptor continuous-time linear system (1.24) is superstable, then (1.295) holds for h satisfying (1.296). Consider the discretization period h equal to

$$h = \frac{p\|\bar{E}\bar{E}^D\|}{\|\bar{E}^D \bar{A}\|}, \tag{1.297}$$

where $p \in \mathbb{R}_+$. Substituting (1.297) into (1.295) we obtain

$$\left\| \frac{p\|\bar{E}\bar{E}^D\|}{\|\bar{E}^D \bar{A}\|} \bar{E}^D \bar{A} + \bar{E}\bar{E}^D \right\| = \|\bar{E}\bar{E}^D\| \left\| \frac{p\bar{E}^D \bar{A}}{\|\bar{E}^D \bar{A}\|} + \frac{\bar{E}\bar{E}^D}{\|\bar{E}\bar{E}^D\|} \right\|. \tag{1.298}$$

Using (1.27d) and (1.298) we have

$$\left\| \frac{p\bar{E}^D \bar{A}}{\|\bar{E}^D \bar{A}\|} + \frac{\bar{E}\bar{E}^D}{\|\bar{E}\bar{E}^D\|} \right\| = \left\| \frac{pT \begin{bmatrix} \hat{E}_d & 0 \\ 0 & 0 \end{bmatrix} T^{-1}}{\left\| T \begin{bmatrix} \hat{E}_d & 0 \\ 0 & 0 \end{bmatrix} T^{-1} \right\|} + \frac{T \begin{bmatrix} \mathbb{I}_{n_1} & 0 \\ 0 & 0 \end{bmatrix} T^{-1}}{\left\| T \begin{bmatrix} \mathbb{I}_{n_1} & 0 \\ 0 & 0 \end{bmatrix} T^{-1} \right\|} \right\|, \tag{1.299}$$

where $\hat{E}_d = J_d^{-1} A_{1d}$. From (1.299) it follows that the value of this norm does not depend on the matrix T. Consider $T = \mathbb{I}_n$. Therefore, from (1.299) we have

$$\left\| \frac{p\hat{E}_d}{\|\hat{E}_d\|} + \mathbb{I}_{n_1} \right\|, \tag{1.300}$$

and the expression (1.300):

1) is equal to 1 for $p = 2$,
2) is greater than 1 for $p > 2$,
3) is less than 1 for $p \in (0, 2)$,

which follows from the proof of Theorem 1.40. Hence, the expression (1.298):

1) is equal to $\|\bar{E}\bar{E}^D\|$ for $p = 2$,
2) is greater than $\|\bar{E}\bar{E}^D\|$ for $p > 2$,
3) is less than $\|\bar{E}\bar{E}^D\|$ for $p \in (0, 2)$.

Therefore, the discrete-time equivalent of the superstable descriptor continuous-time linear system (1.24) is also superstable for h satisfying (1.296). □

Example 1.27. (Continuation of Example 1.13) Consider the superstable descriptor continuous-time linear system (1.24) with the matrices $\bar{E}$ and $\bar{A}$ given by (1.155). We will search for such a range of the discretization period h, for which the discrete-time equivalent (1.182) of the system is also superstable. Taking into account that $\|\bar{E}\bar{E}^D\| = 1$, $\|\bar{E}^D\bar{A}\| = 4$ and using (1.296) we obtain the range of the discretization period

$$0 < h < 0.5. \tag{1.301}$$

Note that the conditions (1.287) and (1.301) are the same. It is easy to check that for $h = 0.5$ we have $\|h\bar{E}^D\bar{A} + \bar{E}\bar{E}^D\| = \|\bar{E}\bar{E}^D\| = 1$, while for h satisfying (1.301) we obtain $\|h\bar{E}^D\bar{A} + \bar{E}\bar{E}^D\| < \|\bar{E}\bar{E}^D\| = 1$. Therefore, the discrete-time equivalent (1.182) of the superstable continuous-time system (1.24) with (1.155) is also superstable if the condition (1.301) is satisfied. Consistent initial conditions of the system are determined by (1.150). The time plots of the state variables and the norm of the state vector for $x_0 = [\,2 \;\; 3 \;\; 0\,]^T$ and two values of the discretization period: $h = 0.3$ satisfying (1.301) and $h = 0.6$ not satisfying (1.301) are presented in Figs. 1.6 and 1.7.

1.2.5.4 Weierstrass-Kronecker Decomposition Method

In this section necessary and sufficient superstability conditions of the descriptor discrete-time linear system (1.195) (or equivalently (1.167)) will be established using the Weierstrass-Kronecker decomposition method. Considerations on the influence of the discretization period h on the superstability of the discrete-time equivalent (1.208) of the continuous-time model (1.55) will be presented.

Note that every superstable system is also asymptotically stable. State vector of the subsystem (1.195b) $\bar{x}_{2,i} = 0$ for $u_i = 0$, $i \in \mathbb{Z}_+$. Using (1.193) and (1.197) we obtain

$$x_i = Q_d \bar{x}_i = Q_d \begin{bmatrix} \bar{x}_{1,i} \\ \bar{x}_{2,i} \end{bmatrix} = Q_d \begin{bmatrix} A_{1d}^i \bar{x}_{10} \\ 0 \end{bmatrix}$$
$$= Q_d \begin{bmatrix} A_{1d}^i & 0 \\ 0 & M_d^i \end{bmatrix} \begin{bmatrix} \bar{x}_{10} \\ 0 \end{bmatrix} \tag{1.302}$$
$$= Q_d \begin{bmatrix} A_{1d}^i & 0 \\ 0 & M_d^i \end{bmatrix} Q_d^{-1} x_0 = A_{Md}^i x_0,$$

where the matrix $M_d \in \mathbb{R}^{n_2 \times n_2}$ is arbitrary and

$$A_{Md} = Q_d \begin{bmatrix} A_{1d} & 0 \\ 0 & M_d \end{bmatrix} Q_d^{-1}. \tag{1.303}$$

From (1.302) we have

$$\|x_i\| = \|A_{Md}^i x_0\| \le \|A_{Md}^i\| \|x_0\| \le \|A_{Md}\|^i \|x_0\| \tag{1.304}$$

If the matrix A_{Md} is superstable (i.e. it satisfies the condition (1.264)), then from inequality (1.304) we obtain

$$\|x_i\| \le \|A_{Md}\|^i \|x_0\| \le \sigma_d^i \|x_0\|, \quad i \in \mathbb{Z}_+, \tag{1.305}$$

and the norm of the state vector decreases monotonically to zero. Therefore, the following theorem has been proved.

Theorem 1.45. The descriptor discrete-time linear system (1.195) (or equivalently (1.167)) is superstable if and only if there exists a matrix $M_d \in \mathbb{R}^{n_2 \times n_2}$ such that the matrix A_{Md} determined by (1.303) satisfies the condition (1.264).

The following theorem can also be used to test the superstability of descriptor discrete-time linear systems.

Theorem 1.46. Let

$$A_{Md,0} = Q_d \begin{bmatrix} A_{1d} & 0 \\ 0 & 0 \end{bmatrix} Q_d^{-1}, \quad Q_I = Q \begin{bmatrix} \mathbb{I}_{n_1} & 0 \\ 0 & 0 \end{bmatrix} Q^{-1}. \tag{1.306}$$

The descriptor discrete-time linear system (1.195) (or equivalently (1.167)) is superstable if and only if

$$\|A_{Md,0}\| < \|Q_I\|. \tag{1.307}$$

Proof. The proof is similar to the proof of Theorem 1.43. □

Now let us consider the descriptor continuous-time linear system (1.55) and assume that it is superstable, i.e. there exists a matrix $M \in \mathbb{R}^{n_2 \times n_2}$ such that the matrix A_M

determined by (1.160) satisfies the condition (1.133). Using (1.206) and (1.209) we obtain

$$\begin{aligned} x_i = Q\tilde{x}_i = Q\begin{bmatrix} \tilde{x}_{1,i} \\ \tilde{x}_{2,i} \end{bmatrix} = Q\begin{bmatrix} (A_1h + \mathbb{I}_{n_1})^i \tilde{x}_{10} \\ 0 \end{bmatrix} \\ = Q\begin{bmatrix} (A_1h + \mathbb{I}_{n_1})^i & 0 \\ 0 & M^i \end{bmatrix}\begin{bmatrix} \tilde{x}_{10} \\ 0 \end{bmatrix} \\ = Q\begin{bmatrix} (A_1h + \mathbb{I}_{n_1})^i & 0 \\ 0 & M^i \end{bmatrix} Q^{-1}x_0 = \bar{A}_M^i x_0, \end{aligned} \tag{1.308}$$

where the matrix $M \in \mathbb{R}^{n_2 \times n_2}$ is arbitrary and

$$\bar{A}_M = Q\begin{bmatrix} A_1h + \mathbb{I}_{n_1} & 0 \\ 0 & M \end{bmatrix} Q^{-1}. \tag{1.309}$$

Let $A_{M,0}$ and $\bar{A}_{M,0}$ be the matrices defined by (1.160) and (1.309) for $M = 0$, respectively. Thus, we have

$$\bar{A}_{M,0} = hA_{M,0} + Q_I, \tag{1.310}$$

where the matrix Q_I is determined by (1.306). Using (1.307) and (1.310) we obtain

$$\|hA_{M,0} + Q_I\| < \|Q_I\|. \tag{1.311}$$

Theorem 1.47. If the descriptor continuous-time linear system (1.55) (or equivalently (1.1)) is superstable, i.e. there exists a matrix $M \in \mathbb{R}^{n_2 \times n_2}$ such that the matrix A_M defined by (1.160) satisfies the condition (1.133), then its discrete-time counterpart (1.208) (or equivalently (1.167)) is also superstable if and only if

$$0 < h < \frac{2\|Q_I\|}{\|A_{M,0}\|}. \tag{1.312}$$

Proof. The proof is similar to the proof of Theorem 1.44. □

Example 1.28. (Continuation of Example 1.14) Consider the superstable descriptor continuous-time linear system (1.55) with the matrix A_1 determined by (1.164). The matrix Q has the form (1.163) and $A_{M,0}$ can be computed from (1.165) for $M = 0$. We will search for such a range of the discretization period h, for which the discrete-time equivalent (1.208) of the system is also superstable. Taking into account that $\|A_{M,0}\| = 4$, $\|Q_I\| = 1$ and using (1.312) we obtain the range of the discretization period

$$0 < h < 0.5. \tag{1.313}$$

Note that the conditions (1.287), (1.301) and (1.313) are the same. It is easy to check that for $h = 0.5$ we have $\|hA_{M,0} + Q_I\| = \|Q_I\| = 1$, while for h satisfying

(1.313) we obtain $\|hA_{M,0} + Q_I\| < \|Q_I\| = 1$. Therefore, the discrete-time equivalent (1.208) of the superstable continuous-time system (1.55) with (1.164) is also superstable if the condition (1.313) is satisfied.

1.3 Concluding Remarks

In this chapter descriptor continuous-time and discrete-time linear systems have been analyzed. In the subsequent sections solutions of the state equation and analytical conditions of the positivity, stability and superstability have been established considering the Laurent series expansion method, Drazin inverse matrix method and Weierstrass-Kronecker decomposition method. In addition, for each of the methods and for the assumed approximation of the derivative (Euler method) formulae linking together the matrices of the continuous-time model and its discrete-time counterpart have been given and the influence of the discretization period on the properties of descriptor systems has been examined.

First of all, it should be noted that each of considered methods requires some preliminary calculations or transformations. The algorithmization of Laurent series expansion method, i.e. computation of its matrices, can be problematic. Another difficult task is to develop an algorithm that performs elementary operations on rows and columns in accordance with the idea of Weierstrass-Kronecker decomposition. The multitude of procedures for determining the Drazin inverse matrix means that this method of analysis has considerable potential for its computer implementation.

In the case of continuous-time systems, two forms of the solutions of the state equation were considered for each method of descriptor systems analysis: the solutions that contain Dirac impulses and their derivatives (Theorems 1.1, 1.3, 1.7) and the solutions without these components (Theorems 1.2, 1.4, 1.8) that have been introduced due to inability of the Dirac delta to occur in physically realizable systems. In the case of discrete-time systems, the solutions of the state equation for each method have been also given (Theorems 1.22, 1.23, 1.24–1.25). It is worth noting that the solution does not depend on the choice of the method of the descriptor systems analysis.

Nevertheless, the considered methods differ in defining the analytical conditions for the positivity, stability and superstability. In the case of the Laurent series expansion method for continuous-time systems the analysis of series coefficients (i.e. the matrices T_k) does not allow to specify the properties of the system. The positivity of these coefficients is not a necessary condition of the system positivity (Theorem 1.9) and the series convergence is not a necessary condition of the stability. The system properties can be determined analyzing the time course of the transition function (Theorems 1.13 and 1.19). Taking all these into account the synthesis of feedback control can be complicated for this method. However, for discrete-time systems the sign of series coefficients (Theorem 1.28) and the series convergence (Theorem 1.35) are also necessary conditions for the positivity and stability.

In this regard, the other methods offer greater possibilities for determining the system properties basing on the matrices in the state equations, i.e. $\bar{E}\bar{E}^D$, $\bar{E}^D\bar{A}$, $\bar{E}^D\bar{B}$ (or their discrete counterparts $\bar{E}_d\bar{E}_d^D$, $\bar{E}_d^D\bar{A}_d$, $\bar{E}_d^D\bar{B}_d$) for the Drazin inverse matrix method and A_1, B_1, B_2 (or their discrete counterparts A_{1d}, B_{1d}, B_{2d}) for the Weierstrass-Kronecker decomposition method. Hence, we can also easily compute the gain matrices in the feedback control and determine their impact on the system properties.

The undoubted advantage of the Drazin inverse matrix method is the possibility to define necessary and sufficient conditions of the positivity (Theorems 1.10 and 1.29), stability (Theorems 1.16 and 1.36) and superstability (Theorems 1.20 and 1.42), which makes it versatile. The disadvantage of this method is that it can be difficult to test the positivity and superstability, which is based on searching for a matrix G (or G_d) such that $\bar{E}^D\bar{A} + G(\mathbb{I}_n - \bar{E}\bar{E}^D)$ (or $\bar{E}_d^D\bar{A}_d + G_d(\mathbb{I}_n - \bar{E}_d\bar{E}_d^D)$) satisfies the conditions (1.88a) and (1.133) (or (1.223a) and (1.264)).

The Weierstrass-Kronecker decomposition method allows us to easily test the stability (Theorems 1.17 and 1.37) since it depends directly on the matrix A_1 (or A_{1d}). However, using this method it can be difficult to determine the positivity and superstability. Note that Theorems 1.11 and 1.31 give necessary and sufficient positivity conditions assuming that the matrix Q (or Q_d) is a monomial matrix, which is in turn a sufficient condition, i.e. the system can be positive although the matrix Q (or Q_d) is not monomial. Superstability analysis (Theorems 1.21 and 1.45) is based on searching for a matrix M (or M_d) such that the matrix A_M (or A_{Md}) defined by (1.160) (or (1.303)) satisfies the condition (1.133) (or (1.264)).

It is also worth comparing the discrete-time equivalents of continuous-time models. Computation of the matrices $T_{d,k}$ based on the matrices T_k for the Laurent series expansion method can be inconvenient for large values of k since by (1.175) they are linear combinations of the matrices T_k for both $k < 0$ and $k \geq 0$. For the Drazin inverse matrix method, we can easily determine the matrices $\bar{E}_d\bar{E}_d^D$, $\bar{E}_d^D\bar{A}_d$, $\bar{E}_d^D\bar{B}_d$ basing on the matrices $\bar{E}\bar{E}^D$, $\bar{E}^D\bar{A}$, $\bar{E}^D\bar{B}$ using (1.186), but the more difficult task is to give an exact formula for computing the matrices $(\bar{A}h + \bar{E})^D$. The least computationally advanced discretization we obtain for the Weierstrass-Kronecker decomposition method, where the matrices A_1, B_1 of the continuous-time system are replaced by $A_1h + \mathbb{I}_n$, B_1h in its discrete-time counterpart. For the subsystem (1.208b) the matrices $\bar{N}_k$ are linear combinations of the matrix N and its powers.

It should be also noted that the choice of the discretization period h can have a significant impact on the positivity (Theorems 1.30 and 1.32), stability (Theorem 1.34) and superstability (Theorems 1.44 and 1.47) of the discrete-time equivalent of the continuous-time system.

Chapter 2
Fractional Descriptor Linear Systems

In this chapter fractional descriptor continuous-time and discrete-time linear systems will be analyzed. In the subsequent sections solutions of the state equation and analitycal conditions of the positivity, stability and superstability will be established considering the Laurent series expansion method, Drazin inverse matrix method and Weierstrass-Kronecker decomposition method.

2.1 Continuous-Time Systems

2.1.1 Euler Gamma Function and Its Properties

Definition 2.1. A function given by the integral [57, 132, 135]

$$\Gamma(x) = \int_0^\infty t^{x-1}e^{-t}dt, \ \ \mathrm{Re}(x) > 0 \tag{2.1}$$

is called the Euler gamma function.
The Euler gamma function can be also defined by

$$\Gamma(x) = \lim_{n\to\infty} \frac{n!n^x}{x(x+1)\dots(x+n)}, \quad x \in \mathbb{C}\backslash\{0, -1, -2, \dots\}. \tag{2.2}$$

We shall show that $\Gamma(x)$ satisfies the equality

$$\Gamma(x+1) = x\Gamma(x). \tag{2.3}$$

Proof. Using (2.1) we obtain

$$\Gamma(x+1) = \int_0^\infty t^x e^{-t}dt = -t^x e^{-t}\big|_0^\infty + x\int_0^\infty t^{x-1}e^{-t}dt = x\Gamma(x). \tag{2.4}$$

□

T. Kaczorek and K. Borawski, *Descriptor Systems of Integer and Fractional Orders*, Studies in Systems, Decision and Control 367,
https://doi.org/10.1007/978-3-030-72480-1_2

Example 2.1. From (2.3) we have for

$$\begin{aligned}
x &= 1 : \Gamma(2) = 1 \cdot \Gamma(1) = 1 \ \text{ since } \Gamma(1) = \int_0^\infty e^{-t} dt = 1, \\
x &= 2 : \Gamma(3) = 2 \cdot \Gamma(2) = 1 \cdot 2 = 2!, \\
x &= 3 : \Gamma(4) = 3 \cdot \Gamma(3) = 3 \cdot 2 \cdot \Gamma(2) = 3!.
\end{aligned}$$

In general case for $x \in \mathbb{N}$ we have

$$\Gamma(n+1) = n\Gamma(n) = n(n-1)\Gamma(n-1) = n(n-1)(n-2)\ldots(1) = n!.$$

The gamma function is also well-defined for x being any real (complex) numbers. For example we have for:

$$\begin{aligned}
x &= 1.5 : \ \Gamma(2.5) = 1.5 \cdot \Gamma(1.5) = 1.5 \cdot 0.5\Gamma(0.5), \\
x &= -0.5 : \Gamma(0.5) = -0.5 \cdot \Gamma(-0.5) = -0.5 \cdot (-1.5)\Gamma(-1.5).
\end{aligned}$$

2.1.2 Mittag-Leffler Function

The Mittag-Leffler function is a generalization of the exponential function $e^{s_i t}$ and it plays important role in solution of the fractional differential equations.

Definition 2.2. A function of the complex variable z defined by [57, 132, 135]

$$E_\alpha(z) = \sum_{k=0}^{\infty} \frac{z^k}{\Gamma(k\alpha + 1)} \tag{2.5}$$

is called the one parameter Mittag-Leffler function.

Example 2.2. For $\alpha = 1$ we obtain

$$E_1(z) = \sum_{k=0}^{\infty} \frac{z^k}{\Gamma(k+1)} = \sum_{k=0}^{\infty} \frac{z^k}{k!} = e^z, \tag{2.6}$$

i.e. the classical exponential function.

An extension of the one parameter Mittag-Leffler function is the following two parameters function.

Definition 2.3. A function of the complex variable z defined by [57, 132, 135]

$$E_{\alpha,\beta}(z) = \sum_{k=0}^{\infty} \frac{z^k}{\Gamma(\alpha k + \beta)} \tag{2.7}$$

is called the two parameters Mittag-Leffler function. For $\beta = 1$ from (2.7) we obtain (2.5).

2.1.3 Caputo Definition of Fractional Derivative-Integral

Definition 2.4. The function defined by [57, 132, 135]

$$ {}_0D_t^{\alpha} f(t) = \frac{1}{\Gamma(N-\alpha)} \int_0^t \frac{f^{(N)}(\tau)}{(t-\tau)^{\alpha+1-N}} d\tau, \quad f^{(N)}(\tau) = \frac{d^N f(\tau)}{d\tau^N} \tag{2.8} $$

is called the Caputo fractional derivative-integral, where $N-1 \le \alpha < N$, $N \in \mathbb{N}$.

Theorem 2.1. The Caputo derivative-integral operator is linear satisfying the relation

$$ {}_0D_t^{\alpha}[\lambda f(t) + \mu g(t)] = \lambda\, {}_0D_t^{\alpha} f(t) + \mu\, {}_0D_t^{\alpha} g(t), \quad \lambda, \mu \in \mathbb{R}. \tag{2.9} $$

Proof. From (2.8) and (2.9) we have

$$ \begin{aligned} {}_0D_t^{\alpha}[\lambda f(t) + \mu g(t)] &= \frac{1}{\Gamma(N-\alpha)} \int_0^t \frac{\frac{d^N}{d\tau^N}[\lambda f(\tau) + \mu g(\tau)]}{(t-\tau)^{\alpha+1-N}} d\tau \\ &= \frac{\lambda}{\Gamma(N-\alpha)} \int_0^t \frac{f^{(N)}(\tau)}{(t-\tau)^{\alpha+1-N}} d\tau \\ &+ \frac{\mu}{\Gamma(N-\alpha)} \int_0^t \frac{g^{(N)}(\tau)}{(t-\tau)^{\alpha+1-N}} d\tau \\ &= \lambda\, {}_0D_t^{\alpha} f(t) + \mu\, {}_0D_t^{\alpha} g(t). \end{aligned} \tag{2.10} $$

□

Theorem 2.2. The Laplace transform of the derivative-integral (2.8) for $N-1 < \alpha < N$ has the form

$$ \mathcal{L}\left[{}_0D_t^{\alpha} f(t)\right] = s^{\alpha} F(s) - \sum_{k=1}^{N} s^{\alpha-k} f^{(k-1)}(0^+). \tag{2.11} $$

Proof. Using Definitions 2.4, C.2 and Eqs. (C.8), (C.12) for $N-1 < \alpha < N$ we obtain

$$\begin{aligned}\mathcal{L}\left[{}_0D_t^\alpha f(t)\right] &= \mathcal{L}\left[\frac{1}{\Gamma(N-\alpha)}\int_0^t (t-\tau)^{N-\alpha-1} f^{(N)}(\tau)d\tau\right] \\ &= \frac{1}{\Gamma(N-\alpha)}\mathcal{L}\left[t^{N-\alpha-1}\right]\mathcal{L}\left[f^{(N)}(t)\right] \\ &= \frac{1}{\Gamma(N-\alpha)}\frac{\Gamma(N-\alpha)}{s^{N-\alpha}}\left[s^N F(s) - \sum_{k=1}^{N} s^{N-k} f^{(k-1)}(0^+)\right] \\ &= s^\alpha F(s) - \sum_{k=1}^{N} s^{\alpha-k} f^{(k-1)}(0^+).\end{aligned} \tag{2.12}$$

□

2.1.4 State Equations of Fractional Descriptor Continuous-Time Linear Systems

Consider the continuous-time linear system described by state equations

$$E\,{}_0D_t^\alpha x(t) = Ax(t) + Bu(t), \quad 0 < \alpha < 1 \tag{2.13a}$$

$$y(t) = Cx(t) + Du(t), \tag{2.13b}$$

where $x(t) \in \mathbb{R}^n, u(t) \in \mathbb{R}^m, y(t) \in \mathbb{R}^p$ are state, input and output vectors, ${}_0D_t^\alpha$ is the Caputo derivative-integral operator defined by (2.8) and $E, A \in \mathbb{R}^{n\times n}$, $B \in \mathbb{R}^{n\times m}$, $C \in \mathbb{R}^{p\times n}$, $D \in \mathbb{R}^{p\times m}$. We assume that $\det E = 0$ and

$$\det[E\lambda - A] \neq 0 \ \text{ for some } \ \lambda \in \mathbb{C}. \tag{2.14}$$

If the matrix E is nonsingular ($\det E \neq 0$), then we can transform (2.13a) to the state equation of a fractional standard continuous-time linear system through premultiplication of (2.13a) by E^{-1}. The system (2.13) with the singular matrix E ($\det E = 0$) will be called a fractional descriptor continuous-time linear system.

2.1.5 Solution to the State Equation of Fractional Descriptor Continuous-Time Linear Systems

Let U be the set of admissible input vectors $u(t) \in U \subset \mathbb{R}^m$ and $X_0 \subset \mathbb{R}^n$ the set of consistent initial conditions $x(0) = x_0 \in X_0$ for which the Eq. (2.13a) has the solution $x(t)$ with $u(t) \in U$.

2.1.5.1 Laurent Series Expansion Method

In this section the solution to the Eq. (2.13a) will be given using the Laurent series expansion method [64].

Theorem 2.3. Solution to the Eq. (2.13a) for $x_0 \in X_0$ and $u(t) \in U$ is given by the formula

$$x(t) = \sum_{k=-n_0}^{\infty} \left\{ \frac{T_k E t^{k\alpha}}{\Gamma(k\alpha + 1)} x_0 + \int_0^t \frac{T_k (t-\tau)^{(k+1)\alpha - 1}}{\Gamma[(k+1)\alpha]} Bu(\tau) d\tau \right\}, \tag{2.15}$$

where n_0 is called the index of the system and the matrices $T_k \in \mathbb{R}^{n \times n}$ can be found by the Laurent series expansion of the matrix $[Es^\alpha - A]^{-1}$.

Proof. Using the Laplace transform and taking into account that for $0 < \alpha < 1$

$$\mathcal{L}[{}_0D_t^\alpha] = s^\alpha X(s) - s^{\alpha-1} x_0 \tag{2.16}$$

we obtain for (2.13a)

$$Es^\alpha X(s) - Es^{\alpha-1} x_0 = AX(s) + BU(s). \tag{2.17}$$

From (2.17) we have

$$X(s) = [Es^\alpha - A]^{-1} [Es^{\alpha-1} x_0 + BU(s)]. \tag{2.18}$$

Let

$$[Es^\alpha - A]^{-1} = \sum_{k=-n_0}^{\infty} T_k s^{-(k+1)\alpha}. \tag{2.19}$$

The matrices T_k, $k = -n_0, -n_0 + 1, \ldots, 0, 1, \ldots$ can be found by the expansion (2.19). Substituting (2.19) into (2.18) we obtain

$$\begin{aligned} X(s) &= \sum_{k=-n_0}^{\infty} T_k s^{-(k+1)\alpha} [Es^{\alpha-1} x_0 + BU(s)] \\ &= \sum_{k=-n_0}^{\infty} T_k E s^{-(k\alpha+1)} x_0 + \sum_{k=-n_0}^{\infty} T_k B s^{-(k+1)\alpha} U(s). \end{aligned} \tag{2.20}$$

Applying to (2.20) the inverse Laplace transform and the convolution theorem we obtain (2.15) since $\mathcal{L}[t^\alpha] = \frac{\Gamma(\alpha+1)}{s^{\alpha+1}}$. □

Remark 2.1. If $E = \mathbb{I}_n$, then $n_0 = 0$ and $T_k = A^k$, $k = 0, 1, \ldots$. In this case from (2.15) we have

$$x(t) = \sum_{k=0}^{\infty} \left(\frac{A^k t^{k\alpha}}{\Gamma(k\alpha + 1)} x_0 + \int_0^t \frac{A^k (t-\tau)^{(k+1)\alpha - 1}}{\Gamma[(k+1)\alpha]} Bu(\tau) d\tau \right)$$
$$= \Phi_0(t) x_0 + \int_0^t \Phi(t-\tau) Bu(\tau) d\tau, \tag{2.21a}$$

where

$$\Phi_0(t) = E_\alpha(At^\alpha) = \sum_{k=0}^{\infty} \frac{A^k t^{k\alpha}}{\Gamma(k\alpha + 1)}, \tag{2.21b}$$

$$\Phi(t) = \sum_{k=0}^{\infty} \frac{A^k t^{(k+1)\alpha - 1}}{\Gamma[(k+1)\alpha]}, \tag{2.21c}$$

where $E_\alpha(At^\alpha)$ is the one parameter Mittag-Leffler function defined by (2.5). The formula (2.21a) is the solution to the state equation of a fractional standard continuous-time linear system [57].

Example 2.3. Consider the fractional descriptor continuous-time linear system (2.13a) with

$$E = \begin{bmatrix} 0.5 & -0.375 & 0 \\ 0 & 0.25 & 0 \\ -0.5 & -0.125 & 0 \end{bmatrix}, \quad A = \begin{bmatrix} -1 & 2 & 0 \\ 0 & -1 & 0 \\ 1 & 0 & -2 \end{bmatrix}$$
$$B = \begin{bmatrix} 1 & 0 \\ 0 & 1 \\ -1 & -2 \end{bmatrix}. \tag{2.22}$$

The matrix pencil (E, A) of (2.22) is regular since

$$\det[Es^\alpha - A] = 0.25(s^\alpha + 2)(s^\alpha + 4) \neq 0. \tag{2.23}$$

The matrix $[Es^\alpha - A]^{-1}$ has the form

$$[Es^\alpha - A]^{-1} = \begin{bmatrix} 0.5s^\alpha + 1 & -0.375s^\alpha - 2 & 0 \\ 0 & 0.25s^\alpha + 1 & 0 \\ -0.5s^\alpha - 1 & -0.125s^\alpha & 2 \end{bmatrix}^{-1}$$
$$= \begin{bmatrix} \frac{2}{s^\alpha + 2} & \frac{3s^\alpha + 16}{(s^\alpha + 2)(s^\alpha + 4)} & 0 \\ 0 & \frac{4}{s^\alpha + 4} & 0 \\ 0.5 & 1 & 0.5 \end{bmatrix}. \tag{2.24}$$

Using (2.19) for $n_0 = -1$, the matrix (2.24) can be written in the form

$$[Es^{\alpha} - A]^{-1} = \sum_{k=-1}^{\infty} T_k s^{-(k+1)\alpha}, \tag{2.25a}$$

where

$$T_k = \begin{cases} \begin{bmatrix} 0 & 0 & 0 \\ 0 & 0 & 0 \\ 0.5 & 1 & 0.5 \end{bmatrix} & \text{for } k = -1, \\ \begin{bmatrix} 2(-2)^k & 5(-2)^k - 2(-4)^k & 0 \\ 0 & 4(-4)^k & 0 \\ 0.5 & 1 & 0.5 \end{bmatrix} & \text{for } k = 0, 1, 2, \dots . \end{cases} \tag{2.25b}$$

Using (2.22) and (2.25b) we have

$$T_k E = \begin{cases} \begin{bmatrix} 0 & 0 & 0 \\ 0 & 0 & 0 \\ 0 & 0 & 0 \end{bmatrix} & \text{for } k = -1, \\ \begin{bmatrix} (-2)^k & 0.5\left[(-2)^k - (-4)^k\right] & 0 \\ 0 & (-4)^k & 0 \\ 0 & 0 & 0 \end{bmatrix} & \text{for } k = 0, 1, 2, \dots, \end{cases} \tag{2.26a}$$

$$T_k B = \begin{cases} \begin{bmatrix} 0 & 0 \\ 0 & 0 \\ 0 & 0 \end{bmatrix} & \text{for } k = -1, \\ \begin{bmatrix} 2(-2)^k & 5(-2)^k - 2(-4)^k \\ 0 & 4(-4)^k \\ 0 & 0 \end{bmatrix} & \text{for } k = 0, 1, 2, \dots . \end{cases} \tag{2.26b}$$

Using (2.15), (2.22) and (2.26) we obtain

$$\begin{aligned} x(t) &= \sum_{k=0}^{\infty} \left(\frac{T_k E t^{k\alpha}}{\Gamma(k\alpha + 1)} x_0 + \int_0^t \frac{T_k (t-\tau)^{(k+1)\alpha - 1}}{\Gamma[(k+1)\alpha]} B u(\tau) d\tau \right) \\ &= \begin{bmatrix} E_\alpha(-2t^\alpha) & 0.5\left[E_\alpha(-2t^\alpha) - E_\alpha(-4t^\alpha)\right] & 0 \\ 0 & E_\alpha(-4t^\alpha) & 0 \\ 0 & 0 & 0 \end{bmatrix} x_0 \\ &+ \int_0^t \begin{bmatrix} 2E_{\alpha,\alpha}\left[-2(t-\tau)^\alpha\right] & 5E_{\alpha,\alpha}\left[-2(t-\tau)^\alpha\right] - 2E_{\alpha,\alpha}\left[-4(t-\tau)^\alpha\right] \\ 0 & 4E_{\alpha,\alpha}\left[-4(t-\tau)^\alpha\right] \\ 0 & 0 \end{bmatrix} \\ &\times (t-\tau)^{\alpha-1} u(\tau) d\tau, \end{aligned} \tag{2.27}$$

where $E_\alpha(z)$ and $E_{\alpha,\beta}(z)$ are Mittag-Leffler functions defined by (2.5) and (2.7).

The equation above shows the solution (2.15) of the state equation of the fractional descriptor linear system (2.13) with (2.22).

2.1.5.2 Drazin Inverse Matrix Method

In this section the solution to the Eq. (2.13a) will be given using the Drazin inverse matrix method [61, 63, 90].

Consider the Eq. (2.13a) and assume that the condition (2.14) is met. There exists a number $c \in \mathbb{C}$ such that $\det[Ec - A] \neq 0$. Premultiplying the Eq. (2.13a) by the matrix $[Ec - A]^{-1}$ we obtain

$$\bar{E}\,{}_0D_t^{\alpha}x(t) = \bar{A}x(t) + \bar{B}u(t), \tag{2.28a}$$

where

$$\bar{E} = [Ec - A]^{-1}E, \quad \bar{A} = [Ec - A]^{-1}A, \quad \bar{B} = [Ec - A]^{-1}B. \tag{2.28b}$$

Note that the Eqs. (2.13a) and (2.28a) have the same solution $x(t)$. Basic definitions and properties concerning Drazin inverses are given in Sect. 1.1.2.2.

Lemma 1.2 can also be used for the fractional descriptor continuous-time linear system (2.28). Therefore, the Eq. (2.28a) is equivalent to the following equations:

$${}_0D_t^{\alpha}x_1(t) = \bar{E}^D\bar{A}x_1(t) + \bar{E}^D\bar{B}u(t), \tag{2.29a}$$

$$\bar{A}^D\bar{E}(\mathbb{I}_n - \bar{E}\bar{E}^D)\,{}_0D_t^{\alpha}x_2(t) = x_2(t) + (\mathbb{I}_n - \bar{E}\bar{E}^D)\bar{A}^D\bar{B}u(t), \tag{2.29b}$$

where

$$x_1(t) = \bar{E}\bar{E}^D x(t), \quad x_2(t) = (\mathbb{I}_n - \bar{E}\bar{E}^D)x(t). \tag{2.29c}$$

Theorem 2.4. [63] Solution to the Eq. (2.28a) for $x_0 \in X_0$ and $u(t) \in U$ has the form

$$\begin{aligned} x(t) &= \Phi_0(t)\bar{E}\bar{E}^D x_0 + \bar{E}^D\int_0^t \Phi(t-\tau)\bar{B}u(\tau)d\tau \\ &\quad + (\bar{E}\bar{E}^D - \mathbb{I}_n)\sum_{k=0}^{q-1}(\bar{E}\bar{A}^D)^k\bar{A}^D\bar{B}u^{(k\alpha)}(t), \end{aligned} \tag{2.30a}$$

where

$$\Phi_0(t) = E_{\alpha}(\bar{E}^D\bar{A}t^{\alpha}) = \sum_{k=0}^{\infty}\frac{(\bar{E}^D\bar{A})^k t^{k\alpha}}{\Gamma(k\alpha + 1)}, \tag{2.30b}$$

$$\Phi(t) = \sum_{k=0}^{\infty} \frac{(\bar{E}^D \bar{A})^k t^{(k+1)\alpha-1}}{\Gamma[(k+1)\alpha]}, \tag{2.30c}$$

$u^{(k\alpha)}(t) = {}_0D_t^{k\alpha} u(t)$, $E_\alpha(\bar{E}^D \bar{A} t^\alpha)$ is the one parameter Mittag-Leffler function defined by (2.5) and q is the index of $\bar{E}$.

It is easy to check that the first two components of (2.30a) are the solution to the Eq. (2.29a) and the third component of (2.30a) is the solution to the Eq. (2.29b).

From (2.30a) for $t = 0$ we have

$$x_0 = \bar{E}\bar{E}^D v + (\bar{E}\bar{E}^D - \mathbb{I}_n) \sum_{k=0}^{q-1} (\bar{E}\bar{A}^D)^k \bar{A}^D \bar{B} u^{(k\alpha)}(0), \tag{2.31}$$

where $v \in \mathbb{R}^n$ is an arbitrary vector.
In the case when $u(t) = 0$ from (1.46) we have $x_0 = \bar{E}\bar{E}^D v$. A homogenous equation $E\,{}_0D_t^\alpha(t) = Ax(t)$ has the solution if and only if $x_0 \in \operatorname{Im} \bar{E}\bar{E}^D$.

By Lemma 1.3 and Theorem 1.5 the matrices $\bar{E}^D\bar{E}$, $\bar{E}^D\bar{A}$, $\bar{E}\bar{A}^D$, $\bar{E}^D\bar{B}$, $\bar{A}^D\bar{B}$, the index q of the matrix $\bar{E}$, the solution (2.30a) of the Eq. (2.13a) and the set of consistent initial conditions X_0 do not depend of the choice of c.

Example 2.4. (Continuation of Example 2.3) Consider the fractional descriptor continuous-time linear system (2.13a) with (2.22). According to (2.23) the matrix pencil (E, A) is regular. We choose $c = 0$ and using (2.28b) we obtain

$$\bar{E} = [-A]^{-1}E = \begin{bmatrix} 0.5 & 0.125 & 0 \\ 0 & 0.25 & 0 \\ 0 & 0 & 0 \end{bmatrix},$$
$$\bar{A} = [-A]^{-1}A = \begin{bmatrix} -1 & 0 & 0 \\ 0 & -1 & 0 \\ 0 & 0 & -1 \end{bmatrix}, \quad \bar{B} = [-A]^{-1}B = \begin{bmatrix} 1 & 2 \\ 0 & 1 \\ 0 & 0 \end{bmatrix}. \tag{2.32}$$

It is easy to see that $\operatorname{rank}\bar{E} = \operatorname{rank}\bar{E}^2$, so $q = 1$. The Drazin inverse of $\bar{E}$ can be computed using (B.1) and (B.2) for $\bar{E} = \bar{E}_1$, where $\bar{E}_{11} = \begin{bmatrix} 0.5 & 0.125 \\ 0 & 0.25 \end{bmatrix}$, $\bar{E}_{12} = \begin{bmatrix} 0 \\ 0 \end{bmatrix}$ and

$$\bar{E}^D = \begin{bmatrix} \bar{E}_{11}^{-1} & \bar{E}_{11}^{-2}\bar{E}_{12} \\ 0 & 0 \end{bmatrix} = \begin{bmatrix} 2 & -1 & 0 \\ 0 & 4 & 0 \\ 0 & 0 & 0 \end{bmatrix}. \tag{2.33}$$

Note that $\bar{A}^D = \bar{A}^{-1} = \bar{A}$. Next, we compute the matrices

$$\bar{E}\bar{E}^D = \begin{bmatrix} 1 & 0 & 0 \\ 0 & 1 & 0 \\ 0 & 0 & 0 \end{bmatrix}, \quad \bar{E}^D\bar{A} = \begin{bmatrix} -2 & 1 & 0 \\ 0 & -4 & 0 \\ 0 & 0 & 0 \end{bmatrix},$$
$$\bar{E}^D\bar{B} = \begin{bmatrix} 2 & 3 \\ 0 & 4 \\ 0 & 0 \end{bmatrix}, \quad (\bar{E}\bar{E}^D - \mathbb{I}_3) = \begin{bmatrix} 0 & 0 & 0 \\ 0 & 0 & 0 \\ 0 & 0 & -1 \end{bmatrix}, \tag{2.34}$$
$$\bar{A}^D\bar{E} = \begin{bmatrix} -0.5 & -0.125 & 0 \\ 0 & -0.25 & 0 \\ 0 & 0 & 0 \end{bmatrix}, \quad \bar{A}^D\bar{B} = \begin{bmatrix} -1 & -2 \\ 0 & -1 \\ 0 & 0 \end{bmatrix}$$

and

$$\Phi_0(t) = \sum_{k=0}^{\infty} \frac{(\bar{E}^D\bar{A})^k t^{k\alpha}}{\Gamma(k\alpha+1)} = \begin{bmatrix} E_\alpha(-2t^\alpha) & 0.5\,[E_\alpha(-2t^\alpha) - E_\alpha(-4t^\alpha)] & 0 \\ 0 & E_\alpha(-4t^\alpha) & 0 \\ 0 & 0 & 0 \end{bmatrix}, \tag{2.35a}$$

$$\Phi_0(t) = \sum_{k=0}^{\infty} \frac{(\bar{E}^D\bar{A})^k t^{(k+1)\alpha-1}}{\Gamma[(k+1)\alpha]} = t^{\alpha-1}\frac{(\bar{E}^D\bar{A})^k t^{k\alpha}}{\Gamma[(k+1)\alpha]}$$
$$= t^{\alpha-1}\begin{bmatrix} E_{\alpha,\alpha}(-2t^\alpha) & 0.5\left[E_{\alpha,\alpha}(-2t^\alpha) - E_{\alpha,\alpha}(-4t^\alpha)\right] & 0 \\ 0 & E_{\alpha,\alpha}(-4t^\alpha) & 0 \\ 0 & 0 & 0 \end{bmatrix}. \tag{2.35b}$$

Taking into account that $q = 1$ and using (2.30a), (2.34) and (2.35) we obtain

$$x(t) = \Phi_0(t)\bar{E}\bar{E}^D x_0 + \int_0^t \Phi(t-\tau)\bar{E}^D\bar{B}u(\tau)d\tau + (\bar{E}\bar{E}^D - \mathbb{I}_3)\bar{A}^D\bar{B}u(t)$$
$$= \begin{bmatrix} E_\alpha(-2t^\alpha) & 0.5\,[E_\alpha(-2t^\alpha) - E_\alpha(-4t^\alpha)] & 0 \\ 0 & E_\alpha(-4t^\alpha) & 0 \\ 0 & 0 & 0 \end{bmatrix} x_0$$
$$+ \int_0^t \begin{bmatrix} 2E_{\alpha,\alpha}[-2(t-\tau)^\alpha] & 5E_{\alpha,\alpha}[-2(t-\tau)^\alpha] - 2E_{\alpha,\alpha}[-4(t-\tau)^\alpha] \\ 0 & 4E_{\alpha,\alpha}[-4(t-\tau)^\alpha] \\ 0 & 0 \end{bmatrix}$$
$$\times (t-\tau)^{\alpha-1}u(\tau)d\tau, \tag{2.36}$$

where $(\bar{E}\bar{E}^D - \mathbb{I}_3)\bar{A}^D\bar{B} = 0$ and $E_\alpha(z)$, $E_{\alpha,\beta}(z)$ are Mittag-Leffler functions defined by (2.5) and (2.7). The equation above shows the solution (2.30a) of the state equation of the fractional descriptor linear system (2.13a) with (2.22). Note that solutions (2.27) and (2.36) are equivalent.

2.1.5.3 Weierstrass-Kronecker Decomposition Method

In this section the solution to the Eq. (2.13a) will be given using the Weierstrass-Kronecker decomposition method [57, 90].

It is well-known [57, 90] that if the condition (2.14) holds, then there exists a pair of nonsingular matrices P, $Q \in \mathbb{R}^{n\times n}$ such that

$$P[Es^{\alpha} - A]Q = \begin{bmatrix} \mathbb{I}_{n_1}s^{\alpha} - A_1 & 0 \\ 0 & Ns^{\alpha} - \mathbb{I}_{n_2} \end{bmatrix}, \quad A_1 \in \mathbb{R}^{n_1\times n_1}, \quad N \in \mathbb{R}^{n_2\times n_2}, \tag{2.37}$$

where $n_1 = \deg\{\det[Es^{\alpha} - A]\}$, $n_2 = n - n_1$ and N is a nilpotent matrix with the nilpotency index μ, i.e. $N^{\mu-1} \neq 0$ and $N^{\mu} = 0$. The matrices P and Q can be computed using one of procedures given in [47, 48, 168].
Premultiplying (2.13a) by the matrix P and introducing the new state vector

$$\bar{x}(t) = \begin{bmatrix} \bar{x}_1(t) \\ \bar{x}_2(t) \end{bmatrix} = Q^{-1}x(t), \quad \bar{x}_1(t) \in \mathbb{R}^{n_1}, \quad \bar{x}_2(t) \in \mathbb{R}^{n_2}, \tag{2.38}$$

we obtain

$$PEQQ^{-1}\, {}_0D_t^{\alpha}x(t) = PAQQ^{-1}x(t) + PBu(t), \tag{2.39a}$$

$$y(t) = CQQ^{-1}x(t) + Du(t) \tag{2.39b}$$

and using (2.37) we get

$${}_0D_t^{\alpha}\bar{x}_1(t) = A_1\bar{x}_1(t) + B_1u(t), \tag{2.40a}$$

$$N\, {}_0D_t^{\alpha}\bar{x}_2(t) = \bar{x}_2(t) + B_2u(t), \tag{2.40b}$$

$$y(t) = C_1\bar{x}_1(t) + C_2\bar{x}_2(t) + Du(t), \tag{2.40c}$$

where

$$PB = \begin{bmatrix} B_1 \\ B_2 \end{bmatrix}, \quad B_1 \in \mathbb{R}^{n_1\times m}, \quad B_2 \in \mathbb{R}^{n_2\times m}, \tag{2.41a}$$

$$CQ = \begin{bmatrix} C_1 & C_2 \end{bmatrix}, \quad C_1 \in \mathbb{R}^{p\times n_1}, \quad C_2 \in \mathbb{R}^{p\times n_2}. \tag{2.41b}$$

Theorem 2.5. [57] Solution to the Eq. (2.40a) for given initial condition $\bar{x}_{10} \in \mathbb{R}^{n_1}$ and input $u(t) \in U$ has the form

$$\bar{x}_1(t) = \Phi_{10}(t)\bar{x}_{10} + \int_0^t \Phi_{11}(t-\tau)B_1u(\tau)d\tau, \tag{2.42}$$

where

$$\Phi_{10}(t) = E_\alpha(A_1 t^\alpha) = \sum_{k=0}^{\infty} \frac{A_1^k t^{k\alpha}}{\Gamma(k\alpha + 1)}, \tag{2.43}$$

$$\Phi_{11}(t) = \sum_{k=0}^{\infty} \frac{A_1^k t^{(k+1)\alpha-1}}{\Gamma[(k+1)\alpha]}, \tag{2.44}$$

where $E_\alpha(A_1 t^\alpha)$ is the one parameter Mittag-Leffler function defined by (2.5).

Theorem 2.6. Solution to the Eq. (2.40b) for given input $u(t) \in U$ and $\bar{x}_{20} = 0$ has the form

$$\bar{x}_2(t) = -\sum_{k=0}^{\mu-1} N^k B_2 u^{(k\alpha)}(t), \tag{2.45}$$

where $u^{(k\alpha)}(t) = {}_0D_t^{k\alpha} u(t)$ and the consistent initial conditions of the subsystem (2.40b) are given by the formula

$$\bar{x}_{20} = -\sum_{k=0}^{\mu-1} N^k B_2 u^{(k\alpha)}(0). \tag{2.46}$$

Proof. From (2.40b) and (2.45) we have

$$\begin{aligned} N\,{}_0D_t^\alpha \bar{x}_2(t) &= -\sum_{k=0}^{\mu-1} N^{k+1} B_2 u^{(k+1)\alpha}(t) = -\sum_{k=0}^{\mu-1} N^k B_2 u^{(k\alpha)}(t) + B_2 u(t) \\ &= \bar{x}_2(t) + B_2 u(t) \end{aligned} \tag{2.47}$$

since $N^\mu = 0$. Substituting $t = 0$ into (2.45) we obtain (2.46). □

Example 2.5. (Continuation of Examples 2.3 and 2.4) Consider the fractional descriptor continuous-time linear system (2.13a) with (2.22). According to (2.23) the matrix pencil (E, A) is regular. In this case

$$P = \begin{bmatrix} 2 & 0 & 0 \\ 0 & 4 & 0 \\ 1 & 2 & 1 \end{bmatrix}, \quad Q = \begin{bmatrix} 1 & 0.75 & 0 \\ 0 & 1 & 0 \\ 0 & 0 & -0.5 \end{bmatrix} \tag{2.48}$$

and

$$\begin{bmatrix} \mathbb{I}_{n_1} & 0 \\ 0 & N \end{bmatrix} = PEQ = \begin{bmatrix} 1 & 0 & 0 \\ 0 & 1 & 0 \\ 0 & 0 & 0 \end{bmatrix},$$

$$\begin{bmatrix} A_1 & 0 \\ 0 & \mathbb{I}_{n_2} \end{bmatrix} = PAQ = \begin{bmatrix} -2 & 2.5 & 0 \\ 0 & -4 & 0 \\ 0 & 0 & 1 \end{bmatrix}, \tag{2.49}$$

$$\begin{bmatrix} B_1 \\ B_2 \end{bmatrix} = PB = \begin{bmatrix} 2 & 0 \\ 0 & 4 \\ 0 & 0 \end{bmatrix}.$$

Next we compute the transition matrices

$$\begin{aligned} \Phi_{10}(t) &= \sum_{k=0}^{\infty} \frac{A_1^k t^{k\alpha}}{\Gamma(k\alpha+1)} \\ &= \begin{bmatrix} E_\alpha(-2t^\alpha) & 1.25\,[E_\alpha(-2t^\alpha) - E_\alpha(-4t^\alpha)] \\ 0 & E_\alpha(-4t^\alpha) \end{bmatrix}. \end{aligned} \tag{2.50a}$$

$$\begin{aligned} \Phi_{11}(t) &= \sum_{k=0}^{\infty} \frac{A_1^k t^{(k+1)\alpha-1}}{\Gamma[(k+1)\alpha]} = t^{\alpha-1} \frac{A_1^k t^{k\alpha}}{\Gamma[(k+1)\alpha]} \\ &= t^{\alpha-1} \begin{bmatrix} E_{\alpha,\alpha}(-2t^\alpha) & 1.25\left[E_{\alpha,\alpha}(-2t^\alpha) - E_{\alpha,\alpha}(-4t^\alpha)\right] \\ 0 & E_{\alpha,\alpha}(-4t^\alpha) \end{bmatrix}. \end{aligned} \tag{2.50b}$$

Using (2.42), (2.49) and (2.50) we obtain

$$\begin{aligned} \bar{x}_1(t) &= \begin{bmatrix} E_\alpha(-2t^\alpha) & 1.25\,[E_\alpha(-2t^\alpha) - E_\alpha(-4t^\alpha)] \\ 0 & E_\alpha(-4t^\alpha) \end{bmatrix} \bar{x}_{10} \\ &+ \int_0^t \begin{bmatrix} 2E_{\alpha,\alpha}\,[-2(t-\tau)^\alpha] & 5\left\{E_{\alpha,\alpha}\,[-2(t-\tau)^\alpha] - E_\alpha\,[-4(t-\tau)^\alpha]\right\} \\ 0 & 4E_{\alpha,\alpha}\,[-4(t-\tau)^\alpha] \end{bmatrix} \\ &\times (t-\tau)^{\alpha-1} u(\tau) d\tau. \end{aligned} \tag{2.51}$$

Taking into account that $\mu = 1$ and using (2.45), (2.49) we obtain

$$\bar{x}_2(t) = -B_2 u(t) = 0. \tag{2.52}$$

From (2.38), (2.51) and (2.52) we have

$$\bar{x}(t) = \begin{bmatrix} E_\alpha(-2t^\alpha) & 1.25\left[E_\alpha(-2t^\alpha) - E_\alpha(-4t^\alpha)\right] & 0 \\ 0 & E_\alpha(-4t^\alpha) & 0 \\ 0 & 0 & 0 \end{bmatrix} \bar{x}_0$$
$$+ \int_0^t \begin{bmatrix} 2E_{\alpha,\alpha}\left[-2(t-\tau)^\alpha\right] & 5\left\{E_{\alpha,\alpha}\left[-2(t-\tau)^\alpha\right] - E_\alpha\left[-4(t-\tau)^\alpha\right]\right\} \\ 0 & 4E_{\alpha,\alpha}\left[-4(t-\tau)^\alpha\right] \\ 0 & 0 \end{bmatrix}$$
$$\times\ (t-\tau)^{\alpha-1}u(\tau)d\tau, \tag{2.53}$$

where $E_\alpha(z)$ and $E_{\alpha,\beta}(z)$ are Mittag-Leffler functions defined by (2.5) and (2.7). Taking into account that $\bar{x}(t) = Q^{-1}x(t)$ we have to premultiply the Eq. (2.53) by Q and substitute $\bar{x}_0 = Q^{-1}x_0$. Thus we obtain the formula (2.27) (or (2.36)), which means that all considered methods of the fractional descriptor systems analysis give the same result.

Remark 2.2. Comparing all methods of the fractional descriptor systems analysis we can notice that the numbers: n_0 in (2.15), q in (2.30) and μ in (2.45) are equivalent.

2.1.6 Positive Fractional Descriptor Continuous-Time Linear Systems

In positive systems inputs, outputs and state variables take only nonnegative values [40, 49]. The following definition of the positivity is suitable for all methods of the fractional descriptor systems analysis.

Definition 2.5. The fractional descriptor continuous-time linear system (2.13) is called (internally) positive if $x(t) \in \mathbb{R}_+^n$ and $y(t) \in \mathbb{R}_+^p$, $t \in [0, +\infty)$ for any consistent initial condition $x_0 \in X_0 \subset \mathbb{R}_+^n$ and all admissible inputs $u(t) \in U \subset \mathbb{R}_+^m$, $t \in [0, +\infty)$ such that $u^{(k\alpha)}(t) \in \mathbb{R}_+^m$, $k = 1, \ldots, l-1$, $t \in [0, +\infty)$, where $u^{(k\alpha)}(t) = {}_0D_t^{k\alpha}u(t)$ and $l = n_0 = q = \mu$.

2.1.6.1 Laurent Series Expansion Method

In this section sufficient positivity conditions of the fractional descriptor continuous-time linear system (2.13) will be established using the Laurent series expansion method.

Theorem 2.7. The fractional descriptor continuous-time linear system (2.13) is (internally) positive if

$$T_k E \in \mathbb{R}_+^{n\times n}, \quad T_k B \in \mathbb{R}_+^{n\times m}, \quad C \in \mathbb{R}_+^{p\times n}, \quad D \in \mathbb{R}_+^{p\times m} \tag{2.54}$$

where $k = -n_0, -n_0+1, \ldots, 0, 1, \ldots$ and the matrices $T_k \in \mathbb{R}^{n\times n}$ can be found by expansion (2.19) of the matrix $[Es^\alpha - A]^{-1}$.

Proof. The proof can be accomplished in a similar way to the proof of Theorem 1.9. □

Example 2.6. Consider the fractional descriptor continuous-time linear system (2.13) with

$$E = \begin{bmatrix} 1 & 0 \\ 0 & 0 \end{bmatrix}, \quad A = \begin{bmatrix} 2 & 1 \\ 0 & -1 \end{bmatrix}, \quad B = \begin{bmatrix} 1 \\ 0 \end{bmatrix}, \quad C = \begin{bmatrix} 2 & 1 \end{bmatrix}, \quad D = 0. \tag{2.55}$$

The matrix pencil (E, A) of (2.55) is regular since

$$\det[Es^\alpha - A] = \begin{vmatrix} s^\alpha - 2 & -1 \\ 0 & 1 \end{vmatrix} = s^\alpha - 2 \neq 0. \tag{2.56}$$

The matrix $[Es^\alpha - A]^{-1}$ has the form

$$[Es^\alpha - A]^{-1} = \begin{bmatrix} s^\alpha - 2 & -1 \\ 0 & 1 \end{bmatrix}^{-1} = \begin{bmatrix} \frac{1}{s^\alpha-2} & \frac{1}{s^\alpha-2} \\ 0 & 1 \end{bmatrix}. \tag{2.57}$$

Using (2.19) the matrix (1.78) can be written in the form

$$[Es^\alpha - A]^{-1} = \sum_{k=-1}^{\infty} T_k s^{-(k+1)\alpha}, \tag{2.58a}$$

where

$$T_k = \begin{cases} \begin{bmatrix} 0 & 0 \\ 0 & 1 \end{bmatrix} & \text{for } k = -1, \\ \begin{bmatrix} 2^k & 2^k \\ 0 & 0 \end{bmatrix} & \text{for } k = 0, 1, 2, \ldots\,. \end{cases} \tag{2.58b}$$

Using (2.55) and (2.58b) we have

$$T_k E = \begin{bmatrix} 2^k & 0 \\ 0 & 0 \end{bmatrix}, \quad T_k B = \begin{bmatrix} 2^k \\ 0 \end{bmatrix} \quad \text{for } k = 0, 1, 2, \ldots\,. \tag{2.59}$$

It is easy to see that $T_k E \in \mathbb{R}_+^{2\times 2}$, $T_k B \in \mathbb{R}_+^{2\times 1}$, $k = 0, 1, 2 \ldots$ and $C \in \mathbb{R}_+^{1\times 2}$, $D \in \mathbb{R}_+^{1\times 1}$. Therefore, by Theorem 2.7 the fractional descriptor continuous-time linear system (2.13) with (2.55) is positive.

2.1.6.2 Drazin Inverse Matrix Method

In this section necessary and sufficient positivity conditions of the fractional descriptor continuous-time linear system (2.28) (or equivalently (2.13)) will be established using the Drazin inverse matrix method.

Lemma 2.1. [57] Let $\bar{E}^D\bar{A} \in \mathbb{M}_n$ and $0 < \alpha < 1$. Then

$$\Phi_0(t) = \sum_{k=0}^{\infty} \frac{(\bar{E}^D\bar{A})^k t^{k\alpha}}{\Gamma(k\alpha + 1)} \in \mathbb{R}_+^{n\times n}, \quad t \in [0, +\infty). \tag{2.60a}$$

$$\Phi(t) = \sum_{k=0}^{\infty} \frac{(\bar{E}^D\bar{A})^k t^{(k+1)\alpha - 1}}{\Gamma[(k+1)\alpha]} \in \mathbb{R}_+^{n\times n}, \quad t \in [0, +\infty). \tag{2.60b}$$

Lemma 2.2. The matrices $\bar{E}^D\bar{A}$ and $\bar{E}^D\bar{A} + G(\mathbb{I}_n - \bar{E}\bar{E}^D)$, where $G \in \mathbb{R}^{n\times n}$ is arbitrary, satisfy the equalities

$$\sum_{k=0}^{\infty} \frac{(\bar{E}^D\bar{A})^k t^{k\alpha}}{\Gamma(k\alpha + 1)} \bar{E}\bar{E}^D = \sum_{k=0}^{\infty} \frac{[\bar{E}^D\bar{A} + G(\mathbb{I}_n - \bar{E}\bar{E}^D)]^k t^{k\alpha}}{\Gamma(k\alpha + 1)} \bar{E}\bar{E}^D, \tag{2.61a}$$

$$\sum_{k=0}^{\infty} \frac{(\bar{E}^D\bar{A})^k t^{(k+1)\alpha-1}}{\Gamma[(k+1)\alpha]} \bar{E}^D = \sum_{k=0}^{\infty} \frac{[\bar{E}^D\bar{A} + G(\mathbb{I}_n - \bar{E}\bar{E}^D)]^k t^{(k+1)\alpha-1}}{\Gamma[(k+1)\alpha]} \bar{E}^D. \tag{2.61b}$$

Proof. From (2.61a) we have

$$\begin{aligned}
&\sum_{k=0}^{\infty} \frac{[\bar{E}^D\bar{A} + G(\mathbb{I}_n - \bar{E}\bar{E}^D)]^k t^{k\alpha}}{\Gamma(k\alpha + 1)} \bar{E}\bar{E}^D \\
&= \left(\mathbb{I}_n + \frac{[\bar{E}^D\bar{A} + G(\mathbb{I}_n - \bar{E}\bar{E}^D)] t^{\alpha}}{\Gamma(\alpha + 1)} + \frac{[\bar{E}^D\bar{A} + G(\mathbb{I}_n - \bar{E}\bar{E}^D)]^2 t^{2\alpha}}{\Gamma(2\alpha + 1)} + \dots\right) \bar{E}\bar{E}^D \\
&= \left(\mathbb{I}_n + \frac{\bar{E}^D\bar{A} t^{\alpha}}{\Gamma(\alpha + 1)} + \frac{(\bar{E}^D\bar{A})^2 t^{2\alpha}}{\Gamma(2\alpha + 1)}\right) \bar{E}\bar{E}^D = \sum_{k=0}^{\infty} \frac{(\bar{E}^D\bar{A})^k t^{k\alpha}}{\Gamma(k\alpha + 1)} \bar{E}\bar{E}^D,
\end{aligned} \tag{2.62}$$

since $(\mathbb{I}_n - \bar{E}\bar{E}^D)\bar{E}\bar{E}^D = 0$. The equality (2.61b) can be proved in a similar way taking into account that $(\mathbb{I}_n - \bar{E}\bar{E}^D)\bar{E}^D = 0$. □

Theorem 2.8. Let

$$\bar{F}_k = (\bar{E}\bar{A}^D)^k \bar{A}^D \bar{B}, \quad k = 0, 1, \dots, q - 1. \tag{2.63}$$

The fractional descriptor continuous-time linear system (2.28) (or equivalently (2.13)) is positive if and only if there exists a matrix $G \in \mathbb{R}^{n\times n}$ [79, 141] such that

$$\bar{E}^D\bar{A} + G(\mathbb{I}_n - \bar{E}\bar{E}^D) \in \mathbb{M}_n \tag{2.64a}$$

and

$$\mathrm{Im}\bar{E}\bar{E}^D \subset \mathbb{R}_+^n,\quad \bar{E}^D\bar{B} \in \mathbb{R}_+^{n\times m},\quad C \in \mathbb{R}_+^{p\times n},\quad D \in \mathbb{R}_+^{p\times m}, \tag{2.64b}$$

$$(\bar{E}\bar{E}^D - \mathbb{I}_n)\bar{F}_k \in \mathbb{R}_+^{n\times m},\quad k = 0, 1, \ldots, q-1. \tag{2.64c}$$

Proof. Using the superposition principle we will consider components of the solution (2.30a) independently:

1. If $\mathrm{Im}\bar{E}\bar{E}^D \subset \mathbb{R}_+^{n\times n}$, then $\bar{E}\bar{E}^D x_0 \subset \mathbb{R}_+^n$. Therefore, for the matrix G such that $\bar{E}^D\bar{A} + G(\mathbb{I}_n - \bar{E}\bar{E}^D) \in \mathbb{M}_n$ according to Lemma 2.1 we have $\sum_{k=0}^{\infty} \frac{[\bar{E}^D\bar{A}+G(\mathbb{I}_n-\bar{E}\bar{E}^D)]^k t^{k\alpha}}{\Gamma(k\alpha+1)}\bar{E}\bar{E}^D \in \mathbb{R}_+^{n\times n}$ and using Lemma 2.2 we obtain $\sum_{k=0}^{\infty} \frac{[\bar{E}^D\bar{A}+G(\mathbb{I}_n-\bar{E}\bar{E}^D)]^k t^{k\alpha}}{\Gamma(k\alpha+1)}\bar{E}\bar{E}^D x_0 = \sum_{k=0}^{\infty} \frac{(\bar{E}^D\bar{A})^k t^{k\alpha}}{\Gamma(k\alpha+1)}\bar{E}\bar{E}^D x_0 \in \mathbb{R}_+^n$, $t \in [0, +\infty)$.
2. We can write the second component in the form $\int_0^t \Phi(t-\tau)\bar{E}^D\bar{B}u(\tau)d\tau$. For $\bar{E}^D\bar{A} + G(\mathbb{I}_n - \bar{E}\bar{E}^D) \in \mathbb{M}_n$, $\bar{E}^D\bar{B} \in \mathbb{R}_+^{n\times m}$ and $u(t) \in U \subset \mathbb{R}_+^m$ we have $\bar{E}^D \int_0^t \Phi(t-\tau)\bar{B}u(\tau)d\tau \in \mathbb{R}_+^n$.
3. If $(\bar{E}\bar{E}^D - \mathbb{I}_n)\bar{F}_k \in \mathbb{R}_+^{n\times m}$, $k = 0, 1, \ldots, q-1$, where the matrices $\bar{F}_k$ are given by (2.63) and $u^{(k\alpha)}(t) \in \mathbb{R}_+^m$, $k = 0, 1, \ldots, q-1$, $t \in [0, +\infty)$, then the last component of the solution (2.30a) is also nonnegative.

Therefore, we have $x(t) \in \mathbb{R}_+^n$, $t \in [0, +\infty)$ if $\mathrm{Im}\bar{E}\bar{E}^D \subset \mathbb{R}_+^{n\times n}$, $\bar{E}^D\bar{A} + G(\mathbb{I}_n - \bar{E}\bar{E}^D) \in \mathbb{M}_n$, $\bar{E}^D\bar{B} \in \mathbb{R}_+^{n\times m}$ and $(\bar{E}\bar{E}^D - \mathbb{I}_n)\bar{F}_k \in \mathbb{R}_+^{n\times m}$, $k = 0, 1, \ldots, q-1$. Substituting (2.30a) into (2.13b) it is easy to see that $y(t) \in \mathbb{R}_+^p$, $t \in [0, +\infty)$ if $C \in \mathbb{R}_+^{p\times n}$ and $D \in \mathbb{R}_+^{p\times m}$. □

Remark 2.3. [79] Note that the fractional descriptor continuous-time linear system (2.28) can be positive even though the matrix $\bar{E}^D\bar{A}$ is not a Metzler matrix. In the special case for $G = 0$ from (2.64a) we obtain the positivity condition $\bar{E}^D\bar{A} \in \mathbb{M}_n$.

Example 2.7. (Continuation of Example 2.6) Consider the fractional descriptor continuous-time linear system with (2.55). According to (2.56) the matrix pencil (E, A) is regular. We choose $c = 0$ and using (2.28b) we obtain

$$\bar{E} = [-A]^{-1}E = \begin{bmatrix} -0.5 & 0 \\ 0 & 0 \end{bmatrix},\quad \bar{A} = [-A]^{-1}A = \begin{bmatrix} -1 & 0 \\ 0 & -1 \end{bmatrix},$$
$$\bar{B} = [-A]^{-1}B = \begin{bmatrix} -0.5 \\ 0 \end{bmatrix}. \tag{2.65}$$

From (B.1) we have $\bar{E} = \bar{E}_1$, where $\bar{E}_{11} = -0.5$, $\bar{E}_{12} = 0$ and the index of $\bar{E}$ is $q = 1$ since $\mathrm{rank}\bar{E} = \mathrm{rank}\bar{E}^2 = 2$.
Using (B.2) we compute

$$\bar{E}^D = \begin{bmatrix} \bar{E}_{11}^{-1} & \bar{E}_{11}^{-2}\bar{E}_{12} \\ 0 & 0 \end{bmatrix} = \begin{bmatrix} -2 & 0 \\ 0 & 0 \end{bmatrix},$$
$$\bar{A}^D = \bar{A}^{-1} = \bar{A} = \begin{bmatrix} -1 & 0 \\ 0 & -1 \end{bmatrix} \tag{2.66}$$

and

$$\mathrm{Im}\bar{E}\bar{E}^D = \mathrm{Im}\begin{bmatrix} 1 & 0 \\ 0 & 0 \end{bmatrix} \subset \mathbb{R}_+^{2\times 2} \text{ for } v = \begin{bmatrix} v_1 \\ 0 \end{bmatrix} \subset \mathbb{R}_+^2, \tag{2.67a}$$

$$\bar{E}^D\bar{A} = \begin{bmatrix} 2 & 0 \\ 0 & 0 \end{bmatrix} \in \mathbb{M}_2, \quad \bar{E}^D\bar{B} = \begin{bmatrix} 1 \\ 0 \end{bmatrix} \in \mathbb{R}_+^{2\times 1}. \tag{2.67b}$$

According to Lemma 1.3 the matrices (2.67) and the index of $\bar{E}$ do not depend of the choice of c. From (2.63) we have

$$\bar{F}_0 = (\bar{E}\bar{A}^D)^0\bar{A}^D\bar{B} = \begin{bmatrix} 0.5 \\ 0 \end{bmatrix}. \tag{2.68}$$

From (2.67a) and (2.68) we have

$$(\bar{E}\bar{E}^D - \mathbb{I}_2)\bar{F}_0 = \begin{bmatrix} 0 \\ 0 \end{bmatrix} \in \mathbb{R}_+^{2\times 1}. \tag{2.69}$$

Therefore, the fractional descriptor continuous-time linear system with (2.55) is positive.

2.1.6.3 Weierstrass-Kronecker Decomposition Method

In this section necessary and sufficient positivity conditions of the fractional descriptor continuous-time linear system (2.40) (or equivalently (2.13)) will be established using the Weierstrass-Kronecker decomposition method with the assumption that the matrix Q determined by (2.37) is a monomial matrix [59, 84].

Let the decomposition (2.37) of the positive fractional descriptor continuous-time linear system (2.13) be possible for a monomial matrix $Q \in \mathbb{R}_+^{n\times n}$. In this case $x(t) = Q\bar{x}(t) \in \mathbb{R}_+^n$ if and only if $\bar{x}(t) \in \mathbb{R}_+^n$. It is also well-known that premultiplying the equation (2.13a) by a nonsingular matrix P does not change the solution $x(t)$.

Theorem 2.9. [19, 59, 84] Let the decomposition (2.37) of the system (2.13) be possible for a monomial matrix $Q \in \mathbb{R}_+^{n\times n}$. The fractional descriptor continuous-time linear system (2.40) (or equivalently (2.13)) is positive if and only if

$$A_1 \in \mathbb{M}_{n_1}, \quad B_1 \in \mathbb{R}_+^{n_1\times m}, \quad C \in \mathbb{R}_+^{p\times n}, \quad D \in \mathbb{R}_+^{p\times m},$$
$$-N^kB_2 \in \mathbb{R}_+^{n_2\times m}, \quad k = 0, 1, \ldots, \mu - 1. \tag{2.70}$$

Proof. For a monomial matrix $Q \in \mathbb{R}_+^{n\times n}$ we have $x(t) \in \mathbb{R}_+^n$ if $\bar{x}(t) \in \mathbb{R}_+^n$. According to Lemma 2.1 for $A_1 \in \mathbb{M}_{n_1}$ we have $\Phi_{10}(t) \in \mathbb{R}_+^{n\times n}$ and $\Phi_{11}(t) \in \mathbb{R}_+^{n\times n}$. If $A_1 \in \mathbb{M}_{n_1}$ and $B_1 \in \mathbb{R}_+^{n_1\times m}$, then from (2.42) we obtain $\bar{x}_1 \in \mathbb{R}_+^{n_1}$ since from Definition 2.5 $\bar{x}_{10} = Q^{-1}x_{10} \in \mathbb{R}_+^{n_1}$ and $u(t) \in \mathbb{R}_+^m$. If $-N^k B_2 \in \mathbb{R}_+^{n_2\times m}$, $k = 0, 1, \dots, \mu - 1$, then from (2.45) we obtain $\bar{x}_2 \in \mathbb{R}_+^{n_2}$ since from Definition 2.5 $u^{(k\alpha)}(t) \in \mathbb{R}_+^m$, $k = 0, 1, \dots, \mu - 1$. From (2.39b) for $C \in \mathbb{R}_+^{p\times n}$ and $D \in \mathbb{R}_+^{p\times m}$ we have $y(t) \in \mathbb{R}_+^p$ since $x(t) \in \mathbb{R}_+^n$ and $u(t) \in \mathbb{R}_+^m$. □

Example 2.8. (Continuation of Example 2.6) Consider the fractional descriptor continuous-time linear system with (2.55). According to (2.56) the matrix pencil (E, A) is regular. In this case

$$P = \begin{bmatrix} 1 & 1 \\ 0 & -1 \end{bmatrix}, \quad Q = \begin{bmatrix} 1 & 0 \\ 0 & 1 \end{bmatrix} \tag{2.71}$$

and

$$\begin{bmatrix} \mathbb{I}_{n_1} & 0 \\ 0 & N \end{bmatrix} = PEQ = \begin{bmatrix} 1 & 0 \\ 0 & 0 \end{bmatrix},$$
$$\begin{bmatrix} A_1 & 0 \\ 0 & \mathbb{I}_{n_2} \end{bmatrix} = PAQ = \begin{bmatrix} 2 & 0 \\ 0 & 1 \end{bmatrix}, \tag{2.72}$$
$$\begin{bmatrix} B_1 \\ B_2 \end{bmatrix} = PB = \begin{bmatrix} 1 \\ 0 \end{bmatrix}.$$

Hence, $n_1 = 1$ and $n_2 = 1$. The matrix $Q \in \mathbb{R}_+^{3\times 3}$ in (2.71) is monomial. From (2.55) and (2.72) we have

$$\begin{gathered} A_1 = 2 \in \mathbb{M}_1, \quad B_1 = 1 \in \mathbb{R}_+^{1\times 1}, \quad -B_2 = 0 \in \mathbb{R}_+^{1\times 1}, \\ C = \begin{bmatrix} 2 & 1 \end{bmatrix} \in \mathbb{R}_+^{1\times 2}, \quad D = 0 \in \mathbb{R}_+^{1\times 1}. \end{gathered} \tag{2.73}$$

By Theorem 2.9 the fractional descriptor continuous-time linear system with (2.55) is positive.

2.1.7 *Stability of Fractional Descriptor Continuous-Time Linear Systems*

In the following considerations it is assumed that $\det A \neq 0$, i.e. the fractional descriptor continuous-time linear system (2.13) has exactly one equilibrium point [47].

2.1.7.1 Laurent Series Expansion Method

In this section necessary and sufficient stability conditions of the fractional descriptor continuous-time linear system (2.13) will be established using the Laurent series expansion method.

Definition 2.6. The fractional descriptor continuous-time linear system (2.13) is called asymptotically stable if

$$\lim_{t\to\infty} x(t) = 0 \tag{2.74}$$

for all consistent initial conditions $x_0 \in X_0$ and $u(t) = 0$.

Let $\Psi(t)$ be the matrix given by

$$\Psi(t) = \sum_{k=-n_0}^{\infty} \frac{T_k E t^{k\alpha}}{\Gamma(k\alpha + 1)}. \tag{2.75}$$

Definition 2.7. The characteristic equation of the matrix pair (E, A) has the form

$$p_{(E,A)}(\lambda) = \det[E\lambda - A] = a_r\lambda^r + a_{r-1}\lambda^{r-1} + \ldots + a_1\lambda + a_0 = 0, \tag{2.76}$$

where $\lambda = s^\alpha$, $r = \operatorname{rank}\Psi(t)$ and $p_{(E,A)}(\lambda)$ is the characteristic polynomial of the matrix pair (E, A).

Theorem 2.10. [57] The matrix pair (E, A) (or equivalently the system (2.13)) is asymptotically stable if and only if its eigenvalues λ_k, $k = 1, \ldots, r$ (roots of the characteristic equation) satisfy the condition

$$|\arg\lambda_k| > \alpha\frac{\pi}{2}, \quad k = 1, \ldots, r. \tag{2.77}$$

Theorem 2.11. The descriptor continuous-time linear system (1.1) is asymptotically stable if and only if

$$\lim_{t\to\infty} \Psi(t) = 0. \tag{2.78}$$

Proof. The proof can be accomplished in a similar way to the proof of Theorem 1.13. □

From the above considerations we have the following theorem.

Theorem 2.12. The descriptor continuous-time linear system (2.13) is asymptotically stable if and only if one of the following equivalent conditions is satisfied:

1. Roots of the characteristic Eq. (2.76) satisfy the condition (2.77).
2. The matrix $\Psi(t)$ defined by (2.75) satisfies the condition (2.78).

Example 2.9. Consider the fractional descriptor continuous-time linear system (2.13) with

$$E = \begin{bmatrix} -0.5 & 0 \\ 0 & 0 \end{bmatrix}, \quad A = \begin{bmatrix} 4 & 0 \\ 1 & -2 \end{bmatrix}. \tag{2.79}$$

The matrix pencil of (2.79) is regular since

$$\det[Es^{\alpha} - A] = \begin{vmatrix} -0.5s^{\alpha} - 4 & 0 \\ -1 & 2 \end{vmatrix} = -s^{\alpha} - 8 \neq 0. \tag{2.80}$$

The matrix $[Es^{\alpha} - A]^{-1}$ has the form

$$[Es^{\alpha} - A]^{-1} = \begin{bmatrix} -0.5s^{\alpha} - 4 & 0 \\ -1 & 2 \end{bmatrix}^{-1} = \begin{bmatrix} -\frac{2}{s^{\alpha}+8} & 0 \\ -\frac{1}{s^{\alpha}+8} & 0.5 \end{bmatrix}. \tag{2.81}$$

Using (2.19) for $n_0 = 1$ the matrix (2.81) can be written in the form

$$[Es^{\alpha} - A]^{-1} = \sum_{k=-1}^{\infty} T_k s^{-(k+1)\alpha}, \tag{2.82a}$$

where

$$T_k = \begin{cases} \begin{bmatrix} 0 & 0 \\ 0 & 0.5 \end{bmatrix} & \text{for } k = -1, \\ \begin{bmatrix} -(-8)^k & 0 \\ -2(-8)^k & 0 \end{bmatrix} & \text{for } k = 0, 1, 2, \ldots. \end{cases} \tag{2.82b}$$

Using (1.107) and (1.110b) we have

$$T_k E = \begin{bmatrix} 0.5(-8)^k & 0 \\ (-8)^k & 0 \end{bmatrix} \quad \text{for } k = 0, 1, 2, \ldots \tag{2.83}$$

and

$$\begin{aligned} \Psi(t) &= \sum_{k=0}^{\infty} \frac{T_k E t^{k\alpha}}{\Gamma(k\alpha + 1)} = \sum_{k=0}^{\infty} \begin{bmatrix} \frac{0.5(-8)^k t^{k\alpha}}{\Gamma(k\alpha+1)} & 0 \\ \frac{(-8)^k t^{k\alpha}}{\Gamma(k\alpha+1)} & 0 \end{bmatrix} \\ &= \begin{bmatrix} 0.5E_{\alpha}(-8t^{\alpha}) & 0 \\ E_{\alpha}(-8t^{\alpha}) & 0 \end{bmatrix}, \end{aligned} \tag{2.84}$$

where $E_{\alpha}(z)$ is the one parameter Mittag-Leffler function defined by (2.5). By the second condition of Theorem 2.12 the fractional descriptor system (2.13) with (2.79) is asymptotically stable for any $\alpha \in (0; 1)$ since $\lim_{t\to\infty} E_{\alpha}(-8t^{\alpha}) = 0$.
The characteristic equation of the system (2.79) is given by

$$p_{(E,A)}(s^{\alpha}) = \det[Es^{\alpha} - A] = \begin{vmatrix} -0.5s^{\alpha} - 4 & 0 \\ -1 & 2 \end{vmatrix} = -s^{\alpha} - 8 = 0. \tag{2.85}$$

From (2.85) we can see that the pair (E, A) has one eigenvalue $s_1^\alpha = -8$ and by Theorem 2.12 the descriptor system (2.13) with (2.79) is asymptotically stable for any $\alpha \in (0; 1)$ since $|\arg s_1^\alpha| = \pi > \alpha\frac{\pi}{2}$.

2.1.7.2 Drazin Inverse Matrix Method

In this section necessary and sufficient stability conditions of the fractional descriptor continuous-time linear system (2.28) (or equivalently (2.13)) will be established using the Drazin inverse matrix method.

Definition 2.6 of the asymptotic stability of the fractional descriptor continuous-time linear system is suitable for this method since the solutions of (2.13) and (2.28) are equivalent.

Definition 2.8. The characteristic equation of the matrix pair $(\bar{E}, \bar{A})$ is given by

$$p_{(\bar{E},\bar{A})}(\lambda) = \det[\bar{E}\lambda - \bar{A}] = \bar{a}_r\lambda^r + \bar{a}_{r-1}\lambda^{r-1} + \ldots + \bar{a}_1\lambda + \bar{a}_0 = 0, \quad (2.86)$$

where $\lambda = s^\alpha$, $r = \operatorname{rank}\bar{E}^D\bar{A}$ and $p_{(\bar{E},\bar{A})}(\lambda)$ is the characteristic polynomial of the matrix pair $(\bar{E}, \bar{A})$.

Lemma 2.3. [11] The characteristic polynomials of the matrix pairs (E, A) and $(\bar{E}, \bar{A})$ are related by

$$p_{(\bar{E},\bar{A})}(\lambda) = \frac{p_{(E,A)}(\lambda)}{\det[Ec - A]}. \quad (2.87)$$

Proof. The proof is similar to the proof of Lemma 1.7. □

Remark 2.4. The characteristic equations of the matrix pairs (E, A) and $(\bar{E}, \bar{A})$ have the same form. Therefore, both pairs have the same set of eigenvalues and Theorem 2.10 can be used for the roots of the Eq. (2.86).

Theorem 2.13. The fractional descriptor continuous-time linear system (2.13) is asymptotically stable if and only if the matrix $\bar{E}^D\bar{A}$ has r stable eigenvalues (satisfying the condition (2.77)) and $n - r$ zero eigenvalues.

Proof. The proof follows immediately from Lemma 1.8. □

Based on the above considerations we have the following Theorem.

Theorem 2.14. The fractional descriptor continuous-time linear system (2.13) is asymptotically stable if and only if one of the following equivalent conditions is satisfied:

1. Roots of the characteristic Eq. (2.86) satisfy the condition (2.77).
2. The matrix $\bar{E}^D\bar{A}$ has r stable eigenvalues and $n - r$ zero eigenvalues.

Example 2.10. (Continuation of Example 2.9) Consider the fractional descriptor continuous-time linear system (2.13) with (2.79). According to (2.80) the matrix pencil (E, A) is regular. We choose $c = 0$ and using (2.28b) we obtain

$$\bar{E} = [-A]^{-1}E = \begin{bmatrix} 0.125 & 0 \\ 0.0625 & 0 \end{bmatrix}, \quad \bar{A} = [-A]^{-1}A = \begin{bmatrix} -1 & 0 \\ 0 & -1 \end{bmatrix}. \tag{2.88}$$

From (B.1) we have $\bar{E} = \bar{E}_2$, where $\bar{E}_{21} = 0.125$, $\bar{E}_{22} = 0.0625$ and the index of $\bar{E}$ is $q = 1$ since $\operatorname{rank}\bar{E} = \operatorname{rank}\bar{E}^2 = 1$. Using (B.2) we compute

$$\bar{E}^D = \begin{bmatrix} \bar{E}_{21}^{-1} & 0 \\ \bar{E}_{22}\bar{E}_{21}^{-2} & 0 \end{bmatrix} = \begin{bmatrix} 8 & 0 \\ 4 & 0 \end{bmatrix}, \quad \bar{A}^D = \bar{A}^{-1} = \bar{A} = \begin{bmatrix} -1 & 0 \\ 0 & -1 \end{bmatrix}. \tag{2.89}$$

From (2.86) and (2.88) we have

$$p_{(\bar{E},\bar{A})}(s^\alpha) = \det[\bar{E}s^\alpha - \bar{A}] = \begin{vmatrix} 0.125s^\alpha + 1 & 0 \\ 0.0625s^\alpha & 1 \end{vmatrix} = 0.125s^\alpha + 1 = 0. \tag{2.90}$$

The pair $(\bar{E}, \bar{A})$ has one eigenvalue $s_1^\alpha = -8$, which is equal to the eigenvalue of the pair (E, A). By Theorem 2.14 the fractional descriptor system (2.13) with (2.79) is asymptotically stable for any $\alpha \in (0; 1)$ since $|\arg s_1^\alpha| = \pi > \alpha\frac{\pi}{2}$.
Using (2.89) and (1.117) we obtain

$$p_{\bar{E}^D\bar{A}}(s^\alpha) = \det[\mathbb{I}_3 s^\alpha - \bar{E}^D\bar{A}] = \begin{vmatrix} s^\alpha + 8 & 0 \\ 4 & s^\alpha \end{vmatrix} = s^\alpha(s^\alpha + 8) = 0. \tag{2.91}$$

It is easy to see that $r = \operatorname{rank}\bar{E}^D\bar{A} = 1$. According to (2.91) the matrix $\bar{E}^D\bar{A}$ has $r = 1$ eigenvalue of the pair $(\bar{E}, \bar{A})$ equal to $s_1^\alpha = -8$ and $n - r = 1$ zero eigenvalue $s_2^\alpha = 0$. Therefore, by Theorem 2.14 the considered system is asymptotically stable for any $\alpha \in (0; 1)$.

2.1.7.3 Weierstrass-Kronecker Decomposition Method

In this section necessary and sufficient stability conditions of the fractional descriptor continuous-time linear system (2.40) (or equivalently (2.13)) using the Weierstrass-Kronecker decomposition method.

Note that the state vector of the subsystem (2.40b) $\bar{x}_2(t) = 0$ for $u(t) = 0$. Therefore, the stability of the system (2.40) (or equivalently (2.13)) depends entirely on the subsystem (2.40a).

Definition 2.9. The fractional descriptor continuous-time linear system (2.40) (or equivalently (2.13)) is called asymptotically stable if

$$\lim_{t \to \infty} \bar{x}_1(t) = 0 \tag{2.92}$$

for all initial conditions $\bar{x}_{10} \in \mathbb{R}^{n_1}$ and $u(t) = 0$.

Definition 2.10. [57] The characteristic equation of the matrix A_1 is given by

$$p_{A_1}(\lambda) = \det[\mathbb{I}_{n_1}\lambda - A_1] = \lambda^{n_1} + \tilde{a}_{n_1-1}\lambda^{n_1-1} + \ldots + \tilde{a}_1\lambda + \tilde{a}_0 = 0, \tag{2.93}$$

where $\lambda = s^\alpha$, $n_1 = \deg\{\det[E\lambda - A]\}$ and $p_{A_1}(\lambda)$ is the characteristic polynomial of the matrix A_1.

Lemma 2.4. [11] The characteristic polynomials of the matrix pair (E, A) and of the matrix A_1 are related by

$$p_{(E,A)}(\lambda) = \frac{(-1)^{n_2}}{\det(PQ)} p_{A_1}(\lambda). \tag{2.94}$$

Proof. The proof is similar to the proof of Lemma 1.9. □

Remark 2.5. The characteristic equations of the matrix pair (E, A) and of the matrix A_1 have the same form. Therefore, the matrix pair (E, A) and the matrix A_1 have the same set of eigenvalues and Theorem 2.10 can be used for the roots of the Eq. (2.93).

Theorem 2.15. [57] The matrix A_1 (or equivalently the system (2.40)) is asymptotically stable if and only if its eigenvalues satisfy the condition (2.77).

Example 2.11. (Continuation of Example 2.9) Consider the fractional descriptor continuous-time linear system (2.13) with (2.79). According to (2.80) the matrix pencil (E, A) is regular. In this case

$$P = \begin{bmatrix} -2 & 0 \\ 0 & 1 \end{bmatrix}, \quad Q = \begin{bmatrix} 1 & 0 \\ 0.5 & -0.5 \end{bmatrix} \tag{2.95}$$

and

$$\begin{gathered} \begin{bmatrix} \mathbb{I}_{n_1} & 0 \\ 0 & N \end{bmatrix} = PEQ = \begin{bmatrix} 1 & 0 \\ 0 & 0 \end{bmatrix}, \\ \begin{bmatrix} A_1 & 0 \\ 0 & \mathbb{I}_{n_2} \end{bmatrix} = PAQ = \begin{bmatrix} -8 & 0 \\ 0 & 1 \end{bmatrix}. \end{gathered} \tag{2.96}$$

Hence, $n_1 = 1$ and $n_2 = 1$. Using (2.93) and (2.96) we obtain

$$p_{A_1}(s^\alpha) = \det[\mathbb{I}_1 s^\alpha - A_1] = s^\alpha + 8 = 0. \tag{2.97}$$

Eigenvalue of the matrix A_1 is $s_1^\alpha = -8$ and is equal to the eigenvalue of the pair (E, A). By Theorem 2.15 the considered fractional descriptor system is asymptotically stable for any $\alpha \in (0; 1)$ since $|\arg s_1^\alpha| = \pi > \alpha\frac{\pi}{2}$.

2.1.8 Superstability of Fractional Descriptor Continuous-Time Linear Systems

The asymptotic stability of a dynamical system ensures that its free response decreases to zero for $t \to \infty$, however its value may increase significantly in the initial part of the state vector trajectory (it is so-called "peak effect"). In superstable systems the norm of the state vector decreases monotonically to zero for $t \to \infty$, which prevents such undesirable effects [70, 137, 138].

2.1.8.1 Properties of Superstable Fractional Continuous-Time Linear Systems

The concept of superstable continuous-time linear systems is given in Sect. 1.1.5.1. Consider the standard fractional continuous-time linear system

$$
{}_0D_t^{\alpha} x(t) = Ax(t), \quad 0 < \alpha < 1, \tag{2.98}
$$

where $x(t) \in \mathbb{R}^n$ is the state vector, ${}_0D_t^{\alpha}$ is the Caputo derivative-integral operator defined by (2.8) and $A \in \mathbb{R}^{n \times n}$. The solution to the Eq. (2.98) can be obtained using (2.21) for $u_i = 0$, $i \in \mathbb{Z}_+$, i.e. it has the form

$$
x_i = \Phi_0(t)x_0 = E_\alpha(At^\alpha)x_0, \tag{2.99}
$$

where $\Phi_0(t)$ is given by (2.21b) and $E_\alpha(At^\alpha)$ is the one parameter Mittag-Leffler function defined by (2.5).

Lemma 2.5. For a superstable matrix A (i.e. satisfying (1.133)) we have

$$
\|E_\alpha(At^\alpha)\| \le E_\alpha(-\sigma t^\alpha), \tag{2.100}
$$

where σ is the superstability degree of the matrix A.

Proof. The proof can be accomplished in a similar way as it has been shown in [137] for the norm of matrix exponential. For a small $t^\alpha = \delta t^\alpha$ we have

$$
E_\alpha(A\delta t^\alpha) \approx \mathbb{I}_n + \frac{A\delta t^\alpha}{\Gamma(\alpha+1)} \tag{2.101}
$$

and

$$\begin{aligned}
\|E_\alpha(A\delta t^\alpha)\| &\approx \max_{1\le i\le n}\left(\left|1+a_{ii}\frac{\delta t^\alpha}{\Gamma(\alpha+1)}\right|+\sum_{\substack{j=1\\ j\neq i}}^{n}\left|a_{ij}\right|\frac{\delta t^\alpha}{\Gamma(\alpha+1)}\right)\\
&= \max_{1\le i\le n}\left[1+\left(a_{ii}+\sum_{\substack{j=1\\ j\neq i}}^{n}\left|a_{ij}\right|\right)\frac{\delta t^\alpha}{\Gamma(\alpha+1)}\right]\le 1-\frac{\sigma\delta t^\alpha}{\Gamma(\alpha+1)}\\
&\le E_\alpha(-\sigma\delta t^\alpha).
\end{aligned} \tag{2.102}$$

Hence, for an arbitrary $t^\alpha = n\delta t^\alpha$ we obtain

$$\begin{aligned}
\|E_\alpha(At^\alpha)\| = \|E_\alpha(An\delta t^\alpha)\| &\le \|E_\alpha(A\delta t^\alpha)\|^n\\
&\le [E_\alpha(-\sigma\delta t^\alpha)]^n = E_\alpha(-\sigma t^\alpha).
\end{aligned} \tag{2.103}$$

□

Theorem 2.16. If the fractional system (2.99) is superstable, then

$$\|x(t)\| \le \|x_0\| E_\alpha(-\sigma t^\alpha), \quad t \in [0, +\infty). \tag{2.104}$$

Proof. The proof follows immediately from Lemma 2.5. □

Therefore, from the above considerations it follows that if the matrix A satisfies the condition (1.133), then the norm of the state vector decreases monotonically to zero for $t \to \infty$.

Example 2.12. Consider the fractional continuous-time linear system (2.98) with $\alpha = 0.9$ and

$$A = \begin{bmatrix} -2 & 1 \\ 0 & -2 \end{bmatrix}. \tag{2.105}$$

It is easy to check that the considered system is asymptotically stable since the eigenvalues of (2.105) are $s_1^{0.9} = s_2^{0.9} = -2$ and the condition (2.77) is satisfied. From (1.133) we have $\sigma = 1 > 0$ and the system is also superstable. In Fig. 2.1 we present time plots of the state variables and the norm of the state vector for $x_0 = [\,1 \quad 1\,]^T$, from which we can see that the norm decreases monotonically to zero.

Now let us consider the fractional continuous-time linear system (2.98) with $\alpha = 0.9$ and

$$A = \begin{bmatrix} -2 & 4 \\ 0 & -2 \end{bmatrix}. \tag{2.106}$$

Note that the eigenvalues of (2.105) and (2.106) are the same. In this case the system is not superstable since by the condition (1.133) we have $\sigma = -2 < 0$. In Fig. 2.2 we show time plots of the state variables and the norm of the state vector for $x_0 = [\,1 \quad 1\,]^T$, from which we can see that the norm does not decrease monotonically to zero.

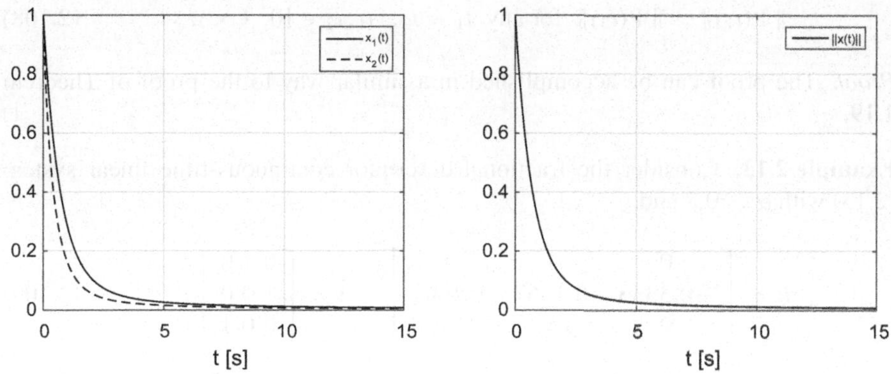

Fig. 2.1 State variables (on the left) and norm of the state vector (on the right) of the fractional continuous-time linear system (2.98) with (2.105) and $\alpha = 0.9$ for $x_0 = [\ 1 \quad 1\ ]^T$

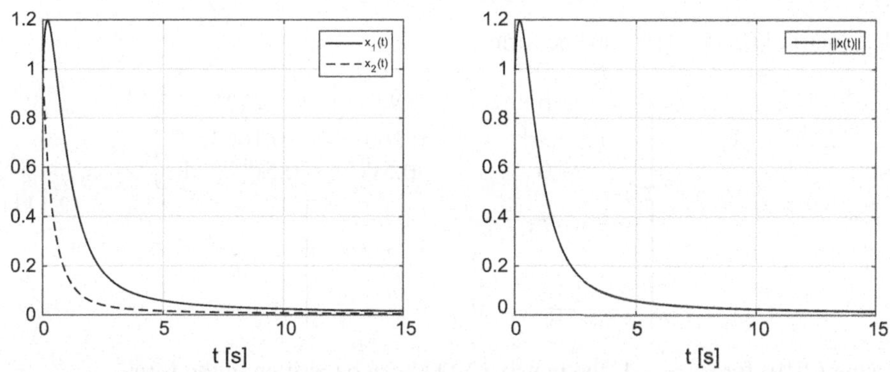

Fig. 2.2 State variables (on the left) and norm of the state vector (on the right) of the fractional continuous-time linear system (2.98) with (2.106) and $\alpha = 0.9$ for $x_0 = [\ 1 \quad 1\ ]^T$

2.1.8.2 Laurent Series Expansion Method

In this section necessary and sufficient superstability conditions of the descriptor continuous-time linear system (2.13) will be established using the Laurent series expansion method.

Consider the fractional descriptor continuous-time linear system (2.13). Solution to the Eq. (2.13a) for $u(t) = 0$ has the form

$$x(t) = \Psi(t)x_0, \tag{2.107}$$

where $\Psi(t)$ is defined by (2.75).

Theorem 2.17. The fractional descriptor continuous-time linear system (2.13) is superstable if and only if

$$\|\Psi(t_1)\| > \|\Psi(t_2)\| \text{ for any } t_1 < t_2, \quad t_1, t_2 \in [0, +\infty). \tag{2.108}$$

Proof. The proof can be accomplished in a similar way to the proof of Theorem 1.19. □

Example 2.13. Consider the fractional descriptor continuous-time linear system (2.13) with $\alpha = 0.8$ and

$$E = \begin{bmatrix} 0 & -1 & -1 \\ -0.3333 & -0.1667 & -0.1667 \\ 0 & 0.5 & 0.5 \end{bmatrix}, \quad A = \begin{bmatrix} 0 & 1 & 0 \\ 1 & 0 & 0 \\ 0 & 0 & 1 \end{bmatrix}. \tag{2.109}$$

The matrix pencil (E, A) of (2.109) is regular since

$$\det[Es^{\alpha} - A] = 0.1667s^{1.6} + 0.8333s^{0.8} + 1 \neq 0. \tag{2.110}$$

The matrix $[Es^{\alpha} - A]^{-1}$ has the form

$$\begin{aligned}[Es^{0.8} - A]^{-1} &= \begin{bmatrix} 0 & -s^{0.8} - 1 & -s^{0.8} \\ -0.3333s^{0.8} - 1 & -0.1667s^{0.8} & -0.1667s^{0.8} \\ 0 & 0.5s^{0.8} & 0.5s^{0.8} - 1 \end{bmatrix}^{-1} \\ &= \begin{bmatrix} \frac{s^{0.8}}{s^{1.6}+5s^{0.8}+6} & -\frac{3}{s^{0.8}+3} & \frac{s^{0.8}}{s^{1.6}+5s^{0.8}+6} \\ \frac{s^{0.8}-2}{s^{0.8}+2} & 0 & \frac{2s^{0.8}}{s^{0.8}+2} \\ -\frac{s^{0.8}}{s^{0.8}+2} & 0 & -\frac{2s^{0.8}+2}{s^{0.8}+2} \end{bmatrix}.\end{aligned} \tag{2.111}$$

Using (2.19) for $n_0 = -1$, the matrix (2.111) can be written in the form

$$[Es^{0.8} - A]^{-1} = \sum_{k=-1}^{\infty} T_k s^{-0.8(k+1)}, \tag{2.112a}$$

where

$$T_k = \begin{cases} \begin{bmatrix} 0 & 0 & 0 \\ 1 & 0 & 2 \\ -1 & 0 & -2 \end{bmatrix} & \text{for } k = -1, \\ \begin{bmatrix} 3(-3)^k - 2(-2)^k & -3(-3)^k & 3(-3)^k - 2(-2)^k \\ -4(-2)^k & 0 & -4(-2)^k \\ 2(-2)^k & 0 & 2(-2)^k \end{bmatrix} & \text{for } k = 0, 1, \ldots. \end{cases} \tag{2.112b}$$

From (2.109) and (2.112b) we have

$$T_k E = \begin{cases} \begin{bmatrix} 0 & 0 & 0 \\ 0 & 0 & 0 \\ 0 & 0 & 0 \end{bmatrix} & \text{for } k = -1, \\ \begin{bmatrix} (-3)^k & (-2)^k - (-3)^k & (-2)^k - (-3)^k \\ 0 & 2(-2)^k & 2(-2)^k \\ 0 & -(-2)^k & -(-2)^k \end{bmatrix} & \text{for } k = 0, 1, 2, \dots \end{cases} \tag{2.113}$$

and

$$\begin{aligned} \Psi(t) &= \sum_{k=0}^{\infty} \frac{T_k E t^{k\alpha}}{\Gamma(k\alpha+1)} \\ &= \sum_{k=0}^{\infty} \begin{bmatrix} \frac{(-3)^k t^{0.8k}}{\Gamma(0.8k+1)} & \frac{[(-2)^k-(-3)^k]t^{0.8k}}{\Gamma(0.8k+1)} & \frac{[(-2)^k-(-3)^k]t^{0.8k}}{\Gamma(0.8k+1)} \\ 0 & \frac{2(-2)^k t^{0.8k}}{\Gamma(0.8k+1)} & \frac{2(-2)^k t^{0.8k}}{\Gamma(0.8k+1)} \\ 0 & -\frac{(-2)^k t^{0.8k}}{\Gamma(0.8k+1)} & -\frac{(-2)^k t^{0.8k}}{\Gamma(0.8k+1)} \end{bmatrix} \\ &= \begin{bmatrix} E_{0.8}(-3t^{0.8}) & \psi_{12}(t) & \psi_{13}(t) \\ 0 & 2E_{0.8}(-2t^{0.8}) & 2E_{0.8}(-2t^{0.8}) \\ 0 & -2E_{0.8}(-2t^{0.8}) & -2E_{0.8}(-2t^{0.8}) \end{bmatrix}, \end{aligned} \tag{2.114}$$

where $\psi_{12}(t) = \psi_{13}(t) = E_{0.8}(-2t^{0.8}) - E_{0.8}(-3t^{0.8})$ and $E_\alpha(z)$ is the one parameter Mittag-Leffler function defined by (2.5). Analyzing (2.114) we can see that

$$\max_t \|\Psi(t)\| = \|\Psi(0)\| = \left\| \begin{bmatrix} 1 & 0 & 0 \\ 0 & 2 & 2 \\ 0 & -2 & -2 \end{bmatrix} \right\| = 4 \tag{2.115}$$

and $\|\Psi(t_1)\| > \|\Psi(t_2)\|$ for any $t_1 < t_2$, $t_1, t_2 \in [0, +\infty)$. By Theorem 2.17 the fractional descriptor continuous-time linear system (2.13) with $\alpha = 0.8$ and (2.109) is superstable. Using (2.15) we can determine consistent initial conditions of the system

$$x_0 = T_0 E v = \begin{bmatrix} 1 & 0 & 0 \\ 0 & 2 & 2 \\ 0 & -1 & -1 \end{bmatrix} \begin{bmatrix} v_1 \\ v_2 \\ v_3 \end{bmatrix} = \begin{bmatrix} v_1 \\ 2(v_2+v_3) \\ -(v_2-v_3) \end{bmatrix}, \tag{2.116}$$

where $v_1, v_2, v_3 \in \mathbb{R}$ are any numbers. In Fig. 2.3 we present time plots of the state variables and the norm of the state vector for $x_0 = [\,0.5 \quad 2 \quad -1\,]^T$, from which we can see that the norm of the state vector decreases monotonically to zero for $t \to \infty$.

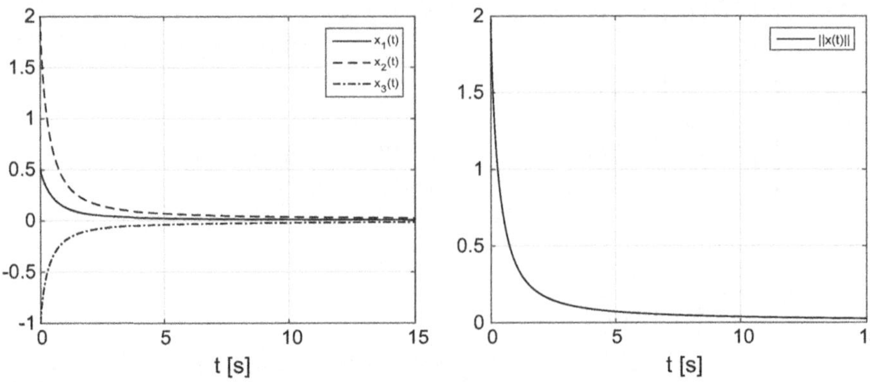

Fig. 2.3 State variables (on the left) and norm of the state vector (on the right) of the fractional descriptor continuous-time linear system (2.13) with $\alpha = 0.8$ and (2.109) for $x_0 = [\ 0.5 \quad 2 \quad -1\ ]^T$

2.1.8.3 Drazin Inverse Matrix Method

In this section necessary and sufficient superstability conditions of the fractional descriptor continuous-time linear system (2.28) (or equivalently (2.13)) will be established using the Drazin inverse matrix method.

Consider the fractional descriptor continuous-time linear system (2.28). Solution to the Eq. (2.28a) for $u(t) = 0$ has the form

$$x(t) = \Phi_0(t)\bar{E}\bar{E}^D x_0 = E_\alpha(\bar{E}^D\bar{A}t^\alpha)x_0, \tag{2.117}$$

where $\Phi_0(t)$ is given by (2.30b) and $E_\alpha(\bar{E}^D\bar{A}t^\alpha)$ is the one parameter Mittag-Leffler function defined by (2.5).

Using Lemma 2.2 from (2.117) we have

$$x(t) = E_\alpha(\bar{E}^D\bar{A}t^\alpha)\bar{E}\bar{E}^D x_0 = E_\alpha\left([\bar{E}^D\bar{A} + G(\mathbb{I}_n - \bar{E}\bar{E}^D)]t^\alpha\right)\bar{E}\bar{E}^D x_0, \tag{2.118}$$

where $G \in \mathbb{R}^{n\times n}$ is an arbitrary matrix.
Taking into account (2.30) and (2.31) for $u(t) = 0$ it is easy to see that if $x_0 \in \operatorname{Im}\bar{E}\bar{E}^D$, then $x_0 = \bar{E}\bar{E}^D x_0$. Therefore, from (2.118) we have

$$\|x(t)\| \le \left\|E_\alpha\left([\bar{E}^D\bar{A} + G(\mathbb{I}_n - \bar{E}\bar{E}^D)]t^\alpha\right)\right\| \|x_0\|. \tag{2.119}$$

If the matrix $\bar{E}^D\bar{A} + G(\mathbb{I}_n - \bar{E}\bar{E}^D)$ is superstable (i.e. it satisfies the condition (1.133)), then by Lemma 2.5 we have $\left\|E_\alpha\left([\bar{E}^D\bar{A} + G(\mathbb{I}_n - \bar{E}\bar{E}^D)]t^\alpha\right)\right\| \le E_\alpha(-\sigma t^\alpha)$. From the inequality (2.119) we obtain

$$\|x(t)\| \le \|x_0\| E_\alpha(-\sigma t^\alpha), \tag{2.120}$$

and the norm of the state vector decreases monotonically to zero. Therefore, the following theorem has been proved.

Theorem 2.18. The fractional descriptor continuous-time linear system (2.28) (or equivalently (2.13)) is superstable if and only if there exists a matrix $G \in \mathbb{R}^{n\times n}$ such that the matrix $\bar{E}^D\bar{A} + G(\mathbb{I}_n - \bar{E}\bar{E}^D)$ satisfies the condition (1.133).

Example 2.14. (Continuation of Example 2.13) Consider the fractional descriptor continuous-time linear system (2.13) with $\alpha = 0.8$ and (2.109). According to (2.110) the matrix pencil (E, A) is regular. Using (2.28b) for $c = 0$ we obtain

$$\bar{E} = [-A]^{-1}E = \begin{bmatrix} 0.3333 & 0.1667 & 0.1667 \\ 0 & 1 & 1 \\ 0 & -0.5 & -0.5 \end{bmatrix},$$
$$\bar{A} = [-A]^{-1}A = \begin{bmatrix} -1 & 0 & 0 \\ 0 & -1 & 0 \\ 0 & 0 & -1 \end{bmatrix}. \tag{2.121}$$

Next, from one of the methods presented in Appendix B we have

$$\bar{E}^D = \begin{bmatrix} 3 & -1 & -1 \\ 0 & 4 & 4 \\ 0 & -2 & -2 \end{bmatrix} \tag{2.122}$$

and

$$\bar{E}^D\bar{A} = \begin{bmatrix} -3 & 1 & 1 \\ 0 & -4 & -4 \\ 0 & 2 & 2 \end{bmatrix}, \quad \bar{E}\bar{E}^D = \begin{bmatrix} 1 & 0 & 0 \\ 0 & 2 & 2 \\ 0 & -1 & -1 \end{bmatrix}. \tag{2.123}$$

The matrix $\bar{E}^D\bar{A}$ given by (2.123) does not satisfy the condition (1.133). However, using (2.123) and

$$G = \begin{bmatrix} 0 & 0 & 0 \\ 0 & 0 & 2 \\ 0 & 2 & 0 \end{bmatrix} \tag{2.124}$$

we obtain

$$\bar{E}^D\bar{A} + G(\mathbb{I}_3 - \bar{E}\bar{E}^D) = \begin{bmatrix} -3 & 1 & 1 \\ 0 & -2 & 0 \\ 0 & 0 & -2 \end{bmatrix}, \tag{2.125}$$

which satisfies (1.133). Therefore, by Theorem 2.18 the considered system is superstable. The time plots of the state variables and the norm of the state vector for $x_0 = [\,0.5 \;\; 2 \;\; -1\,]^T$ are presented in Fig. 2.3.

2.1.8.4 Weierstrass-Kronecker Decomposition Method

In this section necessary and sufficient superstability conditions of the descriptor continuous-time linear system (2.40) (or equivalently (2.13)) using the Weierstrass-Kronecker decomposition method.

Note that every superstable system is also asymptotically stable. State vector of the subsystem (2.40b) $\bar{x}_2 = 0$ for $u(t) = 0$. Using (2.38) and (2.42) we obtain

$$\begin{aligned} x(t) &= Q\bar{x}(t) = Q\begin{bmatrix} \bar{x}_1(t) \\ \bar{x}_2(t) \end{bmatrix} = Q\begin{bmatrix} E_\alpha(A_1 t^\alpha)\bar{x}_{10} \\ 0 \end{bmatrix} \\ &= Q\begin{bmatrix} E_\alpha(A_1 t^\alpha) & 0 \\ 0 & E_\alpha(M t^\alpha) \end{bmatrix}\begin{bmatrix} \bar{x}_{10} \\ 0 \end{bmatrix} \\ &= Q\begin{bmatrix} E_\alpha(A_1 t^\alpha) & 0 \\ 0 & E_\alpha(M t^\alpha) \end{bmatrix} Q^{-1} x_0 \\ &= E_\alpha\left(Q\begin{bmatrix} A_1 & 0 \\ 0 & M \end{bmatrix} Q^{-1}\right) x_0 = E_\alpha(A_M t^\alpha) x_0, \end{aligned} \tag{2.126}$$

where $E_\alpha(A_M t^\alpha)$ is the one parameter Mittag-Leffler function defined by (2.5), the matrix $M \in \mathbb{R}^{n_2 \times n_2}$ is arbitrary and

$$A_M = Q\begin{bmatrix} A_1 & 0 \\ 0 & M \end{bmatrix} Q^{-1}. \tag{2.127}$$

From the Eq. (2.126) we have

$$\|x(t)\| = \|E_\alpha(A_M t^\alpha) x_0\| \leq \|E_\alpha(A_M t^\alpha)\| \|x_0\|. \tag{2.128}$$

If the matrix A_M is superstable (i.e. it satisfies the condition (1.133)), then by Lemma 2.5 we have $\|E_\alpha(A_M t^\alpha)\| \leq E_\alpha(-\sigma t^\alpha)$. From the inequality (2.128) we obtain

$$\|x(t)\| \leq \|x_0\| E_\alpha(-\sigma t^\alpha), \tag{2.129}$$

and the norm of the state vector decreases monotonically to zero. Therefore, the following theorem has been proved.

Theorem 2.19. The fractional descriptor continuous-time linear system (2.40) (or equivalently (2.13)) is superstable if and only if there exists a matrix $M \in \mathbb{R}^{n_2 \times n_2}$ such that the matrix A_M defined by (2.127) satisfies the condition (1.135).

Example 2.15. (Continuation of Example 2.13) Consider the fractional descriptor continuous-time linear system (2.13) with $\alpha = 0.8$ and (2.109). According to (2.110) the matrix pencil (E, A) is regular. In this case

$$P = \begin{bmatrix} -2 & 0 & -2 \\ 1 & -3 & 1 \\ 0.5 & 0 & 1 \end{bmatrix}, \quad Q = \begin{bmatrix} 0 & 1 & 0 \\ 2 & 0 & -2 \\ -1 & 0 & 2 \end{bmatrix} \tag{2.130}$$

and

$$\begin{aligned} \begin{bmatrix} \mathbb{I}_{n_1} & 0 \\ 0 & N \end{bmatrix} &= PEQ = \begin{bmatrix} 1 & 0 & 0 \\ 0 & 1 & 0 \\ 0 & 0 & 0 \end{bmatrix}, \\ \begin{bmatrix} A_{1d} & 0 \\ 0 & \mathbb{I}_{n_2} \end{bmatrix} &= PAQ = \begin{bmatrix} -2 & 0 & 0 \\ 1 & -3 & 0 \\ 0 & 0 & 1 \end{bmatrix}, \end{aligned} \tag{2.131}$$

where $n_1 = 2$ and $n_2 = 1$. Using (2.127) for the matrices (2.130), (2.131) and $M = [m] \in \mathbb{R}^{1\times 1}$ we obtain

$$\begin{aligned} A_M &= \begin{bmatrix} 0 & 1 & 0 \\ 2 & 0 & -2 \\ -1 & 0 & 2 \end{bmatrix} \begin{bmatrix} -2 & 0 & 0 \\ 1 & -3 & 0 \\ 0 & 0 & m \end{bmatrix} \begin{bmatrix} 0 & 1 & 1 \\ 1 & 0 & 0 \\ 0 & 0.5 & 1 \end{bmatrix} \\ &= \begin{bmatrix} -3 & 1 & 1 \\ 0 & -4-m & -4-2m \\ 0 & 2+m & 2+2m \end{bmatrix}. \end{aligned} \tag{2.132}$$

Assuming $m = -2$ the matrix (2.132) satisfies the condition (1.133). Therefore, by Theorem 2.19 the considered system is superstable.

2.2 Discrete-Time Systems

2.2.1 Grünwald-Letnikov Fractional-Order Backward Difference

Definition 2.11. A discrete-time function defined by

$$\begin{gathered} \Delta^n x_i = \Delta^{n-1} x_i - \Delta^{n-1} x_{i-1} = \sum_{k=0}^{n} (-1)^k \binom{n}{k} x_{i-k}, \\ i = 1, 2, 3, \dots, \quad n \in \mathbb{Z}_+, \quad x_i \in \mathbb{R}, \end{gathered} \tag{2.133}$$

where

$$\binom{n}{k} = \frac{n!}{k!(n-k)!} = \frac{n(n-1)\dots(n-k+1)}{k!}, \tag{2.134}$$

is called the n-order (backward) difference of the function x_i.

Definition 2.12. The fractional n-order (backward) difference on the interval $[0, k]$ of the function x_i is defined as follows

$$ {}_0\Delta_k^n x_i = \sum_{j=0}^{k} (-1)^j \binom{n}{j} x_{i-j}. \tag{2.135}$$

From (2.133) it follows that the n-order difference can be written as a linear combination of the values of discrete-time function in $n+1$ points.
The definitions are valid for n being natural numbers and integers. Note that (2.135) is also well defined for fractional and real numbers. In general case n can be also a complex number.

Example 2.16. From (2.133) we have for:

$$\begin{aligned}
n &= 1 : \Delta x_i = x_i - x_{i-1},\\
n &= 2 : \Delta^2 x_i = \Delta x_i - \Delta x_{i-1} = x_i - 2x_{i-1} + x_{i-2},\\
n &= 3 : \Delta^3 x_i = \Delta^2 x_i - \Delta^2 x_{i-1} = x_i - 3x_{i-1} + 3x_{i-2} - x_{i-3}.
\end{aligned}$$

From (2.135) we obtain for:

$$ n = -1 : \quad \begin{aligned} {}_0\Delta_k^{-1} x_k &= \sum_{j=0}^{k} (-1)^j \binom{-1}{j} x_{k-j} = x_k + x_{k-1} + \ldots + x_0 \\ &= \sum_{j=0}^{k} x_{k-j}, \end{aligned}$$

$$ n = -2 : \quad \begin{aligned} {}_0\Delta_k^{-2} x_k &= \sum_{j=0}^{k} (-1)^j \binom{-2}{j} x_{k-j} = x_k + \ldots + (k+1)x_0 \\ &= \sum_{j=0}^{k} (j+1) x_{k-j}. \end{aligned}$$

Definition 2.13. The discrete-time function

$$ \Delta^\alpha x_k = \sum_{j=0}^{k} (-1)^j \binom{\alpha}{j} x_{k-j}, \tag{2.136}$$

where $0 < \alpha < 1$, $\alpha \in \mathbb{R}$ and

$$ \binom{\alpha}{j} = \begin{cases} 1 & \text{for } \; j = 0, \\ \frac{\alpha(\alpha-1)\ldots(\alpha-j+1)}{j!} & \text{for } \; j = 1, 2, 3, \ldots \;, \end{cases} \tag{2.137}$$

is called the Grünwald-Letnikov fractional α-order (backward) difference of the function x_k.

Example 2.17. Using (2.137) for $0 < \alpha < 1$ we obtain for:

$$k = 1 : (-1)^1 \binom{\alpha}{1} = \alpha < 0,$$

$$k = 2 : (-1)^2 \binom{\alpha}{2} = \frac{\alpha(\alpha - 1)}{2!} < 0,$$

$$k = 3 : (-1)^3 \binom{\alpha}{3} = \frac{\alpha(\alpha - 1)(\alpha - 2)}{3!} < 0.$$

2.2.2 State Equations of Descriptor Discrete-Time Linear Systems

Consider the discrete-time linear system described by state equations

$$E_d \Delta^\alpha x_{i+1} = A_d x_i + B_d u_i, \quad 0 < \alpha < 1, \tag{2.138a}$$

$$y_i = C_d x_i + D_d u_i, \quad i \in \mathbb{Z}_+, \tag{2.138b}$$

where $x_i \in \mathbb{R}^n$, $u_i \in \mathbb{R}^m$, $y_i \in \mathbb{R}^p$ are state, input and output vectors and $E_d, A_d \in \mathbb{R}^{n\times n}$, $B_d \in \mathbb{R}^{n\times m}$, $C_d \in \mathbb{R}^{p\times n}$, $D_d \in \mathbb{R}^{p\times m}$. We assume that $\det E_d = 0$ and

$$\det[E_d\lambda - A_d] \neq 0 \text{ for some } \lambda \in \mathbb{C}. \tag{2.139}$$

If the matrix E_d is nonsingular ($\det E_d \neq 0$), then we can transform (2.138a) to the state equation of a fractional standard discrete-time linear system through premultiplication of (2.138a) by E_d^{-1}. The system (2.138) with the singular matrix E_d ($\det E_d = 0$) will be called a fractional descriptor discrete-time linear system.
Substituting the definition of fractional difference (2.136) into (2.138a) we obtain

$$E_d x_{i+1} = A_{d\alpha} x_i + \sum_{j=1}^{i} E_d c_j x_{i-j} + B_d u_i, \quad i \in \mathbb{Z}_+, \tag{2.140a}$$

where

$$A_{d\alpha} = A_d + E_d \alpha. \tag{2.140b}$$

and

$$c_j = (-1)^j \binom{\alpha}{j+1}. \tag{2.140c}$$

Remark 2.6. From (2.140a) it follows that the fractional system is equivalent to the system with increasing number of delays.

Note that coefficients c_j given by (2.140c) strongly decrease for increasing i. Hence, we can limit the upper bound of the sum by some natural number L, which is called the length of the practical implementation [53, 57, 88]. In this case the equations (2.140) can be written in the form

$$E_d x_{i+1} = A_{d\alpha} x_i + \sum_{j=1}^{L} E_d c_j x_{i-j} + B_d u_i, \tag{2.141a}$$

$$y_i = C_d x_i + D_d u_i \tag{2.141b}$$

with $x_{-k} = 0, k = 1, 2, \ldots$.

2.2.3 Solution to the State Equation of Fractional Descriptor Discrete-Time Linear Systems

Let U be the set of admissible input vectors $u_i \in U \subset \mathbb{R}^m$ and $X_0 \subset \mathbb{R}^n$ the set of consistent initial conditions $x(0) = x_0 \in X_0$ for which the equation (2.138a) has the solution x_i with $u_i \in U$.

2.2.3.1 Laurent Series Expansion Method

In this section the solution to the Eq. (2.138a) will be given using the Laurent series expansion method.

Theorem 2.20. Solution to the Eq. (2.138a) for $x_0 \in X_0$ and $u_i \in U, i \in \mathbb{Z}_+$ is given by the formula

$$x_i = T_{d,i} E_d x_0 + \sum_{k=0}^{n_0+i-1} T_{d,i-k-1} B_d u_k, \tag{2.142a}$$

where the matrices $T_{d,k} \in \mathbb{R}^{n \times n}$ are determined by

$$\begin{cases} E_d T_{d,k} - A_{d\alpha} T_{d,k-1} - \sum_{j=1}^{i} c_j E_d T_{d,k-j-1} = \mathbb{I}_n & \text{for } k = 0, \\ E_d T_{d,k} - A_{d\alpha} T_{d,k-1} - \sum_{j=1}^{i} c_j E_d T_{d,k-j-1} = 0 & \text{for } k \neq 0, \\ E_d T_{d,k} = 0 & \text{for } k \leq -n_0 \end{cases} \tag{2.142b}$$

and n_0 is called the index of the system. The matrices $T_{d,k}$ can also be found by the Laurent series expansion of the matrix $\left[E_d z - A_{d\alpha} - \sum_{j=1}^{i} E_d c_j z^{-j}\right]^{-1}$.

Proof. Applying the Z-transform to (2.140a) we obtain

$$E_d z X(z) - E_d z x_0 = A_{d\alpha} X(z) + \sum_{j=1}^{i} E_d c_j z^{-j} X(z) + B_d U(z). \tag{2.143}$$

From (2.143) we have

$$X(z) = \left[E_d z - A_{d\alpha} - \sum_{j=1}^{i} E_d c_j z^{-j}\right]^{-1} [E_d x_0 z + BU(z)]. \tag{2.144}$$

Let

$$\left[E_d z - A_{d\alpha} - \sum_{j=1}^{i} E_d c_j z^{-j}\right]^{-1} = \sum_{k=-n_0}^{\infty} T_{d,k} z^{-(k+1)}. \tag{2.145}$$

The matrices $T_{d,k}$, $k = -n_0, -n_0 + 1, \ldots, 0, 1, \ldots$ can be found by the expansion (2.145). Substituting (2.145) into (2.144) we obtain

$$X(z) = \sum_{k=-n_0}^{\infty} \left[T_{d,k} E_d x_0 z^{-k} + T_{d,k} B_d z^{-(k+1)} U(z)\right]. \tag{2.146}$$

Applying to (2.146) the inverse Z-transform and the convolution theorem we obtain (2.142a). □

From (2.142a) for $i = 0$ we have

$$x_0 = T_{d,0} E v + \sum_{k=0}^{n_0-1} T_{d,-k-1} B_d u_k, \tag{2.147}$$

where $v \in \mathbb{R}^n$ is an arbitrary vector.

Remark 2.7. If the matrix $E_d = \mathbb{I}_n$, then $n_0 = 0$ and $T_{d,i} = \Phi_{d,i}$, where

$$\Phi_{d,i+1} = \Phi_{d,i} A_{d\alpha} + \sum_{j=1}^{i} c_j \Phi_{d,i-j}, \quad \Phi_{d,0} = \mathbb{I}_n, \tag{2.148}$$

$$A_{d\alpha} = A_d + \mathbb{I}_n \alpha. \tag{2.149}$$

Therefore, from (2.142) and (2.148) we have

$$x_i = \Phi_{d,i} x_0 + \sum_{k=0}^{i-1} \Phi_{d,i-k-1} B_d u_k. \tag{2.150}$$

The formula (2.150) is the solution to the state equation of a standard fractional discrete-time linear system [57].

Example 2.18. Consider the fractional descriptor discrete-time linear system (2.141a) with $\alpha = 0.5$, $L = 1$ and

$$\begin{gathered} E_d = \begin{bmatrix} -0 & -0.4 & 0 \\ -1.6 & 0 & 0 \\ 0 & 0 & 0 \end{bmatrix}, \quad A_d = \begin{bmatrix} 0 & 0.2 & 0 \\ 0.8 & 0 & 0 \\ 0 & 0 & 0.2 \end{bmatrix} \\ A_{d\alpha} = A_d + 0.5E_d = \begin{bmatrix} 0 & 0 & 0 \\ 0 & 0 & 0 \\ 0 & 0 & 0.2 \end{bmatrix}, \quad B_d = \begin{bmatrix} -0.2 & 0 \\ 0 & -0.8 \\ 0 & 0 \end{bmatrix}. \end{gathered} \tag{2.151}$$

The matrix pencil (E_d, A_d) of (2.151) is regular since

$$\det[E_d\lambda - A_d] = 0.032(2\lambda + 1)^2 \neq 0. \tag{2.152}$$

The matrix $[E_d z - A_{d\alpha} - E_d c_1 z^{-1}]^{-1}$ has the form

$$[E_d z - A_{d\alpha} - 0.125E_d z^{-1}]^{-1} = \begin{bmatrix} 0 & -\frac{5z}{(8z^2-1)} & 0 \\ -\frac{20z}{(8z^2-1)} & 0 & 0 \\ 0 & 0 & -5 \end{bmatrix}. \tag{2.153}$$

Using (2.145) for $n_0 = 1$, the matrix (2.153) can be written in the form

$$[E_d z - A_{d\alpha} - 0.125E_d z^{-1}]^{-1} = \sum_{k=-1}^{\infty} T_{d,k} z^{-(k+1)}, \tag{2.154a}$$

where

$$T_{d,k} = \begin{cases} \begin{bmatrix} 0 & 0 & 0 \\ 0 & 0 & 0 \\ 0 & 0 & -5 \end{bmatrix} & \text{for } k = -1, \\ \begin{bmatrix} 0 & -0.625(0.125)^{0.5k} & 0 \\ -2.5(0.125)^{0.5k} & 0 & 0 \\ 0 & 0 & 0 \end{bmatrix} & \text{for } k = 0, 2, 4, \ldots, \\ \begin{bmatrix} 0 & 0 & 0 \\ 0 & 0 & 0 \\ 0 & 0 & 0 \end{bmatrix} & \text{for } k = 1, 3, 5, \ldots. \end{cases} \tag{2.154b}$$

Using (2.151) and (2.154b) we have

$$T_{d,k}E_d = \begin{cases} \begin{bmatrix} (0.125)^{0.5k} & 0 & 0 \\ 0 & (0.125)^{0.5k} & 0 \\ 0 & 0 & 0 \end{bmatrix} & \text{for } k = 0, 2, 4, \ldots, \\ \begin{bmatrix} 0 & 0 & 0 \\ 0 & 0 & 0 \\ 0 & 0 & 0 \end{bmatrix} & \text{for } k = 1, 3, 5, \ldots, \end{cases} \tag{2.155a}$$

$$T_{d,k}B_d = \begin{cases} \begin{bmatrix} 0 & 0.5(0.125)^{0.5k} \\ 0.5(0.125)^{0.5k} & 0 \\ 0 & 0 \end{bmatrix} & \text{for } k = 0, 2, 4, \ldots, \\ \begin{bmatrix} 0 & 0 \\ 0 & 0 \\ 0 & 0 \end{bmatrix} & \text{for } k = 1, 3, 5, \ldots . \end{cases} \tag{2.155b}$$

Using (2.142a), (2.151), (2.155) and taking into account that $n_0 = 1$ we obtain

$$\begin{aligned} x_i &= T_{d,i}E_d x_0 + \sum_{k=0}^{i} T_{d,i-k-1}B_d u_k = T_{d,i}E_d x_0 + \sum_{k=0}^{i-1} T_{d,i-k-1}B_d u_k + T_{d,-1}B_d u_i \\ &= \begin{bmatrix} [(-1)^i + 1]^i (0.0625)^i & 0 & 0 \\ 0 & [(-1)^i + 1]^i (0.0625)^i & 0 \\ 0 & 0 & 0 \end{bmatrix} x_0 \\ &+ \sum_{k=0}^{i-1} \begin{bmatrix} 0 & 0.5[(-1)^{\bar{k}} + 1]^{\bar{k}} (0.0625)^{\bar{k}} \\ 0.5[(-1)^{\bar{k}} + 1]^{\bar{k}} (0.0625)^{\bar{k}} & 0 \\ 0 & 0 \end{bmatrix} u_k, \end{aligned} \tag{2.156}$$

where $\bar{k} = i - k - 1$.
The equation above shows the solution (2.142a) of the state equation of the fractional descriptor linear system (2.141a) with $\alpha = 0.5$, $L = 1$ and (2.151).

2.2.3.2 Drazin Inverse Matrix Method

In this section the solution to the Eq. (2.138a) will be given using the Drazin inverse matrix method [61, 63, 90].

Consider the Eq. (2.138a) and assume that the condition (2.139) is met. There exists a number $c \in \mathbb{C}$ such that $\det[E_d c - A_d] \neq 0$. Premultiplying the Eq. (2.138a) by the matrix $[E_d c - A_d]^{-1}$ we obtain

$$\bar{E}_d \Delta^\alpha x_{i+1} = \bar{A}_d x_i + \bar{B}_d u_i, \tag{2.157a}$$

where

$$\bar{E}_d = [E_d c - A_d]^{-1} E_d,\ \ \bar{A}_d = [E_d c - A_d]^{-1} A_d,\ \ \bar{B}_d = [E_d c - A_d]^{-1} B_d. \tag{2.157b}$$

Observe that the Eqs. (2.138a) and (2.157a) have the same solution x_i. Basic definitions and properties concerning Drazin inverses are given in Sect. 1.1.2.2. Substituting the definition of fractional difference (2.136) into (2.157a) we obtain

$$\bar{E}_d x_{i+1} = \bar{A}_{d\alpha} x_i + \sum_{j=1}^{i} \bar{E}_d c_j x_{i-j} + \bar{B}_d u_i, \tag{2.158a}$$

where

$$\bar{A}_{d\alpha} = \bar{A}_d + \bar{E}_d \alpha. \tag{2.159}$$

and the coefficients c_j, $j = 1, 2, \ldots$ are given by (2.140c).

Lemma 1.2 can also be used for the fractional descriptor discrete-time linear system (2.158). Therefore, the Eq. (2.158a) is equivalent to the following equations:

$$\Delta^\alpha x_{1,i+1} = \bar{E}_d^D \bar{A}_d x_{1,i} + \bar{E}_d^D \bar{B}_d u_i, \tag{2.160a}$$

$$\bar{A}_d^D \bar{E}_d (\mathbb{I}_n - \bar{E}_d \bar{E}_d^D) \Delta^\alpha x_{2,i+1} = x_{2,i} + (\mathbb{I}_n - \bar{E}_d \bar{E}_d^D) \bar{A}_d^D \bar{B}_d u_i, \tag{2.160b}$$

where

$$x_{1,i} = \bar{E}\bar{E}^D x_i,\ \ x_{2,i} = (\mathbb{I}_n - \bar{E}\bar{E}^D) x_i. \tag{2.160c}$$

Substitution of (2.136) into (2.160) yields

$$x_{1,i+1} = \bar{E}_d^D \bar{A}_{d\alpha} x_{1,i} + \sum_{j=1}^{i} c_j x_{1,i-j} + \bar{E}_d^D \bar{B}_d u_i, \tag{2.161a}$$

$$\begin{aligned} \bar{A}_d^D \bar{E}_d (\mathbb{I}_n - \bar{E}_d \bar{E}_d^D) x_{2,i+1} &= [\mathbb{I}_n + \alpha \bar{A}_d^D \bar{E}_d (\mathbb{I}_n - \bar{E}_d \bar{E}_d^D)] x_{2,i} \\ &+ \sum_{j=1}^{i} c_j \bar{A}_d^D \bar{E}_d (\mathbb{I}_n - \bar{E}_d \bar{E}_d^D) x_{2,i-j} \\ &+ \bar{A}_d^D \bar{B}_d (\mathbb{I}_n - \bar{E}_d \bar{E}_d^D) u_i, \end{aligned} \tag{2.161b}$$

Theorem 2.21. The solution to the equation (2.158) (or equivalently (2.157a)) for $x_0 \in X_0$ and $u_i \in U$ is given by

$$x_i = \Phi_{d,i}\bar{E}_d\bar{E}_d^D x_0 + \sum_{k=0}^{i-1}\Phi_{d,i-k-1}\bar{E}_d^D\bar{B}_d u_k$$
$$+ (\bar{E}_d\bar{E}_d^D - \mathbb{I}_n)\sum_{k=0}^{q-1}(\bar{E}_d\bar{A}_d^D)^k\bar{A}_d^D\bar{B}_d\psi_{i,k}, \tag{2.162a}$$

where q is the index of $\bar{E}_d$ and

$$\Phi_{d,i+1} = \Phi_{d,i}\bar{E}_d^D\bar{A}_{d\alpha} + \sum_{j=1}^{i} c_j\Phi_{d,i-j}, \quad \Phi_{d,0} = \mathbb{I}_n, \tag{2.162b}$$

$$\psi_{i,0} = u_i,$$
$$\psi_{i,1} = \psi_{i+1,0} - \alpha\psi_{i,0} - \sum_{j=1}^{i} c_j\psi_{i-j,0} = u_{i+1} - \alpha u_i - \sum_{j=1}^{i} c_j u_{i-j},$$
$$\psi_{i,2} = \psi_{i+1,1} - \alpha\psi_{i,1} - \sum_{j=1}^{i} c_j\psi_{i-j,1}$$
$$= u_{i+2} - 2\alpha u_{i+1} + \alpha^2 u_i + 2\alpha\sum_{j=1}^{i} c_j u_{i-j} - \sum_{j=1}^{i+1} c_j u_{i-j+1} \tag{2.162c}$$
$$- \sum_{j=1}^{i} c_j u_{i-j+1} + \sum_{j=1}^{i} c_j \sum_{l=1}^{i-j} c_l u_{i-j-l}$$
$$\vdots$$
$$\psi_{i,q-1} = \psi_{i+1,q-2} - \alpha\psi_{i,q-2} - \sum_{j=1}^{i} c_j\psi_{i-j,q-2}.$$

Proof. We shall show that the solution (2.162) satisfies the equality (2.158a). Using (1.26), (1.27) and (2.162) we obtain

$$
\begin{aligned}
\bar{E}_d x_{i+1} &= \bar{E}_d \Phi_{d,i+1} \bar{E}_d \bar{E}_d^D x_0 + \sum_{k=0}^{i} \bar{E}_d \Phi_{d,i-k} \bar{E}_d^D \bar{B}_d u_k \\
&\quad + \bar{E}_d (\bar{E}_d \bar{E}_d^D - \mathbb{I}_n) \sum_{k=0}^{q-1} (\bar{E}_d \bar{A}_d^D)^k \bar{A}_d^D \bar{B}_d \psi_{i+1,k} \\
&= \left(\Phi_{d,i} \bar{A}_{d\alpha} + \sum_{j=1}^{i} \bar{E}_d c_j \Phi_{d,i-j} \right) \bar{E}_d \bar{E}_d^D x_0 \\
&\quad + \sum_{k=0}^{i-1} \Phi_{d,i-k-1} \bar{A}_{d\alpha} \bar{E}_d^D \bar{B}_d u_k + \bar{E}_d \bar{E}_d^D \bar{B}_d u_i \\
&\quad + \sum_{k=0}^{i-1} \left(\sum_{j=1}^{k} c_j \Phi_{d,k-j} \right) \bar{E}_d \bar{E}_d^D \bar{B}_d u_{i-k-1} \\
&\quad + (\bar{E}_d \bar{E}_d^D - \mathbb{I}_n) \sum_{k=0}^{q-1} (\bar{E}_d \bar{A}_d^D)^{k+1} \bar{B}_d \psi_{i+1,k},
\end{aligned}
\tag{2.163}
$$

$$
\begin{aligned}
\sum_{j=1}^{i} \bar{E}_d c_j x_{i-j} &= \sum_{j=1}^{i} \bar{E}_d c_j \Phi_{d,i-j} \bar{E}_d \bar{E}_d^D v \\
&\quad + \sum_{k=0}^{i-1} \left(\sum_{j=1}^{k} c_j \Phi_{d,k-j} \right) \bar{E}_d \bar{E}_d^D \bar{B}_d u_{i-k-1} \\
&\quad + (\bar{E}_d \bar{E}_d^D - \mathbb{I}_n) \sum_{k=0}^{q-1} (\bar{E}_d \bar{A}_d^D)^{k+1} \bar{B}_d \psi_{i-j,k}
\end{aligned}
\tag{2.164}
$$

and

$$
\begin{aligned}
\bar{A}_{d\alpha} x_i &= \Phi_{d,i} \bar{A}_{d\alpha} \bar{E}_d \bar{E}_d^D x_0 + \sum_{k=0}^{i-1} \Phi_{d,i-k-1} \bar{A}_{d\alpha} \bar{E}_d^D \bar{B}_d u_k \\
&\quad + \alpha (\bar{E}_d \bar{E}_d^D - \mathbb{I}_n) \sum_{k=0}^{q-1} (\bar{E}_d \bar{A}_d^D)^{k+1} \bar{B}_d \psi_{i,k} \\
&\quad + (\bar{E}_d \bar{E}_d^D - \mathbb{I}_n) \sum_{k=0}^{q-1} (\bar{E}_d \bar{A}_d^D)^{k} \bar{B}_d \psi_{i,k}.
\end{aligned}
\tag{2.165}
$$

From (2.163), (2.164) and (2.165) it follows that

$$
\bar{E}_d x_{i+1} - \bar{A}_{d\alpha} x_i - \sum_{j=1}^{i} \bar{E}_d c_j x_{i-j} = \bar{B}_d u_i,
\tag{2.166}
$$

which completes the proof. □

It is easy to check that the first two components of (2.162a) are the solution to the Eq. (2.161a) and the third component of (2.162a) is the solution to the Eq. (2.161b).

From (2.162a) for $i = 0$ we have

$$x_0 = \bar{E}_d \bar{E}_d^D v + (\bar{E}_d \bar{E}_d^D - \mathbb{I}_n) \sum_{k=0}^{q-1} (\bar{E}_d \bar{A}_d^D)^k \bar{A}_d^D \bar{B}_d \psi_{0,k}. \tag{2.167}$$

where $v \in \mathbb{R}^n$ is an arbitrary vector and $\psi_{0,k}$ is defined by (2.162c).
In the case when $u_i = 0, i \in \mathbb{Z}_+$ from (2.167) we have $x_0 = \bar{E}_d \bar{E}_d^D v$. A homogenous equation $E_d \Delta^\alpha x_{i+1} = A_d x_i$ has the solution if and only if $x_0 \in \operatorname{Im} \bar{E}_d \bar{E}_d^D$.

By Lemma 1.3 and Theorem 1.5 the matrices $\bar{E}_d^D \bar{E}_d$, $\bar{E}_d^D \bar{A}_d$, $\bar{E}_d \bar{A}_d^D$, $\bar{E}_d^D \bar{B}_d$, $\bar{A}_d^D \bar{B}_d$, the index q of the matrix $\bar{E}_d$, the solution (2.162a) of the Eq. (2.157a) and the set of consistent initial conditions X_0 do not depend of the choice of c.

Example 2.19. (Continuation of Example 2.18) Consider the fractional descriptor discrete-time linear system (2.141a) with $\alpha = 0.5$, $L = 1$ and (2.151). According to (2.152) the matrix pencil (E_d, A_d) is regular. We choose $c = 0$ and using (2.157b), (2.159) we obtain

$$\begin{gathered} \bar{E}_d = [-A_d]^{-1} E_d = \begin{bmatrix} 2 & 0 & 0 \\ 0 & 2 & 0 \\ 0 & 0 & 0 \end{bmatrix}, \quad \bar{A}_d = [-A_d]^{-1} A_d = \begin{bmatrix} -1 & 0 & 0 \\ 0 & -1 & 0 \\ 0 & 0 & -1 \end{bmatrix} \\ \bar{A}_{d\alpha} = \bar{A}_d + 0.5 \bar{E}_d = \begin{bmatrix} 0 & 0 & 0 \\ 0 & 0 & 0 \\ 0 & 0 & -1 \end{bmatrix}, \quad \bar{B}_d = [-A_d]^{-1} B_d = \begin{bmatrix} 0 & 1 \\ 1 & 0 \\ 0 & 0 \end{bmatrix}. \end{gathered} \tag{2.168}$$

It is easy to see that $\operatorname{rank} \bar{E}_d = \operatorname{rank} \bar{E}_d^2$, so $q = 1$. The Drazin inverse of $\bar{E}_d$ can be computed using (B.1) and (B.2) for $\bar{E}_d = \bar{E}_1$, where $\bar{E}_{11} = \begin{bmatrix} 2 & 0 \\ 0 & 2 \end{bmatrix}$, $\bar{E}_{12} = \begin{bmatrix} 0 \\ 0 \end{bmatrix}$ and

$$\bar{E}_d^D = \begin{bmatrix} \bar{E}_{11}^{-1} & \bar{E}_{11}^{-2} \bar{E}_{12} \\ 0 & 0 \end{bmatrix} = \begin{bmatrix} 0.5 & 0 & 0 \\ 0 & 0.5 & 0 \\ 0 & 0 & 0 \end{bmatrix}. \tag{2.169}$$

Note that $\bar{A}^D = \bar{A}^{-1} = \bar{A}$. Next, we compute the matrices

$$\bar{E}_d\bar{E}_d^D = \begin{bmatrix} 1 & 0 & 0 \\ 0 & 1 & 0 \\ 0 & 0 & 0 \end{bmatrix}, \quad \bar{E}_d^D\bar{A}_{d\alpha} = \begin{bmatrix} 0 & 0 & 0 \\ 0 & 0 & 0 \\ 0 & 0 & 0 \end{bmatrix},$$

$$\bar{E}_d^D\bar{B}_d = \begin{bmatrix} 0 & 0.5 \\ 0.5 & 0 \\ 0 & 0 \end{bmatrix}, \quad (\bar{E}_d\bar{E}_d^D - \mathbb{I}_3) = \begin{bmatrix} 0 & 0 & 0 \\ 0 & 0 & 0 \\ 0 & 0 & -1 \end{bmatrix}, \tag{2.170}$$

$$\bar{A}_d^D\bar{E}_d = \begin{bmatrix} -2 & 0 & 0 \\ 0 & -2 & 0 \\ 0 & 0 & 0 \end{bmatrix}, \quad \bar{A}_d^D\bar{B}_d = \begin{bmatrix} 0 & -1 \\ -1 & 0 \\ 0 & 0 \end{bmatrix}$$

and

$$\Phi_{d,i+1} = \Phi_{d,i}\bar{E}_d^D\bar{A}_{d\alpha} + \sum_{j=1}^{L} c_j\Phi_{d,i-j} = c_1\Phi_{d,i-1} = 0.125\Phi_{d,i-1}, \tag{2.171}$$

where $\Phi_{d,0} = \mathbb{I}_3$, $\Phi_{d,-1} = 0$. From (2.171) it follows that

$$\Phi_{d,i} = \begin{cases} \begin{bmatrix} (0.125)^{0.5i} & 0 & 0 \\ 0 & (0.125)^{0.5i} & 0 \\ 0 & 0 & (0.125)^{0.5i} \end{bmatrix} & \text{for } i = 0, 2, 4, \ldots, \\ \begin{bmatrix} 0 & 0 & 0 \\ 0 & 0 & 0 \\ 0 & 0 & 0 \end{bmatrix} & \text{for } i = 1, 3, 5, \ldots. \end{cases} \tag{2.172}$$

Using (2.162), (2.170) and (2.172) and taking into account that $q = 1$ we obtain

$$\begin{aligned} x_i &= \Phi_{d,i}\bar{E}_d\bar{E}_d^D x_0 + \sum_{k=0}^{i-1}\Phi_{d,i-k-1}\bar{E}_d^D\bar{B}_d u_k + (\bar{E}_d\bar{E}_d^D - \mathbb{I}_n)\bar{A}_d^D\bar{B}_d\psi_{i,0} \\ &= \begin{bmatrix} [(-1)^i + 1]^i(0.0625)^i & 0 & 0 \\ 0 & [(-1)^i + 1]^i(0.0625)^i & 0 \\ 0 & 0 & 0 \end{bmatrix} x_0 \\ &+ \sum_{k=0}^{i-1}\begin{bmatrix} 0 & 0.5[(-1)^{\bar{k}} + 1]^{\bar{k}}(0.0625)^{\bar{k}} \\ 0.5[(-1)^{\bar{k}} + 1]^{\bar{k}}(0.0625)^{\bar{k}} & 0 \\ 0 & 0 \end{bmatrix} u_k, \end{aligned} \tag{2.173}$$

where $\bar{k} = i - k - 1$.

The equation above shows the solution (2.162a) of the state equation of the fractional descriptor linear system (2.141a) with $\alpha = 0.5$, $L = 1$ and (2.151). Note that solutions (2.156) and (2.173) are equivalent.

2.2.3.3 Weierstrass-Kronecker Decomposition Method

In this section the solution to the Eq. (2.138a) will be given using the Weierstrass-Kronecker decomposition method [57, 90].

If the conditon (2.139) holds, then there exists a pair of nonsingular matrices $P_d, Q_d \in \mathbb{R}^{n\times n}$ such that [57, 90]

$$P_d[E_d\lambda - A_d]Q_d = \begin{bmatrix} \mathbb{I}_{n_1}\lambda - A_{1d} & 0 \\ 0 & N\lambda - \mathbb{I}_{n_2} \end{bmatrix}, \quad A_{1d} \in \mathbb{R}^{n_1\times n_1}, \quad N \in \mathbb{R}^{n_2\times n_2}, \tag{2.174}$$

where $n_1 = \deg\{\det[E_d\lambda - A_d]\}$, $n_2 = n - n_1$ and N is a nilpotent matrix with the nilpotency index μ, i.e. $N^{\mu-1} \neq 0$ and $N^\mu = 0$. The matrices P_d and Q_d can be computed using one of procedures given in [47, 48, 168].

Premultiplying (2.138a) by the matrix P_d and introducing the new state vector

$$\bar{x}_i = \begin{bmatrix} \bar{x}_{1,i} \\ \bar{x}_{2,i} \end{bmatrix} = Q_d^{-1}x_i, \quad \bar{x}_{1,i} \in \mathbb{R}^{n_1}, \quad \bar{x}_{2,i} \in \mathbb{R}^{n_2}, \tag{2.175}$$

we obtain

$$P_dE_dQ_dQ_d^{-1}\Delta^\alpha x_{i+1} = P_dA_dQ_dQ_d^{-1}x_i + P_dB_du_i, \tag{2.176a}$$

$$y_i = C_dQ_dQ_d^{-1}x_i + D_du_i \tag{2.176b}$$

and using (2.174) we get

$$\Delta^\alpha \bar{x}_{1,i+1} = A_{1d}\bar{x}_{1,i} + B_{1d}u_i, \tag{2.177a}$$

$$N\Delta^\alpha \bar{x}_{2,i+1} = \bar{x}_{2,i} + B_{2d}u_i, \tag{2.177b}$$

$$y_i = C_{1d}\bar{x}_{1,i} + C_{2d}\bar{x}_{2,i} + D_du_i, \tag{2.177c}$$

where

$$P_dB_d = \begin{bmatrix} B_{1d} \\ B_{2d} \end{bmatrix}, \quad B_{1d} \in \mathbb{R}^{n_1\times m}, \quad B_{2d} \in \mathbb{R}^{n_2\times m}, \tag{2.178a}$$

$$C_dQ_d = \begin{bmatrix} C_{1d} & C_{2d} \end{bmatrix}, \quad C_{1d} \in \mathbb{R}^{p\times n_1}, \quad C_{2d} \in \mathbb{R}^{p\times n_2}. \tag{2.178b}$$

Substituting the definition of fractional difference (2.136) into (2.177) we obtain

$$\bar{x}_{1,i+1} = A_{1d\alpha}\bar{x}_{1,i} + \sum_{j=1}^{i} c_j\bar{x}_{1,i-j} + B_{1d}u_i, \tag{2.179a}$$

$$N\bar{x}_{2,i+1} = (\mathbb{I}_{n_2} + N\alpha)\bar{x}_{2,i} + \sum_{j=1}^{i} c_j N\bar{x}_{2,i-j} + B_{2d}u_i, \tag{2.179b}$$

where

$$A_{1d\alpha} = A_{1d} + \mathbb{I}_{n_1}\alpha \tag{2.179c}$$

and the coefficients c_j, $j = 1, 2, \ldots$ are given by (2.140c).

Theorem 2.22. [57] Solution to the Eq. (2.177a) for given initial condition $\bar{x}_{10} \in \mathbb{R}^{n_1}$ and input $u_i \in U$, $i \in \mathbb{Z}_+$ has the form

$$\bar{x}_{1,i} = \Phi_{d,i}\bar{x}_{10} + \sum_{k=0}^{i-1} \Phi_{d,i-j-1}B_{1d}u_k. \tag{2.180a}$$

where

$$\Phi_{d,i+1} = \Phi_{d,i}A_{1d\alpha} + \sum_{j=1}^{i} c_j\Phi_{d,i-j}, \quad \Phi_{d,0} = \mathbb{I}_{n_1}. \tag{2.180b}$$

Theorem 2.23. Solution to the Eq. (2.177b) for given input $u_i \in U$, $i \in \mathbb{Z}_+$ has the form

$$\bar{x}_{2,i} = -\sum_{k=0}^{\mu-1} N^k B_{2d}\psi_{i,k}, \tag{2.181}$$

and the consistent initial conditions of the subsystem (2.177b) are given by the formula

$$\bar{x}_{20} = -\sum_{k=0}^{\mu-1} N^k B_{2d}\psi_{0,k}, \tag{2.182}$$

where $\psi_{i,k} \in \mathbb{R}^m$ is defined by (2.162c).

Proof. Using (2.179b) and (2.181) we obtain

$$N\bar{x}_{2,i+1} = -\sum_{k=0}^{\mu-1} N^{k+1}B_{2d}\psi_{i+1,k}, \tag{2.183}$$

$$\begin{aligned}(\mathbb{I}_{n_2} + N\alpha)\bar{x}_{2,i} + \sum_{j=1}^{i} c_j N\bar{x}_{2,i-j} &= -\sum_{k=0}^{\mu-1}(N^k + N^{k+1}\alpha)B_{2d}\psi_{i,k} \\ &\quad - \sum_{k=0}^{\mu-1}\sum_{j=1}^{i} c_j N^{k+1}B_{2d}\psi_{i-j,k}\end{aligned} \tag{2.184}$$

Therefore, from (2.183), (2.184) and (2.162c) we have

$$N\bar{x}_{2,i+1} - (\mathbb{I}_{n_2} + N\alpha)\bar{x}_{2,i} - \sum_{j=1}^{i} c_j N\bar{x}_{2,i-j} = B_{2d}u_i, \tag{2.185}$$

which completes the proof. □

Example 2.20. (Continuation of Example 2.18) Consider the fractional descriptor discrete-time linear system (2.141a) with $\alpha = 0.5$, $L = 1$ and (2.151). According to (2.152) the matrix pencil (E_d, A_d) is regular. In this case

$$P = \begin{bmatrix} -2.5 & 0 & 0 \\ 0 & -0.625 & 0 \\ 0 & 0 & 1 \end{bmatrix}, \quad Q = \begin{bmatrix} 0 & 1 & 0 \\ 1 & 0 & 0 \\ 0 & 0 & 5 \end{bmatrix} \tag{2.186}$$

and

$$\begin{bmatrix} \mathbb{I}_{n_1} & 0 \\ 0 & N \end{bmatrix} = PEQ = \begin{bmatrix} 1 & 0 & 0 \\ 0 & 1 & 0 \\ 0 & 0 & 0 \end{bmatrix},$$

$$\begin{bmatrix} A_{1d} & 0 \\ 0 & \mathbb{I}_{n_2} \end{bmatrix} = PAQ = \begin{bmatrix} -0.5 & 0 & 0 \\ 0 & -0.5 & 0 \\ 0 & 0 & 1 \end{bmatrix}, \tag{2.187}$$

$$\begin{bmatrix} B_{1d} \\ B_{2d} \end{bmatrix} = PB = \begin{bmatrix} 0.5 & 0 \\ 0 & 0.5 \\ 0 & 0 \end{bmatrix}.$$

Next we compute

$$A_{1d\alpha} = A_{1d} + 0.5\mathbb{I}_2 = \begin{bmatrix} 0 & 0 \\ 0 & 0 \end{bmatrix} \tag{2.188}$$

and

$$\Phi_{d,i+1} = \Phi_{d,i}A_{1d\alpha} + \sum_{j=1}^{L} c_j\Phi_{d,i-j} = c_1\Phi_{d,i-1} = 0.125\Phi_{d,i-1}, \tag{2.189}$$

where $\Phi_{d,0} = \mathbb{I}_2$, $\Phi_{d,-1} = 0$. From (2.189) it follows that

$$\Phi_{d,i} = \begin{cases} \begin{bmatrix} (0.125)^{0.5i} & 0 \\ 0 & (0.125)^{0.5i} \end{bmatrix} & \text{for } i = 0, 2, 4, \ldots, \\ \begin{bmatrix} 0 & 0 \\ 0 & 0 \end{bmatrix} & \text{for } i = 1, 3, 5, \ldots . \end{cases} \tag{2.190}$$

Using (2.180), (2.187) and (2.190) we obtain

$$\bar{x}_{1,i} = \begin{bmatrix} [(-1)^i + 1]^i (0.0625)^i & 0 \\ 0 & [(-1)^i + 1]^i (0.0625)^i \end{bmatrix} \bar{x}_{10} + \sum_{k=0}^{i-1} \begin{bmatrix} 0.5[(-1)^{\bar{k}} + 1]^{\bar{k}} (0.0625)^{\bar{k}} & 0 \\ 0 & 0.5[(-1)^{\bar{k}} + 1]^{\bar{k}} (0.0625)^{\bar{k}} \end{bmatrix} u_k, \tag{2.191}$$

where $\bar{k} = i - k - 1$.

Taking into account that $\mu = 1$ and using (2.181), (2.187) we obtain

$$\bar{x}_{2,i} = -B_{2d}\psi_{i,0} = -B_{2d}u_i = 0. \tag{2.192}$$

From (2.175), (2.191) and (2.192) we have

$$\bar{x}_i = \begin{bmatrix} [(-1)^i + 1]^i (0.0625)^i & 0 & 0 \\ 0 & [(-1)^i + 1]^i (0.0625)^i & 0 \\ 0 & 0 & 0 \end{bmatrix} \bar{x}_0 + \sum_{k=0}^{i-1} \begin{bmatrix} 0.5[(-1)^{\bar{k}} + 1]^{\bar{k}} (0.0625)^{\bar{k}} & 0 \\ 0 & 0.5[(-1)^{\bar{k}} + 1]^{\bar{k}} (0.0625)^{\bar{k}} \\ 0 & 0 \end{bmatrix} u_k, \tag{2.193}$$

where $\bar{k} = i - k - 1$.

Taking into account that $\bar{x}_i = Q^{-1}x_i$ we have to premultiply the Eq. (2.193) by Q and substitute $\bar{x}_0 = Q^{-1}x_0$. Thus we obtain the formula (2.156) (or (2.173)), which means that all considered methods of the fractional descriptor systems analysis give the same result.

2.2.4 *Positive Fractional Descriptor Discrete-Time Linear Systems*

In positive systems inputs, outputs and state variables take only nonnegative values [40, 49]. The following definition of the positivity is suitable for all methods of the fractional descriptor systems analysis.

Definition 2.14. The fractional descriptor discrete-time linear system (2.138) is called (internally) positive if $x_i \in \mathbb{R}_+^n$ and $y_i \in \mathbb{R}_+^p$, $i \in \mathbb{Z}_+$ for any consistent initial condition $x_0 \in X_0 \subset \mathbb{R}_+^n$ and all admissible inputs $u_i \in U \subset \mathbb{R}_+^m$, $i \in \mathbb{Z}_+$.

2.2.4.1 Laurent Series Expansion Method

In this section necessary and sufficient positivity conditions of the fractional descriptor discrete-time linear system (2.138) will be established using the Laurent series expansion method.

Theorem 2.24. The descriptor discrete-time linear system (2.138) is (internally) positive if and only if

$$\begin{gathered} T_{d,k}E_d \in \mathbb{R}_+^{n\times n}, \quad k=0,1,\ldots, \\ T_{d,k}B_d \in \mathbb{R}_+^{n\times m}, \quad k=-n_0,-n_0+1,\ldots,0,1,\ldots \\ C_d \in \mathbb{R}_+^{p\times n}, \quad D_d \in \mathbb{R}_+^{p\times m}, \end{gathered} \tag{2.194}$$

where the matrices $T_{d,k} \in \mathbb{R}^{n\times n}$ are given by (2.142b).

Proof. The proof can be accomplished in a similar way to the proof of Theorem 1.28. □

Example 2.21. (Continuation of Example 2.18) Consider the fractional descriptor discrete-time linear system (2.141a) with (2.151), $\alpha = 0.5$, $L = 1$ and

$$C_d = \begin{bmatrix} 1 & 0 & 0 \end{bmatrix}, \quad D_d = \begin{bmatrix} 0 & 0 \end{bmatrix}. \tag{2.195}$$

The matrices $T_{d,k}E_d$ and $T_{d,k}B_d$ are given by (2.155). From (2.155) and (2.195) it follows that $T_{d,k}E_d \in \mathbb{R}_+^{3\times 3}$, $T_{d,k}B_d \in \mathbb{R}_+^{3\times 2}$, $k = 0, 1, 2, \ldots$ and $C_d \in \mathbb{R}_+^{1\times 3}$, $D_d \in \mathbb{R}_+^{1\times 2}$. Therefore, by Theorem 2.24 the considered fractional descriptor system is positive.

2.2.4.2 Drazin Inverse Matrix Method

In this section necessary and sufficient positivity conditions of the fractional descriptor discrete-time linear system (2.157) (or equivalently (2.138)) will be established using the Drazin inverse matrix method.

Theorem 2.25. Let

$$\bar{F}_{d,k} = (\bar{E}_d \bar{A}_d^D)^k \bar{A}_d^D \bar{B}_d, \quad k = 0, 1, \ldots, q-1. \tag{2.196}$$

The fractional descriptor discrete-time linear system (2.157) (or equivalently (2.138)) is positive if and only if there exists a matrix $G_d \in \mathbb{R}^{n\times n}$ [79, 141] such that

$$\bar{E}_d^D \bar{A}_{d\alpha} + G_d(\mathbb{I}_n - \bar{E}_d \bar{E}_d^D) \in \mathbb{R}_+^{n\times n}, \tag{2.197a}$$

and

$$\operatorname{Im}\bar{E}_d\bar{E}_d^D \subset \mathbb{R}_+^n, \quad \bar{E}_d^D\bar{B}_d \in \mathbb{R}_+^{n\times m}, \quad C_d \in \mathbb{R}_+^{p\times n}, \quad D_d \in \mathbb{R}_+^{p\times m} \tag{2.197b}$$

$$(\bar{E}_d\bar{E}_d^D - \mathbb{I}_n)\bar{F}_{d,k} \in \mathbb{R}_+^{n\times m}, \quad k = 0, 1, \ldots, q-1. \tag{2.197c}$$

Proof. Using the superposition principle we consider components of the solution (2.162a) independently:

1) If $\operatorname{Im}\bar{E}_d\bar{E}_d^D \subset \mathbb{R}_+^{n\times n}$, then $\bar{E}_d\bar{E}_d^D x_0 \subset \mathbb{R}_+^n$. Therefore, for the matrix G_d such that $\bar{E}_d^D\bar{A}_{d\alpha} + G_d(\mathbb{I}_n - \bar{E}_d\bar{E}_d^D) \in \mathbb{R}_+^{n\times n}$ we have $[\bar{E}_d^D\bar{A}_{d\alpha} + G_d(\mathbb{I}_n - \bar{E}_d\bar{E}_d^D)]^i \in \mathbb{R}_+^{n\times n}$. Taking into account that $[\bar{E}_d^D\bar{A}_{d\alpha} + G_d(\mathbb{I}_n - \bar{E}_d\bar{E}_d^D)]^i\bar{E}_d\bar{E}_d^D = (\bar{E}_d^D\bar{A}_{d\alpha})^i\bar{E}_d\bar{E}_d^D$ we obtain $[\bar{E}_d^D\bar{A}_{d\alpha} + G_d(\mathbb{I}_n - \bar{E}_d\bar{E}_d^D)]^i\bar{E}_d\bar{E}_d^D x_0 = (\bar{E}_d^D\bar{A}_{d\alpha})^i\bar{E}_d\bar{E}_d^D x_0 \in \mathbb{R}_+^n$, $i \in \mathbb{Z}_+$. Taking into account that for $\alpha \in (0; 1)$ coefficients $c_j > 0$, $j = 1, 2, \ldots$ we have $\Phi_{d,i}\bar{E}_d\bar{E}_d^D x_0 \in \mathbb{R}_+^{n\times n}$, $i \in \mathbb{Z}_+$.
2) If 1) holds and $\bar{E}_d^D\bar{B}_d \in \mathbb{R}_+^{n\times n}$, then $\sum_{k=0}^{i-1}\Phi_{d,i-k-1}\bar{E}_d^D\bar{B}_d u_k \in \mathbb{R}_+^n$ for $u_i \in U \subset \mathbb{R}_+^m$, $i \in \mathbb{Z}_+$.
3) If $(\bar{E}_d\bar{E}_d^D - \mathbb{I}_n)\bar{F}_{d,k} \in \mathbb{R}_+^{n\times m}$, $k = 0, 1, \ldots, q-1$, where the matrices $\bar{F}_{d,k}$ are given by (2.196), then the last component of the solution (2.162a) is also non-negative.

Therefore, we have $x_i \in \mathbb{R}_+^n$, $i \in \mathbb{Z}_+$ if and only if the conditions (2.197) hold. Substituting (2.162a) into (2.138b) it is easy to see that $y_i \in \mathbb{R}_+^p$, $i \in \mathbb{Z}_+$ if $C_d \in \mathbb{R}_+^{p\times n}$ and $D_d \in \mathbb{R}_+^{p\times m}$. □

Remark 2.8. Note that the fractional descriptor discrete-time linear system (2.157) can be positive even though the matrix $\bar{E}_d^D\bar{A}_{d\alpha}$ contains negative entries. In the special case for $G_d = 0$ from (2.197a) we obtain the positivity condition $\bar{E}_d^D\bar{A}_{d\alpha} \in \mathbb{R}_+^{n\times n}$.

Example 2.22. (Continuation of Example 2.19) Consider the fractional descriptor discrete-time linear system (2.141a) with $\alpha = 0.5$, $L = 1$ and (2.151), (2.195). From (2.170) we have

$$\operatorname{Im}\bar{E}_d\bar{E}_d^D = \operatorname{Im}\begin{bmatrix} 1 & 0 & 0 \\ 0 & 1 & 0 \\ 0 & 0 & 0 \end{bmatrix} \subset \mathbb{R}_+^{3\times 3} \text{ for } v = \begin{bmatrix} v_1 \\ v_2 \\ 0 \end{bmatrix} \subset \mathbb{R}_+^3 \tag{2.198}$$

and $\bar{E}_d^D\bar{A}_{d\alpha} \in \mathbb{R}_+^{3\times 3}$, $\bar{E}_d^D\bar{B}_d \in \mathbb{R}_+^{3\times 2}$, $(\bar{E}_d\bar{E}_d^D - \mathbb{I}_3)\bar{A}_d^D\bar{B}_d \in \mathbb{R}_+^{3\times 2}$. From (2.195) it follows that $C_d \in \mathbb{R}_+^{1\times 3}$, $D_d \in \mathbb{R}_+^{1\times 2}$. Therefore, by Theorem 2.25 the considered fractional descriptor system is positive.

2.2.4.3 Weierstrass-Kronecker Decomposition Method

In this section necessary and sufficient positivity conditions of the fractional descriptor discrete-time linear system (2.177) (or equivalently (2.138)) will be established using the Weierstrass-Kronecker decomposition method with the assumption that the matrix Q_d determined by (2.174) is a monomial matrix [59, 66].

Theorem 2.26. [19, 59, 84] Let the decomposition (2.174) of the system (2.138) be possible for a monomial matrix $Q_d \in \mathbb{R}_+^{n\times n}$. The fractional descriptor discrete-time linear system (2.177) (or equivalently (2.138)) is positive if and only if

$$\begin{aligned} &A_{1d\alpha} \in \mathbb{R}_+^{n_1\times n_1}, \quad B_{1d} \in \mathbb{R}_+^{n_1\times m}, \quad C_d \in \mathbb{R}_+^{p\times n}, \quad D_d \in \mathbb{R}_+^{p\times m}, \\ &-N^k B_{2d} \in \mathbb{R}_+^{n_2\times m}, \quad k = 0, 1, \ldots, \mu - 1. \end{aligned} \tag{2.199}$$

Proof. For a monomial matrix $Q_d \in \mathbb{R}_+^{n\times n}$ we have $x_i \in \mathbb{R}_+^n$, $i \in \mathbb{Z}_+$ if $\bar{x}_i \in \mathbb{R}_+^n$, $i \in \mathbb{Z}_+$. Taking into account that for $\alpha \in (0; 1)$ coefficients $c_j > 0$, $j = 1, 2, \ldots$ and $A_{1d\alpha} \in \mathbb{R}_+^{n_1\times n_1}$, $B_{1d} \in \mathbb{R}_+^{n_1\times m}$, then from (2.180a) we obtain $\bar{x}_{1,i} \in \mathbb{R}_+^{n_1}$, $i \in \mathbb{Z}_+$ since by Definition 2.14 $\bar{x}_{10} = Q^{-1}x_{10} \in \mathbb{R}_+^{n_1}$ and $u_i \in \mathbb{R}_+^m$, $i \in \mathbb{Z}_+$. If $-N^k B_{2d} \in \mathbb{R}_+^{n_2\times m}$, $k = 0, 1, \ldots, \mu - 1$, then from (2.181) we obtain $\bar{x}_{2,i} \in \mathbb{R}_+^{n_2}$ since $u_i \in \mathbb{R}_+^m$, $i \in \mathbb{Z}_+$. From (2.176b) for $C_d \in \mathbb{R}_+^{p\times n}$ and $D_d \in \mathbb{R}_+^{p\times m}$ we have $y_i \in \mathbb{R}_+^p$, $i \in \mathbb{Z}_+$ since $x_i \in \mathbb{R}_+^n$ and $u_i \in \mathbb{R}_+^m$, $i \in \mathbb{Z}_+$. □

Example 2.23. (Continuation of Example 2.20) Consider the fractional descriptor discrete-time linear system (2.141a) with $\alpha = 0.5$, $L = 1$ and (2.151), (2.195). From (2.187) and (2.188) we have $A_{1d\alpha} \in \mathbb{R}_+^{2\times 2}$, $B_{1d} \in \mathbb{R}_+^{2\times 2}$ and $-N^0 B_{2d} = -B_{2d} \in \mathbb{R}_+^{1\times 2}$. From (2.195) it follows that $C_d \in \mathbb{R}_+^{1\times 3}$, $D_d \in \mathbb{R}_+^{1\times 2}$. Taking into account that the matrix $Q \in \mathbb{R}_+^{3\times 3}$ is monomial, by Theorem 2.26 the considered fractional descriptor system is positive.

2.2.5 *Stability of Fractional Descriptor Discrete-Time Linear Systems*

2.2.5.1 Laurent Series Expansion Method

In this section necessary and sufficient stability conditions of the fractional descriptor discrete-time linear system (2.138a) will be established using the Laurent series expansion method.

Definition 2.15. [53, 57, 88] The fractional descriptor discrete-time linear system (2.138a) is called practically stable for given length L of practical implementation if the fractional linear system (2.141a) is asymptotically stable.
The fractional descriptor discrete-time linear system (2.138a) is called asymptotically stable if the fractional linear system (2.141a) is asymptotically stable for $L \to \infty$.

To test the stability of the system (2.138a) well-known methods for fractional discrete-time linear systems can be applied, see e.g. [25, 152].

Theorem 2.27. The fractional descriptor discrete-time linear system (2.138a) is practically stable for given length L of the practical implementation if and only if all roots of the characteristic equation

$$\det\left[E_d z - A_{d\alpha} - \sum_{j=1}^{L} E_d c_j z^{-j}\right] = 0 \tag{2.200}$$

lie inside the unit circle.

Proof. The proof can be accomplished in a similar way as it has been shown in [152]. The characteristic Eq. (2.200) is obtained through application of the Z-transform to the Eq. (2.141a) for $u_i = 0, i \in \mathbb{Z}_+$. □

Taking into account that [25]

$$\sum_{j=1}^{\infty} c_j z^{-j} = z - \alpha - (z-1)^{\alpha} z^{1-\alpha} \tag{2.201}$$

we obtain the following theorem.

Theorem 2.28. The fractional descriptor discrete-time linear system (2.138a) is asymptotically stable if and only if all roots of the characteristic equation

$$\det\left[E_d (z-1)^{\alpha} z^{1-\alpha} - A_d\right] = 0 \tag{2.202}$$

lie inside the unit circle.

Example 2.24. Consider the fractional descriptor discrete-time linear system (2.138a) with $\alpha = 0.4$ and

$$E_d = \begin{bmatrix} 1 & 0 \\ 0.5 & 0 \end{bmatrix}, \quad A_d = \begin{bmatrix} 0.2 & 0 \\ 0.1 & 0.3 \end{bmatrix},$$
$$A_{d\alpha} = A_d + 0.4E_d = \begin{bmatrix} 0.6 & 0 \\ 0.3 & 0.3 \end{bmatrix}. \tag{2.203}$$

The matrix pencil (E_d, A_d) of (2.203) is regular since

$$\det[E_d \lambda - A_d] = -0.3z + 0.06 \neq 0. \tag{2.204}$$

To test the stability of the system (2.138a) we have to consider the system (2.141a). The characteristic equation of (2.141a) with $\alpha = 0.4$, $L = 1$ and (2.203) has the form

$$\det[E_d z - A_{d\alpha} - 0.12E_d z^{-1}] = -0.3z + 0.036z^{-1} + 0.18 = 0. \tag{2.205}$$

The system (2.141a) with $L = 1$ is asymptotically stable since roots of (2.205) are $z_1 = -0.1583$, $z_2 = 0.7583$. Therefore, by Theorem 2.27 the fractional descriptor discrete-time linear system (2.138a) is practically stable for length of the practical implementation $L = 1$.
Now let us consider the system (2.141a) with $\alpha = 0.4$, $L \to \infty$ and (2.203). Using (2.202) we obtain

$$\det[E_d(z-1)^{0.4}z^{0.6} - A_d] = -0.3(z-1)^{0.4}z^{0.6} + 0.06 = 0. \tag{2.206}$$

The system (2.141a) with $L \to \infty$ is unstable since the root of (2.206) is $z_1 = 1.0174$. Therefore, by Theorem 2.28 the fractional descriptor discrete-time linear system (2.138a) is unstable.

2.2.5.2 Drazin Inverse Matrix Method

In this section necessary and sufficient stability conditions of the fractional descriptor discrete-time linear system (2.157) (or equivalently (2.138)) will be established using the Drazin inverse matrix method.

The upper bound of the sum in (2.158a) can be limited by some natural number L (length of practical implementation), like in the case of (2.141a), i.e. the state equation has the form

$$\bar{E}_d x_{i+1} = \bar{A}_{d\alpha} x_i + \sum_{j=1}^{L} \bar{A}_d c_j x_{i-j} + \bar{B}_d u_i, \tag{2.207}$$

where $\bar{A}_{d\alpha}$ is given by (2.159) and $x_{-k} = 0$, $k = 1, 2, \ldots$.

Definition 2.16. The fractional descriptor discrete-time linear system (2.157) (or equivalently (2.138)) is called practically stable for given length L of practical implementation if the fractional linear system (2.207) is asymptotically stable.
The fractional descriptor discrete-time linear system v (or equivalently (2.138)) is called asymptotically stable if the fractional linear system (2.207) is asymptotically stable for $L \to \infty$.

Lemma 2.6. The characteristic equation

$$\det\left[\bar{E}_d z - \bar{A}_{d\alpha} - \sum_{j=1}^{L} \bar{E}_d c_j z^{-j}\right] = 0 \tag{2.208}$$

of (2.207) has the same set of roots as the characteristic equation (2.200) of (2.141a).

Proof. From (2.157b) and (2.208) we have

$$\begin{aligned}
&\det\left[\bar{E}_d z - \bar{A}_{d\alpha} - \sum_{j=1}^{L} \bar{E}_d c_j z^{-j}\right] \\
&= \det\left\{[E_d c - A_d]^{-1}\left(E - A_{d\alpha} - \sum_{j=1}^{L} E_d c_j z^{-j}\right)\right\} \\
&= \{\det[E_d c - A_d]\}^{-1} \det\left[E_d z - A_{d\alpha} - \sum_{j=1}^{L} E_d c_j z^{-j}\right] = 0,
\end{aligned} \tag{2.209}$$

which is equivalent to (2.200). □

Therefore, by Lemma 2.6 and the approach given in Sect. 2.2.5.1 we obtain the following theorem.

Theorem 2.29. The fractional descriptor discrete-time linear system (2.138a) is:

1) practically stable for given length L of the practical implementation if and only if all roots of the characteristic equation (2.208) lie inside the unit circle;
2) asymptotically stable if and only if all roots of the characteristic equation

$$\det\left[\bar{E}_d (z-1)^{\alpha} z^{1-\alpha} - \bar{A}_d\right] = 0 \tag{2.210}$$

lie inside the unit circle.

Example 2.25. (Continuation of Example 2.24) Consider the fractional descriptor discrete-time linear system (2.138a) with (2.203). According to (2.204) the matrix pencil (E_d, A_d) is regular. We choose $c = 0$ and using (2.157b) we obtain

$$\begin{aligned}
\bar{E}_d &= [-A_d]^{-1} E_d = \begin{bmatrix} -5 & 0 \\ 0 & 0 \end{bmatrix}, \quad \bar{A}_d = [-A_d]^{-1} A_d = \begin{bmatrix} -1 & 0 \\ 0 & -1 \end{bmatrix}, \\
\bar{A}_{d\alpha} &= \bar{A}_d + 0.4\bar{E}_d = \begin{bmatrix} -3 & 0 \\ 0 & -1 \end{bmatrix}.
\end{aligned} \tag{2.211}$$

To test the stability of the system (2.157) (or equivalently (2.138a)) we have to consider the system (2.207).

The characteristic equation of (2.207) with $\alpha = 0.4$, $L = 1$ and (2.211) has the form

$$\det[\bar{E}_d z - \bar{A}_{d\alpha} - 0.12\bar{E}_d z^{-1}] = -5z + 0.6z^{-1} + 3 = 0. \tag{2.212}$$

The system (2.207) with $L = 1$ is asymptotically stable since roots of (2.212) are $z_1 = -0.1583$, $z_2 = 0.7583$. Therefore, by Theorem 2.29 the fractional descriptor discrete-time linear system (2.157) (or equivalently (2.138a)) is practically stable for length of the practical implementation $L = 1$.

Now let us consider the system (2.207) with $\alpha = 0.4$, $L \to \infty$ and (2.211). Using (2.210) we obtain

$$\det[\bar{E}_d(z-1)^{0.4}z^{0.6} - \bar{A}_d] = -5(z-1)^{0.4}z^{0.6} + 1 = 0. \tag{2.213}$$

The system (2.207) with $L \to \infty$ is unstable since the root of (2.213) is $z_1 = 1.0174$. Therefore, by Theorem 2.29 the fractional descriptor discrete-time linear system (2.157) (or equivalently (2.138a)) is unstable.

2.2.5.3 Weierstrass-Kronecker Decomposition Method

In this section necessary and sufficient stability conditions of the fractional descriptor discrete-time linear system (2.177) (or equivalently (2.138)) using the Weierstrass-Kronecker decomposition method.

Note that the state vector of the subsystem (2.177b) $\bar{x}_2(t) = 0$ for $u(t) = 0$. Therefore, the stability of the system (2.177) (or equivalently (2.138)) depends entirely on the subsystem (2.177a).

The upper bound of the sum in (2.179a) can be limited by some natural number L (length of practical implementation), like in the case of (2.141a), i.e. the state equation has the form

$$\bar{x}_{1,i+1} = A_{1d\alpha}\bar{x}_{1,i} + \sum_{j=1}^{L} c_j \bar{x}_{1,i-j} + B_{1d}u_i, \tag{2.214}$$

where $A_{1d\alpha}$ is given by (2.179c) and $\bar{x}_{1,-k} = 0$, $k = 1, 2, \ldots$.

Definition 2.17. The fractional descriptor discrete-time linear system (2.177) (or equivalently (2.138)) is called practically stable for given length L of practical implementation if the fractional linear system (2.214) is asymptotically stable.
The fractional descriptor discrete-time linear system (2.177) (or equivalently (2.138)) is called asymptotically stable if the fractional linear system (2.214) is asymptotically stable for $L \to \infty$.

Lemma 2.7. The characteristic equation

$$\det\left[\mathbb{I}_{n_1} z - A_{1d\alpha} - \sum_{j=1}^{L} \mathbb{I}_{n_1} c_j z^{-j}\right] = 0 \tag{2.215}$$

of (2.214) has the same set of roots as the characteristic equation (2.200) of (2.141a).

Proof. Using (2.174) and (2.200) we obtain

$$
\begin{aligned}
&\det\left\{P\left[E_d z - A_{d\alpha} - \sum_{j=1}^{L} E_d c_j z^{-j}\right]Q\right\} \\
&= \det\begin{bmatrix} \mathbb{I}_{n_1} z - A_{1d\alpha} - \sum_{j=1}^{L} \mathbb{I}_{n_1} c_j z^{-j} & 0 \\ 0 & Nz - (\mathbb{I}_{n_2} + N\alpha) - \sum_{j=1}^{L} c_j N z^{-j} \end{bmatrix} \\
&= (-1)^{n_2} \det\left[\mathbb{I}_{n_1} z - A_{1d\alpha} - \sum_{j=1}^{L} \mathbb{I}_{n_1} c_j z^{-j}\right] = 0
\end{aligned}
\tag{2.216}
$$

since

$$
\det\left[Nz - (\mathbb{I}_{n_2} + N\alpha) - \sum_{j=1}^{L} c_j N z^{-j}\right] = (-1)^{n_2}. \tag{2.217}
$$

Therefore, from (2.216) it follows that the characteristic equations (2.200) and (2.215) have the same set of roots. □

Using Lemma 2.7 and the approach given in Sect. 2.2.5.1 we obtain the following theorem.

Theorem 2.30. The fractional descriptor discrete-time linear system (2.138a) is:

1) practically stable for given length L of the practical implementation if and only if all roots of the characteristic equation (2.215) lie inside the unit circle;
2) asymptotically stable if and only if all roots of the characteristic equation

$$
\det\left[\mathbb{I}_{n_1}(z-1)^{\alpha} z^{1-\alpha} - A_{1d}\right] = 0 \tag{2.218}
$$

lie inside the unit circle.

Example 2.26. (Continuation of Example 2.24) Consider the fractional descriptor discrete-time linear system (2.138a) with (2.203). According to (2.204) the matrix pencil (E_d, A_d) is regular. In this case

$$
P = \begin{bmatrix} 1 & 0 \\ -0.5 & 1 \end{bmatrix}, \quad Q = \begin{bmatrix} 1 & 0 \\ 0 & 3.3333 \end{bmatrix} \tag{2.219}
$$

and

$$
\begin{aligned}
\begin{bmatrix} \mathbb{I}_{n_1} & 0 \\ 0 & N \end{bmatrix} &= PEQ = \begin{bmatrix} 1 & 0 \\ 0 & 0 \end{bmatrix}, \\
\begin{bmatrix} A_{1d} & 0 \\ 0 & \mathbb{I}_{n_2} \end{bmatrix} &= PAQ = \begin{bmatrix} 0.2 & 0 \\ 0 & 1 \end{bmatrix},
\end{aligned}
\tag{2.220a}
$$

$$A_{1d\alpha} = A_{1d} + 0.4\mathbb{I}_1 = 0.6. \tag{2.220b}$$

To test the stability of the system (2.177) (or equivalently (2.138a)) we have to consider the system (2.214).

The characteristic equation of (2.214) with $\alpha = 0.4$, $L = 1$ and (2.220) has the form

$$\det[\mathbb{I}_{n_1} z - \bar{A}_{1d\alpha} - 0.12\mathbb{I}_{n_1} z^{-1}] = z - 0.12z^{-1} + 0.6 = 0. \tag{2.221}$$

The system (2.214) with $L = 1$ is asymptotically stable since roots of (2.221) are $z_1 = -0.1583$, $z_2 = 0.7583$. Therefore, by Theorem 2.30 the fractional descriptor discrete-time linear system (2.177) (or equivalently (2.138a)) is practically stable for length of the practical implementation $L = 1$.

Now let us consider the system (2.214) with $\alpha = 0.4$, $L \to \infty$ and (2.220). Using (2.218) we obtain

$$\det[\mathbb{I}_{n_1}(z-1)^{0.4} z^{0.6} - A_{1d}] = (z-1)^{0.4} z^{0.6} - 0.2 = 0. \tag{2.222}$$

The system (2.214) with $L \to \infty$ is unstable since the root of (2.222) is $z_1 = 1.0174$. Therefore, by Theorem 2.30 the fractional descriptor discrete-time linear system (2.177) (or equivalently (2.138a)) is unstable.

2.2.6 *Superstability of Fractional Descriptor Discrete-Time Linear Systems*

The asymptotic stability of a dynamical system ensures that its free response decreases to zero for $i \to \infty$, however its value may increase significantly in the initial part of the state vector trajectory (it is so-called “peak effect”). In superstable systems the norm of the state vector decreases monotonically to zero for $i \to \infty$, which prevents such undesirable effects [70, 137, 138].

2.2.6.1 Properties of Superstable Fractional Discrete-Time Linear Systems

The concept of superstable discrete-time linear systems is given in Sect. 1.2.5.1. Consider the standard fractional discrete-time linear system

$$\Delta^\alpha x_{i+1} = A_d x_i, \quad i \in \mathbb{Z}_+, \quad 0 < \alpha < 1, \tag{2.223}$$

where $x_i \in \mathbb{R}^n$ is the state vector and $A_d \in \mathbb{R}^{n \times n}$. Substituting the definition of fractional difference (2.136) into (2.223) and limiting the upper bound of the sum by some natural number L we obtain

$$x_{i+1} = A_{d\alpha} x_i + \sum_{j=1}^{L} c_j x_{i-j}, \tag{2.224}$$

where $A_{d\alpha}$ is given by (2.149) and $x_{-k} = 0, k = 1, 2, \ldots$. The solution to the equation (2.224) can be obtained using (2.150) for $u_i = 0, i \in \mathbb{Z}_+$, i.e. it has the form

$$x_i = \Phi_{d,i} x_0, \tag{2.225}$$

where the matrices $\Phi_{d,i}$ are determined by (2.148).

Definition 2.18. The fractional discrete-time linear system (2.223) is called practically superstable for given length L of practical implementation if the fractional linear system (2.224) is superstable.
The fractional discrete-time linear system (2.223) is called superstable if the fractional linear system (2.224) is superstable for $L \to \infty$.

Theorem 2.31. The fractional discrete-time linear system (2.224) is superstable if

$$\|A_{d\alpha}\| \in \left(\frac{d - \sqrt{d^2 - 4c_L}}{2}; \frac{d + \sqrt{d^2 - 4c_L}}{2} \right) \quad \text{for } L \geq 1 \tag{2.226a}$$

or

$$\|A_{d\alpha}\| < 1 \quad \text{for } L = 0, \tag{2.226b}$$

where

$$d = 1 - \sum_{j=1}^{L-1} c_j. \tag{2.226c}$$

Proof. According to Theorem 1.38 the fractional discrete-time linear system (2.223) will be superstable if $\|x_{i+1}\| < \|x_i\|$ and therefore

$$\frac{\|\Phi_{d,i+1}\|}{\|\Phi_{d,i}\|} < 1, \quad i \in \mathbb{Z}_+ \tag{2.227}$$

since $\|x_i\| \leq \|\Phi_{d,i}\| \|x_0\|$. Given that

$$\|\Phi_{d,i+1}\| \leq \|\Phi_{d,i}\| \|A_{d\alpha}\| + \left\| \sum_{j=1}^{L} c_j \bar{\Phi}_{i-j} \right\| \tag{2.228}$$

the inequality (2.227) can be rewritten in the form

$$
\begin{aligned}
\frac{\|\Phi_{d,i+1}\|}{\|\Phi_{d,i}\|} &\le \|A_{d\alpha}\| + \frac{\left\|\sum_{j=1}^{L} c_j \Phi_{d,i-j}\right\|}{\|\Phi_{d,i}\|} \\
&\le \|A_{d\alpha}\| + \frac{\left\|\begin{bmatrix} c_1\mathbb{I}_n & c_2\mathbb{I}_n & \dots & c_L\mathbb{I}_n \end{bmatrix}\begin{bmatrix}\Phi_{d,i-1} \\ \Phi_{d,i-2} \\ \vdots \\ \Phi_{d,i-L}\end{bmatrix}\right\|}{\left\|\begin{bmatrix} A_{d\alpha} & c_1\mathbb{I}_n & \dots & c_{L-1}\mathbb{I}_n \end{bmatrix}\begin{bmatrix}\Phi_{d,i-1} \\ \Phi_{d,i-2} \\ \vdots \\ \Phi_{d,i-L}\end{bmatrix}\right\|} \\
&\le \|A_{d\alpha}\| + \frac{\sum_{j=1}^{L} c_j}{\|A_{d\alpha}\| + \sum_{j=1}^{L-1} c_j} < 1.
\end{aligned}
\tag{2.229}
$$

From (2.229) we have

$$
\|A_{d\alpha}\|^2 + \|A_{d\alpha}\|\left(\sum_{j=1}^{L-1} c_j - 1\right) + \sum_{j=1}^{L} c_j - \sum_{j=1}^{L-1} c_j < 0. \tag{2.230}
$$

Introducing (2.226c) and solving (2.230) with respect to $\|A_{d\alpha}\|$ we obtain (2.226a). For a system with no delays ($L = 0$) from (2.230) we obtain the condition (2.226b). □

Theorem 2.32. The fractional discrete-time linear system (2.224) is superstable for $L \to \infty$ if

$$
\|A_{d\alpha}\| < \alpha. \tag{2.231}
$$

Proof. In [88] it has been shown that

$$
\sum_{j=1}^{\infty} c_j = 1 - \alpha. \tag{2.232}
$$

From (2.232) and (2.230) for $L \to \infty$ we have

$$\|A_{d\alpha}\|^2 - \|A_{d\alpha}\|\alpha < 0 \tag{2.233}$$

Solving (2.233) with respect to $\|A_{d\alpha}\|$ we obtain (2.231). □

Therefore, the following theorem has been proved.

Theorem 2.33. The fractional discrete-time linear system (2.223) is:

1) practically superstable for given length L of practical implementation if the condition (2.226) holds;
2) superstable if the condition (2.231) holds.

Example 2.27. Consider the fractional discrete-time linear system (2.223) with $\alpha = 0.5$ and

$$A_d = \begin{bmatrix} -0.45 & 0 \\ 0.72 & -0.4 \end{bmatrix}. \tag{2.234}$$

In this case we have

$$\|A_{d\alpha}\| = \|A_d + \mathbb{I}_2\alpha\| = \left\|\begin{bmatrix} 0.05 & 0 \\ 0.72 & 0.1 \end{bmatrix}\right\| = 0.82. \tag{2.235}$$

From (2.226) and (2.231) we obtain the desired values of the norm of the matrix (2.235) for superstable systems:

$$\begin{aligned}
&\|A_{d\alpha}\| \in (0.1464; 0.8536) && \text{for} \quad L = 1, \\
&\|A_{d\alpha}\| \in (0.0785; 0.7965) && \text{for} \quad L = 2, \\
&\quad\vdots \\
&\|A_{d\alpha}\| \in (0; 0.5) && \text{for} \quad L \to \infty.
\end{aligned}$$

Therefore, from the above values and (2.235) we conclude that the considered system is practically superstable for the length of practical implementation $L = 1$. Taking into account that the presented approach gives only sufficient superstability conditions, we can not determine if the system is practically superstable for higher lengths of practical implementation ($L > 1$) or superstable (for $L \to \infty$).
In Fig. 2.4 we present time plots of the state variables and the norm of the state vector for $x_0 = [\,2 \quad 1\,]^T$ and $L = 1$. We can see that the norm of the state vector decreases monotonically for $i \to \infty$ and the system is practically superstable for $L = 1$.

In Fig. 2.5 we show time plots of the norm of the state vector for $x_0 = [\,2 \quad 1\,]^T$ and different values of the length of practical implementation L, from which we can conclude that the system is not practically superstable for $L > 1$ and therefore is also not superstable.

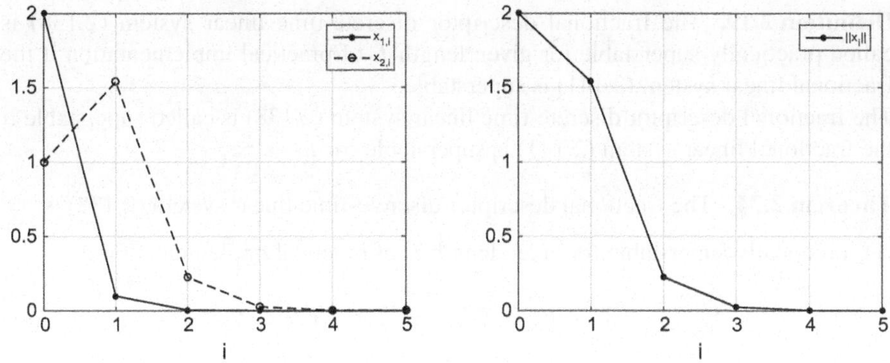

Fig. 2.4 State variables (on the left) and norm of the state vector (on the right) of the fractional discrete-time linear system (2.223) with (2.234) and $\alpha = 0.5$ for $x_0 = [\ 2 \quad 1\ ]^T$ and $L = 1$

Fig. 2.5 Norm of the state vector of the fractional discrete-time linear system (2.223) with (2.234) and $\alpha = 0.5$ for $x_0 = [\ 2 \quad 1\ ]^T$ and different values of L

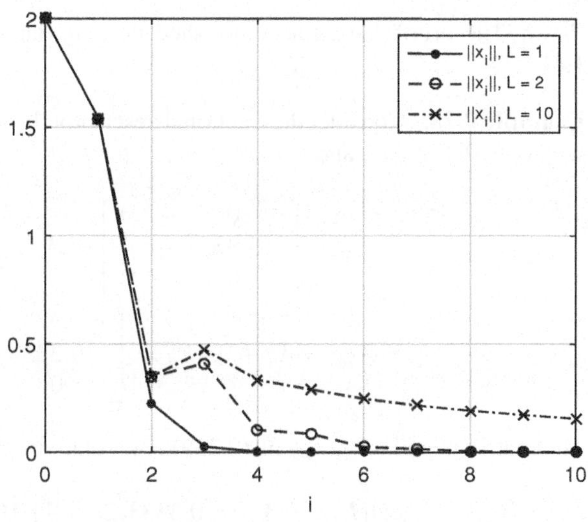

2.2.6.2 Laurent Series Expansion Method

In this section sufficient superstability conditions of the fractional descriptor discrete-time linear system (2.138) will be established using the Laurent series expansion method.

Consider the fractional descriptor discrete-time linear system (2.138). Solution to the Eq. (2.138a) for $u_i = 0$, $i \in \mathbb{Z}_+$ is given by

$$x_i = T_{d,i} E_d x_0, \tag{2.236}$$

where the matrices $T_{d,k} \in \mathbb{R}^{n \times n}$ are determined by (2.142b).

Definition 2.19. The fractional descriptor discrete-time linear system (2.138) is called practically superstable for given length L of practical implementation if the fractional linear system (2.141) is superstable.
The fractional descriptor discrete-time linear system (2.138) is called superstable if the fractional linear system (2.141) is superstable for $L \to \infty$.

Theorem 2.34. The fractional descriptor discrete-time linear system (2.138) is:

1) practically superstable for given length L of practical implementation if

$$\|T_{d,i}E\| > \|T_{d,i+1}E\| \text{ for } i \in \mathbb{Z}_+; \tag{2.237}$$

2) superstable if

$$\|T_{d,i}E\| > \|T_{d,i+1}E\| \text{ for } i \in \mathbb{Z}_+ \text{ and } L \to \infty. \tag{2.238}$$

Proof. The proof can be accomplished in a similar way to the proof of Theorem 1.41. □

Example 2.28. Consider the fractional descriptor discrete-time linear system (2.138) with $\alpha = 0.4$, $L = 1$ and

$$E_d = \begin{bmatrix} 0 & -2 & 0 \\ -3.3333 & -5 & 0 \\ 0 & -1 & 0 \end{bmatrix}, \quad A_d = \begin{bmatrix} 0 & 1 & 0 \\ 1 & 0 & 0 \\ 0 & 0 & 1 \end{bmatrix}$$
$$A_{d\alpha} = A_d + 0.4E_d = \begin{bmatrix} 0 & 0.2 & 0 \\ -0.3333 & -2 & 0 \\ 0 & -0.4 & 1 \end{bmatrix}. \tag{2.239}$$

The matrix pencil (E_d, A_d) of (2.239) is regular since

$$\det[E_d\lambda - A_d] = 0.3333(2\lambda + 1)(10\lambda + 3) \neq 0. \tag{2.240}$$

The matrix $[E_d z - A_{d\alpha} - E_d c_1 z^{-1}]^{-1}$ has the form

$$[E_d z - A_{d\alpha} - 0.12E_d z^{-1}]^{-1} = \begin{bmatrix} \frac{1875z^3-750z^2-225z}{2500z^4-625z^2+36} & \frac{15z}{-50z^2+5z+6} & 0 \\ -\frac{25z}{50z^2+5z-6} & 0 & 0 \\ \frac{25z^2-10z-3}{50z^2+5z-6} & 0 & 1 \end{bmatrix}. \tag{2.241}$$

Using (2.145) for $n_0 = 1$, the matrix (2.241) can be written in the form

$$[E_d z - A_{d\alpha} - 0.12E_d z^{-1}]^{-1} = \sum_{k=-1}^{\infty} T_{d,k} z^{-(k+1)}, \tag{2.242a}$$

where

$$T_{d,k} = \begin{cases} \begin{bmatrix} 0 & 0 & 0 \\ 0 & 0 & 0 \\ 0.5 & 0 & -1 \end{bmatrix} & \text{for } k = -1, \\ \begin{bmatrix} t_{d,11} & t_{d,12} & 0 \\ t_{d,21} & 0 & 0 \\ t_{d,31} & 0 & 0 \end{bmatrix} & \text{for } k = 0, 1, 2, \ldots, \end{cases} \tag{2.242b}$$

$$\begin{aligned} t_{d,11} &= 0.4286(-0.4)^k - 0.6429(0.4)^k - 0.4821(-0.3)^k + 0.8036(0.3)^k, \\ t_{d,12} &= 0.0857(0.4)^k - 0.1286(-0.3)^k, \\ t_{d,21} &= -0.2857(-0.4)^k - 0.2143(0.3)^k, \\ t_{d,31} &= -0.1429(-0.4)^k - 0.1071(0.3)^k. \end{aligned} \tag{2.242c}$$

Using (2.239) and (2.242) we have

$$T_{d,k}E_d = \begin{cases} \begin{bmatrix} 0 & 0 & 0 \\ 0 & 0 & 0 \\ 0 & 0 & 0 \end{bmatrix} & \text{for } k = -1, \\ \begin{bmatrix} \bar{t}_{d,11} & \bar{t}_{d,12} & 0 \\ 0 & \bar{t}_{d,22} & 0 \\ 0 & \bar{t}_{d,32} & 0 \end{bmatrix} & \text{for } k = 0, 1, 2, \ldots, \end{cases} \tag{2.243a}$$

$$\begin{aligned} \bar{t}_{d,11} &= 0.2857(0.4)^k + 0.4286(-0.3)^k, \\ \bar{t}_{d,12} &= 0.8572(0.4)^k - 0.8572(-0.4)^k + 1.6071(-0.3)^k - 1.6071(0.3)^k, \\ \bar{t}_{d,22} &= 0.5714(-0.4)^k + 0.4286(0.3)^k, \\ \bar{t}_{d,32} &= 0.2857(-0.4)^k + 0.2143(0.3)^k. \end{aligned} \tag{2.243b}$$

From (2.243) it follows that the condition (2.237) is satisfied. Therefore, by Theorem 2.34 the considered system is practically superstable for the length of practical implementation $L = 1$. Performing similar analysis for $L = 2$, i.e. expanding in the Laurent series the matrix $[E_d z - A_{d\alpha} - c_1 E_d z^{-1} - c_2 E_d z^{-2}]^{-1}$ it can be shown that the system is not practically superstable for $L > 1$ and therefore is also not superstable (the condition (2.238) is not satisfied).

In Fig. 2.6 we present time plots of the state variables and the norm of the state vector for $x_0 = [\,1 \quad 2 \quad 1\,]^T$ and $L = 1$. We can see that the norm of the state vector decreases monotonically for $i \to \infty$ and the system is practically superstable for $L = 1$.

In Fig. 2.7 we show time plots of the norm of the state vector for $x_0 = [\,1 \quad 2 \quad 1\,]^T$ and different values of the length of practical implementation L, from which we can conclude that the system is not practically superstable for $L > 1$ and therefore is also not superstable.

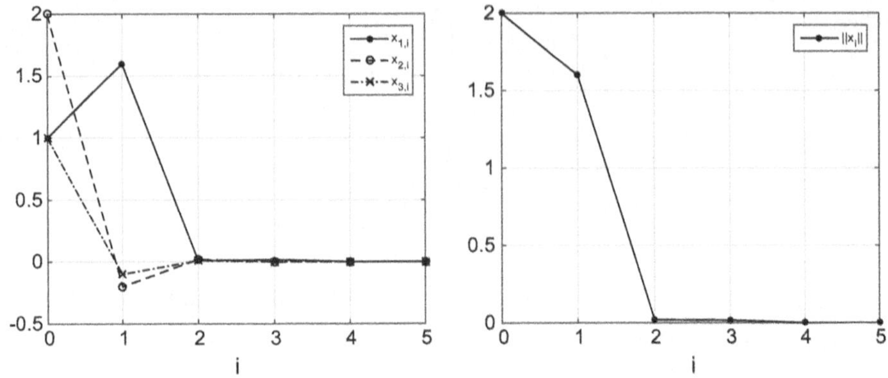

Fig. 2.6 State variables (on the left) and norm of the state vector (on the right) of the fractional descriptor discrete-time linear system (2.138) with (2.239) and $\alpha = 0.4$ for $x_0 = [\,1 \;\; 2 \;\; 1\,]^T$ and $L = 1$

Fig. 2.7 Norm of the state vector of the fractional descriptor discrete-time linear system (2.138) with (2.239) and $\alpha = 0.4$ for $x_0 = [\,1 \;\; 2 \;\; 1\,]^T$ and different values of L

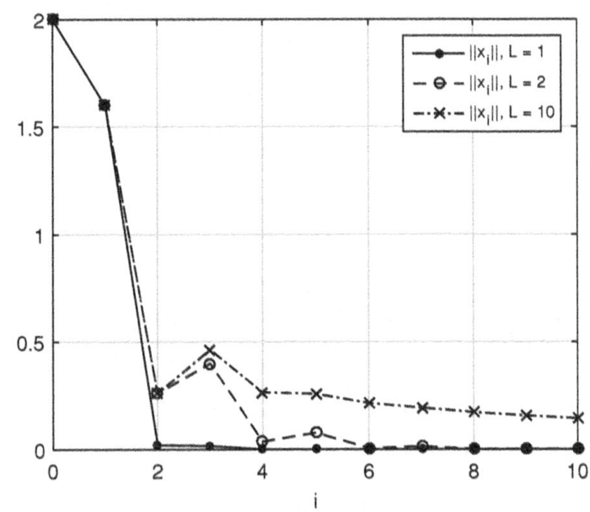

2.2.6.3 Drazin Inverse Matrix Method

In this section sufficient superstability conditions of the fractional descriptor discrete-time linear system (2.157) (or equivalently (2.138)) will be established using the Drazin inverse matrix method.

Consider the fractional descriptor discrete-time linear system (2.157). From (1.26b) and (2.162a) for $u_i = 0$, $i \in \mathbb{Z}_+$ it follows that

$$x_i = \bar{E}_d \bar{E}_d^D x_i, \quad i \in \mathbb{Z}_+, \tag{2.244}$$

Premultiplying (2.207) by $\bar{E}^D$ and taking into account (2.244) we obtain

$$x_{i+1} = \bar{E}_d^D \bar{A}_{d\alpha} x_i + \sum_{j=1}^{L} c_j x_{i-j}, \tag{2.245}$$

where $\bar{A}_{d\alpha}$ is given by (2.159). The Eq. (2.245) can also be written in the form

$$x_{i+1} = \bar{H}_d x_i + \sum_{j=1}^{L} c_j x_{i-j}, \tag{2.246}$$

where

$$\bar{H}_d = \bar{E}_d^D \bar{A}_{d\alpha} + G_d(\mathbb{I}_n - \bar{E}_d \bar{E}_d^D) \tag{2.247}$$

since (2.244) holds and $(\mathbb{I}_n - \bar{E}_d \bar{E}_d^D)\bar{E}_d \bar{E}_d^D = 0$, $G_d \in \mathbb{R}^{n\times n}$ is an arbitrary matrix. Taking into consideration (2.162a) the solution to the equation (2.246) can be expressed by

$$x_i = \bar{\Phi}_{d,i} x_0, \tag{2.248a}$$

where

$$\begin{gathered} \bar{\Phi}_{d,i+1} = \bar{\Phi}_{d,i} \bar{H}_d + \sum_{j=1}^{L} c_j \bar{\Phi}_{d,i-j}, \\ \bar{\Phi}_{d,0} = \mathbb{I}_n, \quad \bar{\Phi}_{d,-k} = 0, \quad k = 1, \dots, h. \end{gathered} \tag{2.248b}$$

Definition 2.20. The fractional descriptor discrete-time linear system (2.157) (or equivalently (2.138)) is called practically superstable for given length L of practical implementation if the fractional discrete-time linear system (2.246) is superstable. The fractional descriptor discrete-time linear system (2.157) (or equivalently (2.138)) is called superstable if the fractional linear system (2.246) is superstable for $L \to \infty$.

Theorem 2.35. The fractional discrete-time linear system (2.246) is superstable if there exists a matrix $G_d \in \mathbb{R}^{n\times n}$ such that

$$\|\bar{H}_d\| \in \left(\frac{d - \sqrt{d^2 - 4c_L}}{2}; \frac{d + \sqrt{d^2 - 4c_L}}{2} \right) \quad \text{for } L \geq 1 \tag{2.249a}$$

or

$$\|\bar{H}_d\| < 1 \text{ for } L = 0, \tag{2.249b}$$

where the coefficient d is given by (2.226c).

Proof. The proof is similar to the proof of Theorem 2.31. □

Theorem 2.36. The fractional discrete-time linear system (2.246) is superstable for $L \to \infty$ if there exists a matrix $G_d \in \mathbb{R}^{n \times n}$ such that

$$\|\bar{H}_d\| < \alpha. \tag{2.250}$$

Proof. The proof can be accomplished in a similar way to the proof of Theorem 2.32. □

Therefore, from the above considerations we have the following theorem.

Theorem 2.37. The fractional descriptor discrete-time linear system (2.157) (or equivalently (2.138)) is:

1) practically superstable for given length L of practical implementation if there exists a matrix $G_d \in \mathbb{R}^{n \times n}$ such that the condition (2.249) holds;
2) superstable if there exists a matrix $G_d \in \mathbb{R}^{n \times n}$ such that the condition (2.250) holds.

Example 2.29. (Continuation of Example 2.28) Consider the fractional descriptor discrete-time linear system (2.138) with $\alpha = 0.4$ and (2.239). According to (2.240) the matrix pencil (E_d, A_d) is regular. Using (2.157b) for $c = 0$ we obtain

$$\begin{gathered}
\bar{E}_d = [-A_d]^{-1} E_d = \begin{bmatrix} 3.3333 & 5 & 0 \\ 0 & 2 & 0 \\ 0 & 1 & 0 \end{bmatrix}, \\
\bar{A}_d = [-A_d]^{-1} A_d = \begin{bmatrix} -1 & 0 & 0 \\ 0 & -1 & 0 \\ 0 & 0 & -1 \end{bmatrix}, \\
\bar{A}_{d\alpha} = \bar{A}_d + 0.4\bar{E}_d = \begin{bmatrix} 0.3333 & 2 & 0 \\ 0 & -0.2 & 0 \\ 0 & 0.4 & -1 \end{bmatrix}.
\end{gathered} \tag{2.251}$$

Next, from (B.1) and (B.2) we have

$$\bar{E}_d^D = \begin{bmatrix} 0.3 & -0.75 & 0 \\ 0 & 0.5 & 0 \\ 0 & 0.25 & 0 \end{bmatrix} \tag{2.252}$$

and

$$\bar{E}_d^D \bar{A}_{d\alpha} = \begin{bmatrix} 0.1 & 0.75 & 0 \\ 0 & -0.1 & 0 \\ 0 & -0.05 & 0 \end{bmatrix}. \tag{2.253}$$

Note that for $G_d = 0$ the norm of the matrix

$$\|\bar{H}_d\| = \|\bar{E}_d^D \bar{A}_{d\alpha} + G_d(\mathbb{I}_3 - \bar{E}_d \bar{E}_d^D)\| = \left\| \begin{bmatrix} 0.1 & 0.75 & 0 \\ 0 & -0.1 & 0 \\ 0 & -0.05 & 0 \end{bmatrix} \right\| = 0.85 \qquad (2.254)$$

takes its minimal value. From (2.249) and (2.250) we obtain the desired values of the norm of the matrix (2.254) for superstable systems

$$\begin{aligned} \|\bar{H}_d\| &\in (0.1394; 0.8606) \quad \text{for} \quad L = 1, \\ \|\bar{H}_d\| &\in (0.08; 0.8) \quad \text{for} \quad L = 2, \\ &\vdots \\ \|\bar{H}_d\| &\in (0; 0.4) \quad \text{for} \quad L \to \infty. \end{aligned}$$

Therefore, from the above values and (2.235) we conclude that the considered system is practically superstable for the length of practical implementation $L = 1$. Taking into account that the presented approach gives only sufficient superstability conditions, we can not determine if the system is practically superstable for higher lengths of practical implementation ($L > 1$) or superstable (for $L \to \infty$). The time plots of the state variables and the norm of the state vector for $x_0 = [\,1 \;\; 2 \;\; 1\,]^T$ and $L = 1$ are presented in Fig. 2.6. The time plots of the norm of the state vector for $x_0 = [\,1 \;\; 2 \;\; 1\,]^T$ and different values of the length of practical implementation L are presented in Fig. 2.7. From these we can conclude that the considered system is not practically superstable for $L > 1$ and therefore is also not superstable.

2.2.6.4 Weierstrass-Kronecker Decomposition Method

In this section sufficient superstability conditions of the fractional descriptor discrete-time linear system (2.177) (or equivalently (2.138)) will be established using the Weierstrass-Kronecker decomposition method.

Note that every superstable system is also asymptotically stable. State vector of the subsystem (2.177b) $\bar{x}_{2,i} = 0$ for $u_i = 0$, $i \in \mathbb{Z}_+$. Using (2.175) and (2.180) we obtain

$$\begin{aligned} x_i = Q_d \bar{x}_i = Q_d \begin{bmatrix} \bar{x}_{1,i} \\ \bar{x}_{2,i} \end{bmatrix} &= Q_d \begin{bmatrix} \Phi_{d,i} \bar{x}_{10} \\ 0 \end{bmatrix} \\ &= Q_d \begin{bmatrix} \Phi_{d,i} & 0 \\ 0 & \Phi_{M,i} \end{bmatrix} \begin{bmatrix} \bar{x}_{10} \\ 0 \end{bmatrix} \\ &= Q_d \begin{bmatrix} \Phi_{d,i} & 0 \\ 0 & \Phi_{M,i} \end{bmatrix} Q_d^{-1} x_0 = \bar{\Phi}_{d,i} x_0, \end{aligned} \qquad (2.255)$$

where

$$\bar{\Phi}_{d,i+1} = \bar{\Phi}_{d,i} A_{Md} + \sum_{j=1}^{L} c_j \bar{\Phi}_{d,i-j}, \qquad (2.256a)$$

$$A_{Md} = Q_d \begin{bmatrix} A_{1d\alpha} & 0 \\ 0 & M_d \end{bmatrix} Q_d^{-1} \tag{2.256b}$$

and the matrix $M_d \in \mathbb{R}^{n_2 \times n_2}$ is arbitrary.

Definition 2.21. The fractional descriptor discrete-time linear system (2.177) (or equivalently (2.138)) is called practically superstable for given length L of practical implementation if the fractional discrete-time linear system (2.214) is superstable. The fractional descriptor discrete-time linear system (2.177) (or equivalently (2.138)) is called superstable if the fractional linear system (2.214) is superstable for $L \to \infty$.

Taking into account (2.255), (2.256) and the approach given in Sect. 2.2.6.1 we obtain the following theorems.

Theorem 2.38. The fractional discrete-time linear system (2.214) is superstable if there exists a matrix $M_d \in \mathbb{R}^{n \times n}$ such that

$$\|A_{Md}\| \in \left(\frac{d - \sqrt{d^2 - 4c_L}}{2}; \frac{d + \sqrt{d^2 - 4c_L}}{2} \right) \quad \text{for } L \geq 1 \tag{2.257a}$$

or

$$\|A_{Md}\| < 1 \text{ for } L = 0, \tag{2.257b}$$

where the coefficient d is given by (2.226c).

Theorem 2.39. The fractional discrete-time linear system (2.214) is superstable for $L \to \infty$ if there exists a matrix $M_d \in \mathbb{R}^{n \times n}$ such that

$$\|A_{Md}\| < \alpha. \tag{2.258}$$

From Theorems 2.38 and 2.39 we have the resulting proposition.

Theorem 2.40. The fractional descriptor discrete-time linear system (2.177) (or equivalently (2.138)) is:

1) practically superstable for given length L of practical implementation if there exists a matrix $M_d \in \mathbb{R}^{n \times n}$ such that the condition (2.257) holds;
2) superstable if there exists a matrix $M_d \in \mathbb{R}^{n \times n}$ such that the condition (2.258) holds.

Example 2.30. (Continuation of Example 2.28) Consider the fractional descriptor discrete-time linear system (2.138) with $\alpha = 0.4$ and (2.239). According to (2.240) the matrix pencil (E_d, A_d) is regular. In this case

$$P_d = \begin{bmatrix} -0.5 & 0 & 0 \\ 0.75 & -0.3 & 0 \\ -0.5 & 0 & 1 \end{bmatrix}, \quad Q_d = \begin{bmatrix} 0 & 1 & 0 \\ 1 & 0 & 0 \\ 0.5 & 0 & 1 \end{bmatrix} \tag{2.259}$$

and

$$
\begin{bmatrix} \mathbb{I}_{n_1} & 0 \\ 0 & N \end{bmatrix} = P_d E_d Q_d = \begin{bmatrix} 1 & 0 & 0 \\ 0 & 1 & 0 \\ 0 & 0 & 0 \end{bmatrix},
$$

$$
\begin{bmatrix} A_{1d} & 0 \\ 0 & \mathbb{I}_{n_2} \end{bmatrix} = P_d A_d Q_d = \begin{bmatrix} -0.5 & 0 & 0 \\ 0.75 & -0.3 & 0 \\ 0 & 0 & 1 \end{bmatrix}, \tag{2.260}
$$

where $n_1 = 2$ and $n_2 = 1$. Next we compute

$$
A_{1d\alpha} = A_{1d} + 0.4\mathbb{I}_2 = \begin{bmatrix} -0.1 & 0 \\ 0.75 & 0.1 \end{bmatrix}. \tag{2.261}
$$

Using (2.256b) for (2.259), (2.261) and $M_d = [m_d] \in \mathbb{R}^{1\times 1}$ we obtain

$$
\begin{aligned}
A_{Md} &= \begin{bmatrix} 0 & 1 & 0 \\ 1 & 0 & 0 \\ 0.5 & 0 & 1 \end{bmatrix} \begin{bmatrix} -0.1 & 0 & 0 \\ 0.75 & 0.1 & 0 \\ 0 & 0 & m_d \end{bmatrix} \begin{bmatrix} 0 & 1 & 0 \\ 1 & 0 & 0 \\ 0 & -0.5 & 1 \end{bmatrix} \\
&= \begin{bmatrix} 0.1 & 0.75 & 0 \\ 0 & -0.1 & 0 \\ 0 & -0.5m_d - 0.05 & m_d \end{bmatrix}.
\end{aligned} \tag{2.262}
$$

From (2.257) and (2.258) we obtain the desired values of the norm of the matrix (2.262) for superstable systems

$$
\begin{aligned}
\|A_{Md}\| &\in (0.1394; 0.8606) \quad \text{for} \quad L = 1, \\
\|A_{Md}\| &\in (0.08; 0.8) \quad \text{for} \quad L = 2, \\
&\vdots \\
\|A_{Md}\| &\in (0; 0.4) \quad \text{for} \quad L \to \infty.
\end{aligned}
$$

From the above values for $m_d = 0$ and $\|A_{Md}\| = 0.85$ we conclude that the considered system is practically superstable for the length of practical implementation $L = 1$.

2.3 Concluding Remarks

In this chapter fractional descriptor continuous-time and discrete-time linear systems have been analyzed. In the subsequent sections solutions of the state equation and analytical conditions of the positivity, stability and superstability have been established considering the Laurent series expansion method, Drazin inverse matrix method and Weierstrass-Kronecker decomposition method. The Caputo definition of fractional derivative-integral and the Grünwald-Letnikov fractional (backward) difference operators have been used for continuous-time and discrete-time systems, respectively.

Many of the conclusions given in Sect. 1.3 can also be applied here. The solution of the state equation does not depend on the choice of the method of the descriptor systems analysis. However, the considered methods differ in defining the analytical conditions for the positivity, stability and superstability.

In the case of fractional continuous-time systems, the main difference compared to integer-order systems (which have been considered in Sect. 1.1) is the Mittag-Leffler function occurring in the solution of the state equation (Theorems 2.3–2.5). The Mittag-Leffler function is a generalization of the exponential function and it plays important role in the analysis of fractional differential equations. The conditions for the positivity (Theorems 2.7–2.9), stability (Theorems 2.10-2.15) and superstability (Theorems 2.17–2.19) are analogous to those presented in Sect. 1.1.

We can see more differences in the case of fractional discrete-time systems. Applying the fractional difference operator we obtain the system with increasing number of delays in the state vector. The computational complexity of the Laurent series expansion method grows with the higher length of the practical implementation (i.e. with more delays in the state space model) and it is not recommended to use in the analysis of the fractional discrete-time linear systems, even though it is the only method for which necessary and sufficient conditions for the positivity (Theorem 2.24) and superstability (Theorem 2.34) have been established. Using the other two methods necessary and sufficient conditions for the positivity (Theorems 2.25 and 2.26) and sufficient conditions for superstability (Theorems 2.37 and 2.40) have been obtained. Note that for the Weierstrass-Kronecker decomposition method additional positivity condition is needed, i.e. the matrix Q (or Q_d) has to be a monomial matrix.

Chapter 3
Stability of Positive Descriptor Systems

In this chapter stability of positive descriptor systems will be analyzed. In the subsequent sections the following topics will be considered: stability tests for positive linear systems, stability of positive interval systems, stability of nonlinear systems with positive linear parts.

3.1 Stability Tests for Positive Linear Systems

To test the stability of descriptor systems an approach based on the location of the roots of the characteristic equation can be used, which is discussed in Sects. 1.1.4, 1.2.4, 2.1.7, 2.2.5. However, if the system is a positive one, some other methods of stability analysis can be used [47, 57]. These methods are the subject of research in this section.

3.1.1 Continuous-Time Systems

Consider the positive descriptor continuous-time linear system

$$E\dot{x}(t) = Ax(t), \tag{3.1}$$

where $x(t) \in \mathbb{R}^n_+$ is the state vector and $E, A \in \mathbb{R}^{n\times n}$. It is assumed that $\det E = 0$ and $\det[Es - A] \neq 0$ for some $s \in \mathbb{C}$. Following the considerations presented in Sect. 1.1 the system (3.1) can be transformed into the following equivalent forms.

1) Using the Drazin inverse matrix method we obtain

$$\begin{aligned} \dot{x}_1(t) &= \bar{E}^D \bar{A} x_1(t), \\ x_2(t) &= 0, \end{aligned} \tag{3.2}$$

T. Kaczorek and K. Borawski, *Descriptor Systems of Integer and Fractional Orders*, Studies in Systems, Decision and Control 367,
https://doi.org/10.1007/978-3-030-72480-1_3

where $x_1(t), x_2(t) \in \mathbb{R}^n_+$ are given by (1.28) and the matrices $\bar{E}, \bar{A} \in \mathbb{R}^{n\times n}$ have the form (1.24b). Theorem 1.10 gives the positivity conditions of the system (3.2).

2) Using the Weierstrass-Kronecker decomposition method we have

$$\begin{aligned}\dot{\bar{x}}_1(t) &= A_1\bar{x}_1(t),\\ \bar{x}_2(t) &= 0,\end{aligned} \tag{3.3}$$

where $\bar{x}_1(t) \in \mathbb{R}^{n_1}_+, \bar{x}_2(t) \in \mathbb{R}^{n_2}_+$ are given by (1.53) and the matrix $A_1 \in \mathbb{R}^{n_1\times n_1}$ is determined by (1.52). Theorem 1.11 gives the positivity conditions of the system (3.3).

Theorem 3.1. The positive descriptor continuous-time linear system (3.1) is asymptotically stable if and only if all coefficients of the following characteristic equations

1) $p_{(E,A)}(s) = 0$ given by (1.100),
2) $p_{(\bar{E},\bar{A})}(s) = 0$ given by (1.114),
3) $p_{\bar{E}^D\bar{A}}(s) = 0$ given by (1.117),
4) $p_{A_1}(s) = 0$ given by (1.125).

are positive.

Proof. First we shall show the proof for the equation (1.125).

Necessity. The eigenvalues $s_1, s_2, \dots, s_{n_1}$ of A_1 are real or complex conjugate since the coefficients $\tilde{a}_k, k = 1, \dots, n_1$ of $p_{A_1}(s)$ are real. Hence if $\text{Re}s_k < 0, k = 1, \dots, n_1$ then all coefficients of the polynomial $p_{A_1}(s) = (s - s_1)(s - s_2)\dots(s - s_{n_1})$ are positive, i.e. $\tilde{a}_k > 0$ for $k = 1, \dots, n_1$.

Sufficiency. This will be proved by contradiction. It is well-known [49] that if A_1 is a Metzler matrix, then $\bar{s} = \max_k \text{Re}s_k$ is its eigenvalue and $\text{Re}s_k < 0$ if $\bar{s} < 0$. For $\tilde{a}_k > 0, k = 1, \dots, n_1$ and real s we have $p_{A_1}(s) = s^{n_1} + a_{n_1-1}s^{n_1-1} + \dots + a_1 s + a_0 > 0$ and A_1 has no real nonnegative eigenvalue. Thus, we get the contradiction and $\bar{s} < 0$.
By Lemmas 1.7–1.9 the characteristic polynomials $p_{(E,A)}(s)$, $p_{(\bar{E},\bar{A})}(s)$, $p_{\bar{E}^D\bar{A}}(s)$ and $p_{A_1}(s)$ are related by (1.115), (1.117) and (1.126), respectively. Therefore, if one of the foregoing characteristic equations have only positive coefficients, then also every considered characteristic equation has only positive coefficients. □

Theorem 3.2. The positive descriptor continuous-time linear system (3.1) is asymptotically stable if and only if there exists a strictly positive vector $\bar{\lambda} = \begin{bmatrix} \bar{\lambda}_1 & \dots & \bar{\lambda}_n \end{bmatrix}$, $\bar{\lambda}_k > 0, k = 1, \dots, n$ such that

$$[\bar{E}^D\bar{A} + G(\mathbb{I}_n - \bar{E}\bar{E}^D)]\bar{\lambda} < 0 \text{ for an arbitrary matrix } G \in \mathbb{R}^{n\times n} \tag{3.4}$$

or there exists a strictly positive vector $\lambda = \begin{bmatrix} \lambda_1 & \dots & \lambda_{n_1} \end{bmatrix}$, $\lambda_k > 0, k = 1, \dots, n_1$ such that

$$A_1\lambda < 0. \tag{3.5}$$

Proof. First we consider the condition (3.5). We shall show that if the system (3.3) is asymptotically stable then there exists a strictly positive vector $\lambda \in \mathbb{R}_+^{n_1}$ satisfying (3.5). Integrating in the interval $[0, \infty)$ the first equation of (3.3) we obtain

$$\int_0^\infty \dot{\bar{x}}_1(t)dt = A_1 \int_0^\infty \bar{x}_1(t)dt \tag{3.6}$$

and

$$\bar{x}_1(\infty) - \bar{x}_1(0) = A_1 \int_0^\infty \bar{x}_1(t)dt. \tag{3.7}$$

For an asymptotically stable positive system we have

$$\bar{x}_1(\infty) = 0, \;\; \bar{x}_1(0) > 0, \;\; \int_0^\infty \bar{x}_1(t)dt > 0. \tag{3.8}$$

Taking into account (3.7) and (3.8) we obtain (3.5) for

$$\lambda = \int_0^\infty \bar{x}_1(t)dt. \tag{3.9}$$

It is well-known that the system (3.3) is asymptotically stable if and only if the corresponding transposed system

$$\dot{\bar{x}}_1(t) = A_1^T \bar{x}_1(t) \tag{3.10}$$

is asymptotically stable. As a Lyapunov function for the positive system (3.10) we choose the function

$$V(\bar{x}_1(t)) = \bar{x}_1^T(t)\lambda, \tag{3.11}$$

which is positive for any nonzero $\bar{x}_1(t) \in \mathbb{R}_+^n$ and $V(\bar{x}_1(t)) = 0$ for $\bar{x}_1(t) = 0$. Using (3.10) and (3.11) we obtain

$$\dot{V}(\bar{x}_1(t)) = \dot{\bar{x}}_1^T(t)\lambda = \bar{x}_1^T(t)A\lambda. \tag{3.12}$$

Therefore, if the condition (3.5) is satisfied then from (3.12) we have $\dot{V}(\bar{x}_1) < 0$ and the system (3.3) is asymptotically stable.

The condition (3.4) can be proved in a similar way. The matrix $G \in \mathbb{R}^{n\times n}$ and the term $G(\mathbb{I}_n - \bar{E}\bar{E}^D)$ eliminates from the matrix $\bar{E}^D\bar{A}$ unimportant entires that are further canceled through multiplication by $x_1(t) \in \operatorname{Im}\bar{E}\bar{E}^D$. Therefore, we have to find both: an arbitrary matrix G and strictly positive vector λ such that (3.4) holds. □

Example 3.1. Consider the descriptor continuous-time linear system (3.1) with

$$E = \begin{bmatrix} 0 & 1 & 0 \\ 0.125 & 1 & 0 \\ 0 & 0 & 0 \end{bmatrix}, \quad A = \begin{bmatrix} 0 & -3 & 1 \\ -0.25 & -2.75 & 1 \\ 0 & 0 & 0.5 \end{bmatrix}. \tag{3.13}$$

The matrix pencil (E, A) of (3.13) is regular since

$$\begin{aligned} \det[Es - A] &= \begin{vmatrix} 0 & s+3 & -1 \\ 0.125s + 0.25 & s + 2.75 & -1 \\ 0 & 0 & -0.5 \end{vmatrix} \\ &= 0.0625(s+2)(s+3) \neq 0. \end{aligned} \tag{3.14}$$

The characteristic equation of (3.13) is given by

$$p_{(E,A)}(s) = \det[Es - A] = 0.0625(s^2 + 5s + 6) = 0. \tag{3.15}$$

From (3.15) it follows that the pair (E, A) has two eigenvalues $s_1 = -2$, $s_2 = -3$ and by condition (1.101) the considered system is asymptotically stable. To check the positivity of the system we use two approaches as follows.

1) Drazin inverse matrix method case

We choose $c = 0$ and using (1.24b) we obtain

$$\begin{aligned} \bar{E} &= [-A]^{-1}E = \begin{bmatrix} 0.5 & 0.3333 & 0 \\ 0 & 0.3333 & 0 \\ 0 & 0 & 0 \end{bmatrix}, \\ \bar{A} &= [-A]^{-1}A = \begin{bmatrix} -1 & 0 & 0 \\ 0 & -1 & 0 \\ 0 & 0 & -1 \end{bmatrix}. \end{aligned} \tag{3.16}$$

Next, from (B.1) and (B.2) we have

$$\bar{E}^D = \begin{bmatrix} 2 & -2 & 0 \\ 0 & 3 & 0 \\ 0 & 0 & 0 \end{bmatrix} \tag{3.17}$$

and

$$\bar{E}^D\bar{A} = \begin{bmatrix} -2 & 2 & 0 \\ 0 & -3 & 0 \\ 0 & 0 & 0 \end{bmatrix}. \tag{3.18}$$

By Theorem 1.10 the matrix $\bar{E}^D\bar{A} \in \mathbb{M}_3$ and the system (3.13) is positive. Therefore, Theorems 3.1 and 3.2 can be used to test the stability of the system. From (3.16) and (3.18) we have

$$\begin{aligned} p_{(\bar{E},\bar{A})}(s) &= \det[\bar{E}s - \bar{A}] = 0.1667(s^2 + 5s + 6) = 0, \\ p_{\bar{E}^D\bar{A}}(s) &= \det[\mathbb{I}_3 s - \bar{E}^D\bar{A}] = s(s^2 + 5s + 6) = 0. \end{aligned} \tag{3.19}$$

Note that all coefficients of the characteristic equations $p_{(E,A)}(s)$, $p_{(\bar{E},\bar{A})}(s)$, $p_{\bar{E}^D\bar{A}}(s)$ given by (3.15) and (3.19) are positive and by Theorem 3.1 the positive descriptor continuous-time linear system with (3.13) is asymptotically stable.

We obtain the same result using Theorem 3.2 since for

$$G = \begin{bmatrix} 0 & 0 & 0 \\ 0 & 0 & 0 \\ 0 & 0 & -1 \end{bmatrix} \tag{3.20}$$

and $\bar{\lambda} = [\,2 \quad 1 \quad 1\,]^T$ we have

$$\begin{aligned} [\bar{E}^D\bar{A} + G(\mathbb{I}_3 - \bar{E}\bar{E}^D)]\bar{\lambda} &= \begin{bmatrix} -2 & 2 & 0 \\ 0 & -3 & 0 \\ 0 & 0 & -1 \end{bmatrix} \begin{bmatrix} 2 \\ 1 \\ 1 \end{bmatrix} \\ &= -\begin{bmatrix} 2 \\ 3 \\ 1 \end{bmatrix} < \begin{bmatrix} 0 \\ 0 \\ 0 \end{bmatrix} \end{aligned} \tag{3.21}$$

and the condition (3.4) is satisfied.

2) Weierstrass-Kronecker decomposition method case

In this case

$$P = \begin{bmatrix} 1 & 0 & -2 \\ -8 & 8 & 0 \\ 0 & 0 & 1 \end{bmatrix}, \quad Q = \begin{bmatrix} 0 & 1 & 0 \\ 1 & 0 & 0 \\ 0 & 0 & 2 \end{bmatrix} \tag{3.22}$$

and

$$\begin{aligned} \begin{bmatrix} \mathbb{I}_{n_1} & 0 \\ 0 & N \end{bmatrix} &= PEQ = \begin{bmatrix} 1 & 0 & 0 \\ 0 & 1 & 0 \\ 0 & 0 & 0 \end{bmatrix}, \\ \begin{bmatrix} A_1 & 0 \\ 0 & \mathbb{I}_{n_2} \end{bmatrix} &= PAQ = \begin{bmatrix} -3 & 0 & 0 \\ 2 & -2 & 0 \\ 0 & 0 & 1 \end{bmatrix}, \end{aligned} \tag{3.23}$$

where $n_1 = 2, n_2 = 1$. By Theorem 1.11 the matrix $A_1 \in \mathbb{M}_2$ and $Q \in \mathbb{R}_+^{3\times 3}$ is monomial. Therefore, the system (3.13) is positive. From (3.23) we have

$$p_{A_1}(s) = \det[\mathbb{I}_2 s - A_1] = s^2 + 5s + 6 = 0. \tag{3.24}$$

Note that all coefficients of the characteristic equation $p_{A_1}(s)$ given by (3.24) are positive and by Theorem 3.1 the positive descriptor continuous-time linear system with (3.13) is asymptotically stable.

We obtain the same result using Theorem 3.2 since for $\lambda = [\,1 \quad 2\,]^T$ we have

$$A_1\lambda = \begin{bmatrix} -3 & 0 \\ 2 & -2 \end{bmatrix}\begin{bmatrix} 1 \\ 2 \end{bmatrix} = -\begin{bmatrix} 3 \\ 2 \end{bmatrix} < \begin{bmatrix} 0 \\ 0 \end{bmatrix} \tag{3.25}$$

and the condition (3.5) is satisfied.

3.1.2 Discrete-Time Systems

Consider the positive descriptor discrete-time linear system

$$E_d x_{i+1} = A_d x_i, \quad i \in \mathbb{Z}_+, \tag{3.26}$$

where $x_i \in \mathbb{R}_+^n$ is the state vector and $E_d, A_d \in \mathbb{R}^{n\times n}$. It is assumed that $\det E_d = 0$ and $\det[E_d z - A_d] \neq 0$ for some $z \in \mathbb{C}$. Following the considerations presented in Sect. 1.2 the system (3.26) can be transformed into the following equivalent forms.

1) Using the Drazin inverse matrix method we obtain

$$\begin{aligned} x_{1,i+1} &= \bar{E}_d^D \bar{A}_d x_{1,i}, \\ x_{2,i} &= 0, \end{aligned} \tag{3.27}$$

where $x_{1,i}, x_{2,i} \in \mathbb{R}_+^n$ are given by (2.160c) and the matrices $\bar{E}, \bar{A} \in \mathbb{R}^{n\times n}$ have the form (1.182b). Theorem 1.29 gives the positivity conditions of the system (3.27).

2) Using the Weierstrass-Kronecker decomposition method we have

$$\begin{aligned} \bar{x}_{1,i+1} &= A_{1d}\bar{x}_{1,i}, \\ \bar{x}_{2,i} &= 0, \end{aligned} \tag{3.28}$$

where $\bar{x}_{1,i} \in \mathbb{R}_+^{n_1}$, $\bar{x}_{2,i} \in \mathbb{R}_+^{n_2}$ are given by (1.193) and the matrix $A_{1d} \in \mathbb{R}^{n_1\times n_1}$ is determined by (1.192). Theorem 1.31 gives the positivity conditions of the system (3.28).

Theorem 3.3. The positive descriptor discrete-time linear system (3.26) is asymptotically stable if and only if all coefficients of the following characteristic equations

1) $p_{(E_d,A_d)}(z+1) = p_{(E_d,A_d-E_d)}(z) = \det[E_d(z+1) - A_d] = 0,$
2) $p_{(\bar{E}_d,\bar{A}_d)}(z+1) = p_{(\bar{E}_d,\bar{A}_d-\bar{E}_d)}(z) = \det[\bar{E}_d(z+1) - \bar{A}_d] = 0,$
3) $p_{\bar{E}_d^D\bar{A}_d}(z+1) = (z+1)^{n-r} p_{(\bar{E}_d,\bar{A}_d)}(z+1) = 0,$
4) $p_{A_{1d}}(z+1) = p_{A_{1d}-\mathbb{I}_{n_1}}(z) = \det[\mathbb{I}_{n_1}(z+1) - A_{1d}] = 0$

are positive, where $r = \operatorname{rank}\bar{E}_d^D\bar{A}_d$.

Proof. First we shall show the proof for the fourth equation. From the equality

$$p_{A_{1d}-\mathbb{I}_{n_1}}(z) = \det[\mathbb{I}_{n_1} z - A_{1d} + \mathbb{I}_{n_1}] = \det[\mathbb{I}_{n_1}(z+1) - A_{1d}] = p_{A_{1d}}(z+1) \tag{3.29}$$

it follows that if $z_1, z_2, \ldots, z_{n_1}$ is a spectrum of A_{1d} then $z_1 - 1, z_2 - 1, \ldots, z_{n_1} - 1$ is the spectrum of $A_{1d} - \mathbb{I}_{n_1}$. The real eigenvalues of A_{1d} of the positive system (3.26) are located inside the unit circle if and only if the roots of the characteristic equation $p_{A_{1d}-\mathbb{I}_{n_1}}(z) = 0$ have negative real parts. Therefore, the positive descriptor discrete-time linear system (3.26) is asymptotically stable if and only if all coefficients of the equation $p_{A_{1d}-\mathbb{I}_{n_1}}(z) = 0$ are positive.

By Lemmas 1.12–1.14 the characteristic polynomials $p_{(E_d,A_d)}(z)$, $p_{(\bar{E}_d,\bar{A}_d)}(z)$, $p_{\bar{E}_d^D\bar{A}_d}(z)$ and $p_{A_{1d}}(z)$ are related by (1.252), (1.253) and (1.257), respectively. Therefore, if one of the foregoing characteristic equations have only positive coefficients, then also every considered characteristic equation has only positive coefficients. □

Theorem 3.4. The positive descriptor discrete-time linear system (3.26) is asymptotically stable if and only if there exists a strictly positive vector $\bar{\lambda} = \begin{bmatrix} \bar{\lambda}_1 & \ldots & \bar{\lambda}_n \end{bmatrix}$, $\bar{\lambda}_k > 0$, $k = 1, \ldots, n$ such that

$$[\bar{E}_d^D\bar{A}_d + G_d(\mathbb{I}_n - \bar{E}_d\bar{E}_d^D) - \mathbb{I}_n]\bar{\lambda} < 0 \text{ for an arbitrary matrix } G_d \in \mathbb{R}^{n\times n} \tag{3.30}$$

or there exists a strictly positive vector $\lambda = \begin{bmatrix} \lambda_1 & \ldots & \lambda_{n_1} \end{bmatrix}$, $\lambda_k > 0$, $k = 1, \ldots, n_1$ such that

$$(A_{1d} - \mathbb{I}_{n_1})\lambda < 0. \tag{3.31}$$

Proof. First we consider the condition (3.31). We shall show that if the system (3.28) is asymptotically stable then there exists a strictly positive vector $\lambda \in \mathbb{R}_+^{n_1}$ satisfying (3.31). From (3.28) we have

$$\sum_{i=0}^{\infty} \bar{x}_{1,i+1} = A_{1d} \sum_{i=0}^{\infty} \bar{x}_{1,i}. \tag{3.32}$$

For an asymptotically stable positive system we obtain

$$\bar{x}_{1,\infty} = 0, \quad \bar{x}_{10} > 0, \quad \sum_{i=0}^{\infty} \bar{x}_{1,i} > 0. \tag{3.33}$$

Taking into account (3.32) and (3.33) we obtain (3.31) for

$$\lambda = \sum_{i=0}^{\infty} \bar{x}_{1,i} \tag{3.34}$$

since from (3.32) and (3.34) we have

$$\lambda - \bar{x}_{10} = A_{1d}\lambda. \tag{3.35}$$

It is well-known that the system (3.28) is asymptotically stable if and only if the corresponding transposed system

$$\bar{x}_{1,i+1} = A_{1d}^T \bar{x}_{1,i} \tag{3.36}$$

is asymptotically stable. As a Lyapunov function for the positive system (3.36) we choose the function

$$V(\bar{x}_{1,i}) = \bar{x}_{1,i}^T \lambda, \tag{3.37}$$

which is positive for any nonzero $\bar{x}_{1,i} \in \mathbb{R}_+^n$ and $V(\bar{x}_{1,i}) = 0$ for $\bar{x}_{1,i} = 0$. Using (3.36) and (3.37) we obtain

$$\Delta V(\bar{x}_{1,i}) = V(\bar{x}_{1,i+1}) - V(\bar{x}_{1,i}) = \bar{x}_{1,i+1}^T \lambda - \bar{x}_{1,i}^T \lambda = \bar{x}_{1,i}^T (A_{1d} - \mathbb{I}_n)\lambda. \tag{3.38}$$

Therefore, if the condition (3.31) is satisfied then from (3.38) we have $\Delta V(\bar{x}_{1,i}) < 0$ and the system (3.28) is asymptotically stable.

The condition (3.30) can be proved in a similar way. The matrix $G_d \in \mathbb{R}^{n\times n}$ and the term $G_d(\mathbb{I}_n - \bar{E}_d \bar{E}_d^D)$ eliminates from the matrix $\bar{E}_d^D \bar{A}_d$ unimportant entires that are further canceled through multiplication by $x_{1,i} \in \operatorname{Im} \bar{E}_d \bar{E}_d^D$. Therefore, we have to find both: an arbitrary matrix G_d and strictly positive vector λ such that (3.30) holds. □

Example 3.2. Consider the descriptor discrete-time linear system (3.26) with

$$E_d = \begin{bmatrix} 1 & 0 & 0 \\ 2 & 2 & 0 \\ -0.8 & -0.8 & 0 \end{bmatrix}, \quad A_d = \begin{bmatrix} 0.5 & 0.1 & 0 \\ 1 & 0.6 & 0 \\ -0.4 & -0.24 & 2 \end{bmatrix}. \tag{3.39}$$

The matrix pencil (E_d, A_d) of (3.39) is regular since

$$\begin{aligned} \det[E_d z - A_d] &= \begin{vmatrix} z-0.5 & -0.1 & 0 \\ 2z-1 & 2z-0.6 & 0 \\ -0.8z+0.4 & -0.8z+0.24 & -2 \end{vmatrix} \\ &= -4z^2 + 2.8z - 0.4 \neq 0. \end{aligned} \tag{3.40}$$

The characteristic equation of (3.39) is given by

$$p_{(E_d, A_d)}(z) = \det[E_d z - A_d] = -4z^2 + 2.8z - 0.4 = 0. \tag{3.41}$$

From (3.41) it follows that the pair (E_d, A_d) has two eigenvalues $z_1 = 0.5$, $z_2 = 0.2$ and by condition (1.239) the considered system is asymptotically stable. To check the positivity of the system we use two approaches as follows.

1) Drazin inverse matrix method case

Using (1.182b) for $c = 0$ we obtain

$$\bar{E}_d = [-A_d]^{-1}E_d = \begin{bmatrix} -2 & 1 & 0 \\ 0 & -5 & 0 \\ 0 & 0 & 0 \end{bmatrix},$$
$$\bar{A}_d = [-A_d]^{-1}A_d = \begin{bmatrix} -1 & 0 & 0 \\ 0 & -1 & 0 \\ 0 & 0 & -1 \end{bmatrix}. \tag{3.42}$$

Next, from (B.1) and (B.2) we have

$$\bar{E}_d^D = \begin{bmatrix} -0.5 & -0.1 & 0 \\ 0 & -0.2 & 0 \\ 0 & 0 & 0 \end{bmatrix} \tag{3.43}$$

and

$$\bar{E}_d^D\bar{A}_d = \begin{bmatrix} 0.5 & 0.1 & 0 \\ 0 & 0.2 & 0 \\ 0 & 0 & 0 \end{bmatrix}. \tag{3.44}$$

By Theorem 1.29 the matrix $\bar{E}_d^D\bar{A}_d \in \mathbb{R}_+^{3\times 3}$ and the system (3.39) is positive. Therefore, Theorems 3.3 and 3.4 can be used to test the stability of the system. From (3.39), (3.42) and (3.44) we have

$$\begin{aligned} p_{(E_d,A_d)}(z+1) &= \det[E_d(z+1) - A_d] = -4z^2 - 5.2z - 1.6 = 0 \\ p_{(\bar{E}_d,\bar{A}_d)}(z+1) &= \det[\bar{E}_d(z+1) - \bar{A}_d] = 10z^2 + 13z + 4 = 0, \\ p_{\bar{E}_d^D\bar{A}_d}(z+1) &= \det[\mathbb{I}_3(z+1) - \bar{E}_d^D\bar{A}_d] = (z+1)(z^2 + 1.3z + 0.4) = 0. \end{aligned} \tag{3.45}$$

Note that all coefficients of the characteristic equations $p_{(\bar{E}_d,\bar{A}_d)}(z+1)$, $p_{\bar{E}_d^D\bar{A}_d}(z+1)$ given by (3.45) are positive. Although the characteristic equation $p_{(E_d,A_d)}(z+1)$ has only negative coefficients, we can multiply it by (-1) to obtain positive coefficients. Therefore, by Theorem 3.3 the positive descriptor discrete-time linear system with (3.39) is asymptotically stable.

We obtain the same result using Theorem 3.4 since for $G_d = 0$ and $\bar{\lambda} = [\,1 \quad 1 \quad 1\,]^T$ we have

$$\begin{aligned} [\bar{E}_d^D\bar{A}_d + G_d(\mathbb{I}_3 - \bar{E}_d\bar{E}_d^D) - \mathbb{I}_3]\bar{\lambda} &= \begin{bmatrix} -0.5 & 0.1 & 0 \\ 0 & -0.8 & 0 \\ 0 & 0 & -1 \end{bmatrix}\begin{bmatrix} 1 \\ 1 \\ 1 \end{bmatrix} \\ &= -\begin{bmatrix} 0.4 \\ 0.8 \\ 1 \end{bmatrix} < \begin{bmatrix} 0 \\ 0 \\ 0 \end{bmatrix} \end{aligned} \tag{3.46}$$

and the condition (3.30) is satisfied.

2) Weierstrass-Kronecker decomposition method case

In this case

$$P_d = \begin{bmatrix} 1 & 0 & 0 \\ -1 & 0.5 & 0 \\ 0 & 0.2 & 0.5 \end{bmatrix}, \quad Q_d = \begin{bmatrix} 1 & 0 & 0 \\ 0 & 1 & 0 \\ 0 & 0 & 1 \end{bmatrix} \tag{3.47}$$

and

$$\begin{bmatrix} \mathbb{I}_{n_1} & 0 \\ 0 & N \end{bmatrix} = P_d E_d Q_d = \begin{bmatrix} 1 & 0 & 0 \\ 0 & 1 & 0 \\ 0 & 0 & 0 \end{bmatrix},$$
$$\begin{bmatrix} A_{1d} & 0 \\ 0 & \mathbb{I}_{n_2} \end{bmatrix} = P_d A_d Q_d = \begin{bmatrix} 0.5 & 0.1 & 0 \\ 0 & 0.2 & 0 \\ 0 & 0 & 1 \end{bmatrix}, \tag{3.48}$$

where $n_1 = 2$, $n_2 = 1$. By Theorem 1.31 the matrix $A_{1d} \in \mathbb{R}_+^{2\times 2}$ and $Q_d \in \mathbb{R}_+^{3\times 3}$ is monomial. Therefore, the system (3.39) is positive. From (3.48) we have

$$p_{A_{1d}}(z+1) = \det[\mathbb{I}_{n_1}(z+1) - A_{1d}] = z^2 + 1.3z + 0.4 = 0 \tag{3.49}$$

Note that all coefficients of the characteristic equation $p_{A_{1d}}(z+1)$ given by (3.49) are positive and by Theorem 3.3 the positive descriptor discrete-time linear system with (3.39) is asymptotically stable.

We obtain the same result using Theorem 3.4 since for $\lambda = [\,1 \quad 1\,]^T$ we have

$$(A_1 - \mathbb{I}_2)\lambda = \begin{bmatrix} -0.5 & 0.1 \\ 0 & -0.8 \end{bmatrix} \begin{bmatrix} 1 \\ 1 \end{bmatrix} = -\begin{bmatrix} 0.4 \\ 0.8 \end{bmatrix} < \begin{bmatrix} 0 \\ 0 \end{bmatrix} \tag{3.50}$$

and the condition (3.31) is satisfied.

3.1.3 Fractional Continuous-Time Systems

Consider the fractional positive descriptor continuous-time linear system

$$E\,{}_0D_t^{\alpha} x(t) = Ax(t), \quad 0 < \alpha < 1, \tag{3.51}$$

where $x(t) \in \mathbb{R}_+^n$, is the state vector, ${}_0D_t^{\alpha}$ is the Caputo derivative-integral operator defined by (2.8) and $E, A \in \mathbb{R}^{n\times n}$. It is assumed that $\det E = 0$ and $\det[Es^{\alpha} - A] \neq 0$ for some $s^{\alpha} \in \mathbb{C}$. Following the considerations presented in Sect. 2.1 the system (3.51) can be transformed into the following equivalent forms.

1) Using the Drazin inverse matrix method we obtain

$$\begin{aligned} {}_0D_t^\alpha x_1(t) &= \bar{E}^D\bar{A}x_1(t), \\ x_2(t) &= 0, \end{aligned} \tag{3.52}$$

where $x_1(t), x_2(t) \in \mathbb{R}_+^n$ are given by (2.29c) and the matrices $\bar{E}, \bar{A} \in \mathbb{R}^{n\times n}$ have the form (2.28b). Theorem 2.8 gives the positivity conditions of the system (3.52).

2) Using the Weierstrass-Kronecker decomposition method we have

$$\begin{aligned} {}_0D_t^\alpha \bar{x}_1(t) &= A_1\bar{x}_1(t), \\ \bar{x}_2(t) &= 0, \end{aligned} \tag{3.53}$$

where $\bar{x}_1(t) \in \mathbb{R}_+^{n_1}, \bar{x}_2(t) \in \mathbb{R}_+^{n_2}$ are given by (2.38) and the matrix $A_1 \in \mathbb{R}^{n_1\times n_1}$ is determined by (2.37). Theorem 2.9 gives the positivity conditions of the system (3.3).

Theorem 3.5. The fractional positive descriptor continuous-time linear system (3.51) is asymptotically stable if and only if all coefficients of the following characteristic equations

1) $p_{(E,A)}(s^\alpha) = 0$ given by (2.76),
2) $p_{(\bar{E},\bar{A})}(s^\alpha) = 0$ given by (2.86),
3) $p_{\bar{E}^D\bar{A}}(s^\alpha) = 0$ given by (1.117),
4) $p_{A_1}(s^\alpha) = 0$ given by (2.93).

are positive.

Proof. The proof can be accomplished in a similar way to the proof of Theorem 3.1. Consider the Eq. (2.93). By condition (2.77) the stability region of a fractional continuous-time system can be extended to complex eigenvalues with positive real parts for $\alpha < 1$. However, if A_1 is a Metzler matrix, then its rightmost eigenvalue is real [49]. Therefore, for a fractional positive system the stability region is the same as for an integer-order system ($\alpha = 1$) and A_1 has no real nonnegative eigenvalue. By Lemmas 2.3–2.4 the characteristic polynomials $p_{(E,A)}(s^\alpha)$, $p_{(\bar{E},\bar{A})}(s^\alpha)$, $p_{\bar{E}^D\bar{A}}(s^\alpha)$ and $p_{A_1}(s^\alpha)$ are related by (2.87), (1.117) and (2.94), respectively. Therefore, if one of the foregoing characteristic equations have only positive coefficients, then also every considered characteristic equation has only positive coefficients. □

Theorem 3.6. The fractional positive descriptor continuous-time linear system (3.51) is asymptotically stable if and only if there exists a strictly positive vector $\bar{\lambda} = [\,\bar{\lambda}_1 \;\; \ldots \;\; \bar{\lambda}_n\,]$, $\bar{\lambda}_k > 0$, $k = 1, \ldots, n$ such that

$$[\bar{E}^D\bar{A} + G(\mathbb{I}_n - \bar{E}\bar{E}^D)]\bar{\lambda} < 0 \text{ for an arbitrary matrix } G \in \mathbb{R}^{n\times n} \tag{3.54}$$

or there exists a strictly positive vector $\lambda = [\,\lambda_1 \;\; \ldots \;\; \lambda_{n_1}\,]$, $\lambda_k > 0$, $k = 1, \ldots, n_1$ such that

$$A_1\lambda < 0. \tag{3.55}$$

Proof. The proof can be accomplished in a similar way to the proof of Theorem 3.2. Consider the inequality (3.55). Applying to the first equation of (3.53) the fractional integral defined by [57, 132, 135]

$$ {}_0I_t^\alpha f(t) = \frac{1}{\Gamma(\alpha)} \int_0^t (t-\tau)^{\alpha-1} f(\tau) d\tau \tag{3.56}$$

we obtain

$$ \bar{x}_1(\infty) - \bar{x}_1(0) = A_1 \, {}_0I_t^\alpha \bar{x}_1(t). \tag{3.57}$$

For an asymptotically stable positive system we have

$$ \bar{x}_1(\infty) = 0, \;\; \bar{x}_1(0) > 0, \;\; {}_0I_t^\alpha \bar{x}_1(t) > 0. \tag{3.58}$$

From (3.57) and (3.58) we obtain (3.55) for

$$ \lambda = {}_0I_t^\alpha \bar{x}_1(t). \tag{3.59}$$

Taking into account that the system (3.53) is asymptotically stable if and only if the corresponding transposed system

$$ {}_0D_t^\alpha \bar{x}_1(t) = A_1^T \bar{x}_1(t) \tag{3.60}$$

is asymptotically stable, the Lyapunov function

$$ V(\bar{x}_1(t)) = \bar{x}_1^T(t)\lambda \tag{3.61}$$

is chosen, which is positive for any nonzero $\bar{x}_1(t) \in \mathbb{R}_+^n$ and $V(\bar{x}_1(t)) = 0$ for $\bar{x}_1(t) = 0$. From (3.60) and (3.61) we have

$$ {}_0D_t^\alpha V(\bar{x}_1(t)) = {}_0D_t^\alpha \bar{x}_1^T(t)\lambda = \bar{x}_1(t)^T A\lambda. \tag{3.62}$$

Therefore, if the condition (3.55) is satisfied then from (3.62) we have ${}_0D_t^\alpha V(\bar{x}_1(t)) < 0$ and the system (3.53) is asymptotically stable. The condition (3.54) can be proved similarly. □

Example 3.3. Consider the fractional descriptor continuous-time linear system (3.51) with

$$ E = \begin{bmatrix} -1 & -0.75 & -0.75 \\ -2 & -0.25 & -0.25 \\ 0 & -0.75 & -0.75 \end{bmatrix}, \; A = \begin{bmatrix} 2 & 1 & 0 \\ 4 & 0 & 1 \\ 0 & 2 & 1 \end{bmatrix}. \tag{3.63}$$

The matrix pencil (E, A) of (3.63) is regular since

$$\det[Es^{\alpha} - A] = \begin{vmatrix} -s-2 & -0.75s-1 & -0.75s \\ -2s-4 & -0.25s & -0.25s-1 \\ 0 & -0.75s-2 & -0.75-1 \end{vmatrix} \tag{3.64}$$
$$= s^{2\alpha} + 6s^{\alpha} + 8 \neq 0.$$

The characteristic equation of (3.63) is given by

$$p_{(E,A)}(s^{\alpha}) = \det[Es^{\alpha} - A] = s^{2\alpha} + 6s^{\alpha} + 8 = 0. \tag{3.65}$$

From (3.15) it follows that the pair (E, A) has two eigenvalues $s_1^{\alpha} = -2, s_2^{\alpha} = -4$ and by condition (2.77) the considered system is asymptotically stable for any $\alpha \in (0; 1)$. To check the positivity of the system we use the Drazin inverse matrix method. From (2.28b) for $c = 0$ we have

$$\bar{E} = [-A]^{-1}E = \begin{bmatrix} 0.5 & 0.125 & 0.125 \\ 0 & 0.5 & 0.5 \\ 0 & -0.25 & -0.25 \end{bmatrix},$$
$$\bar{A} = [-A]^{-1}A = \begin{bmatrix} -1 & 0 & 0 \\ 0 & -1 & 0 \\ 0 & 0 & -1 \end{bmatrix}. \tag{3.66}$$

Next, using one of the methods presented in Appendix B we obtain

$$\bar{E}^D = \begin{bmatrix} 2 & -1 & -1 \\ 0 & 8 & 8 \\ 0 & -4 & -4 \end{bmatrix} \tag{3.67}$$

and

$$\bar{E}^D\bar{A} = \begin{bmatrix} -2 & 1 & 1 \\ 0 & -8 & -8 \\ 0 & 4 & 4 \end{bmatrix}. \tag{3.68}$$

By Theorem 2.8 there exists a matrix

$$G = \begin{bmatrix} 0 & 0 & 0 \\ 0 & -4 & 0 \\ 0 & 0 & -3 \end{bmatrix} \tag{3.69}$$

such that

$$\bar{E}^D\bar{A} + G(\mathbb{I}_3 - \bar{E}\bar{E}^D) = \begin{bmatrix} -2 & 1 & 1 \\ 0 & -4 & 0 \\ 0 & 1 & -2 \end{bmatrix} \in \mathbb{M}_3 \tag{3.70}$$

and the system (3.63) is positive. Therefore, Theorems 3.5 and 3.6 can be used to test the stability of the system. From (3.66) and (3.68) we have

$$\begin{aligned} p_{(\bar{E},\bar{A})}(s^\alpha) &= \det[\bar{E}s^\alpha - \bar{A}] = 0.125s^{2\alpha} + 0.75s^\alpha + 1 = 0, \\ p_{\bar{E}^D\bar{A}}(s^\alpha) &= \det[\mathbb{I}_3 s^\alpha - \bar{E}^D\bar{A}] = s^\alpha(s^{2\alpha} + 6s^\alpha + 8) = 0. \end{aligned} \tag{3.71}$$

Note that all coefficients of the characteristic equations $p_{(E,A)}(s^\alpha)$, $p_{(\bar{E},\bar{A})}(s^\alpha)$, $p_{\bar{E}^D\bar{A}}(s^\alpha)$ given by (3.65) and (3.71) are positive and by Theorem 3.5 the fractional positive descriptor continuous-time linear system with (3.63) is asymptotically stable for any $\alpha \in (0; 1)$.

We obtain the same result using Theorem 3.6 since for G given by (3.69) and $\lambda = [\,2 \;\; 1 \;\; 1\,]^T$ we obtain

$$\begin{aligned} [\bar{E}^D\bar{A} + G(\mathbb{I}_3 - \bar{E}\bar{E}^D)]\lambda &= \begin{bmatrix} -2 & 1 & 1 \\ 0 & -4 & 0 \\ 0 & 1 & -2 \end{bmatrix} \begin{bmatrix} 2 \\ 1 \\ 1 \end{bmatrix} \\ &= -\begin{bmatrix} 2 \\ 4 \\ 1 \end{bmatrix} < \begin{bmatrix} 0 \\ 0 \\ 0 \end{bmatrix} \end{aligned} \tag{3.72}$$

and the condition (3.54) is satisfied.

3.1.4 Fractional Discrete-Time Systems

Consider the fractional positive descriptor discrete-time linear system

$$E_d \Delta^\alpha x_{i+1} = A_d x_i, \quad i \in \mathbb{Z}_+, \quad 0 < \alpha < 1, \tag{3.73}$$

where $x_i \in \mathbb{R}^n_+$, is the state vector, Δ^α is the Grünwald-Letnikov fractional difference operator defined by (2.136) and $E_d, A_d \in \mathbb{R}^{n\times n}$. It is assumed that $\det E_d = 0$ and $\det[Ez - A] \neq 0$ for some $z \in \mathbb{C}$. Substituting (2.136) into (3.73) we get

$$E_d x_{i+1} = A_{d\alpha} x_i + \sum_{j=1}^{i} E_d c_j x_{i-j}, \tag{3.74}$$

where $A_{d\alpha}$ is given by (2.140b) and the coefficients c_j are determined by (2.140c). Following the considerations presented in Sect. 2.2 the system (3.73) can be transformed into the following equivalent forms.

1) Using the Drazin inverse matrix method we obtain

$$\begin{aligned} \Delta^\alpha x_{1,i+1} &= \bar{E}_d^D \bar{A}_d x_{1,i}, \\ x_{2,i} &= 0, \end{aligned} \tag{3.75}$$

where $x_{1,i}, x_{2,i} \in \mathbb{R}_+^n$ are given by (2.160c) and the matrices $\bar{E}_d, \bar{A}_d \in \mathbb{R}^{n\times n}$ have the form (2.157b). Substituting (2.136) into (3.75) we get

$$\begin{aligned} x_{1,i+1} &= \bar{E}_d^D \bar{A}_{d\alpha} x_{1,i} + \sum_{j=1}^{L} c_j x_{1,i-j}, \\ x_{2,i} &= 0, \end{aligned} \tag{3.76}$$

where $\bar{A}_{d\alpha}$ is given by (2.159), L is the length of practical implementation and $x_{1,-k} = 0$, $k = 1, 2, \ldots$. Theorem 2.25 gives the positivity conditions of the system (3.76).

2) Using the Weierstrass-Kronecker decomposition method we have

$$\begin{aligned} \Delta^{\alpha} \bar{x}_{1,i+1} &= A_{1d} \bar{x}_{1,i}, \\ \bar{x}_{2,i} &= 0, \end{aligned} \tag{3.77}$$

where $\bar{x}_{1,i} \in \mathbb{R}_+^{n_1}$, $\bar{x}_{2,i} \in \mathbb{R}_+^{n_2}$ are given by (2.175) and the matrix $A_{1d} \in \mathbb{R}^{n_1\times n_1}$ is determined by (2.174). Substituting (2.136) into (3.77) we get

$$\begin{aligned} \bar{x}_{1,i+1} &= A_{1d\alpha} \bar{x}_{1,i} + \sum_{j=1}^{L} c_j \bar{x}_{1,i-j}, \\ \bar{x}_{2,i} &= 0, \end{aligned} \tag{3.78}$$

where $A_{1d\alpha}$ is given by (2.179c), L is the length of practical implementation and $\bar{x}_{1,-k} = 0$, $k = 1, 2, \ldots$. Theorem 2.26 gives the positivity conditions of the system (3.78).

Remark 3.1. Note that the positive fractional system (3.76) (or (3.78)) is a system with increasing number of delays. It is well-known [23] that the asymptotic stability of the positive discrete-time linear systems with delays is independent of the numbers and values of the delays and it depends only on the sum of state matrices of the system.

From Remark 3.1 we obtain the following theorems [23, 57, 88].

Theorem 3.7. The positive fractional system (3.76) is asymptotically stable if and only if the positive system

$$\begin{aligned} x_{1,i+1} &= \bar{D}_1 x_{1,i}, \\ x_{2,i} &= 0 \end{aligned} \tag{3.79}$$

with

$$\bar{D}_1 = \bar{E}_d^D \bar{A}_{d\alpha} + \sum_{j=1}^{L} c_j \mathbb{I}_n \in \mathbb{R}^{n\times n} \tag{3.80}$$

is asymptotically stable.

Theorem 3.8. The positive fractional system (3.78) is asymptotically stable if and only if the positive system

$$\begin{aligned}\bar{x}_{1,i+1} &= D_1 x_{1,i},\\ \bar{x}_{2,i} &= 0\end{aligned} \tag{3.81}$$

with

$$D_1 = A_{1d\alpha} + \sum_{j=1}^{L} c_j \mathbb{I}_{n_1} \in \mathbb{R}^{n_1 \times n_1} \tag{3.82}$$

is asymptotically stable.

Theorem 3.9. The fractional positive descriptor discrete-time linear system (3.73) is practically stable for given length L of practical implementation if and only if all coefficients of the following characteristic equations

1) $p_{(E_d,D)}(z+1) = p_{(E_d,D-E_d)}(z) = \det[E_d(z+1) - D] = 0,$
2) $p_{(\bar{E}_d,\bar{D})}(z+1) = p_{(\bar{E}_d,\bar{D}-\bar{E}_d)}(z) = \det[\bar{E}_d(z+1) - \bar{D}] = 0,$
3) $p_{\bar{D}_1}(z+1) = (z+1)^{n-r} p_{(\bar{E}_d,\bar{D})}(z+1) = 0,$
4) $p_{D_1}(z+1) = p_{D_1-\mathbb{I}_{n_1}}(z) = \det[\mathbb{I}_{n_1}(z+1) - D_1] = 0$

are positive, where

$$D = A_{d\alpha} + \sum_{j=1}^{L} c_j E_d \in \mathbb{R}^{n\times n}, \tag{3.83a}$$

$$\bar{D} = \bar{A}_{d\alpha} + \sum_{j=1}^{L} c_j \bar{E}_d \in \mathbb{R}^{n\times n}, \tag{3.83b}$$

$\bar{D}_1$, D_1 are given by (3.80) and (3.82), respectively and $r = \operatorname{rank} \bar{E}_d^D \bar{A}_{d\alpha}$.

Proof. The proof follows immediately from Theorem 3.3. The equivalence of the above characteristic equations can be shown using Lemmas 1.8, 2.6 and 2.7. □

Theorem 3.10. The positive descriptor discrete-time linear system (3.26) is practically stable for given length L of practical implementation if and only if there exists a strictly positive vector $\bar{\lambda} = \begin{bmatrix} \bar{\lambda}_1 & \dots & \bar{\lambda}_n \end{bmatrix}$, $\bar{\lambda}_k > 0$, $k = 1, \dots, n$ such that

$$[\bar{D}_1 + G_d(\mathbb{I}_n - \bar{E}_d \bar{E}_d^D) - \mathbb{I}_n]\bar{\lambda} < 0 \text{ for an arbitrary matrix } G_d \in \mathbb{R}^{n\times n} \tag{3.84}$$

or there exists a strictly positive vector $\lambda = \begin{bmatrix} \lambda_1 & \dots & \lambda_{n_1} \end{bmatrix}$, $\lambda_k > 0$, $k = 1, \dots, n_1$ such that

$$(D_1 - \mathbb{I}_{n_1})\lambda < 0, \tag{3.85}$$

where $\bar{D}_1$, D_1 are given by (3.80) and (3.82), respectively.

Proof. The proof follows immediately from Theorem 3.4. □

To test the asymptotic stability of the positive fractional system (3.73) we assume that $L \to \infty$.

Lemma 3.1. For $L \to \infty$ the matrices (3.80), (3.82) and (3.83) have the forms

$$\begin{aligned} D_\infty &= A_d + E_d, \\ \bar{D}_\infty &= \bar{A}_d + \bar{E}_d, \\ \bar{D}_{1,\infty} &= \bar{E}_d^D \bar{A}_d + \mathbb{I}_n - (\mathbb{I}_n - \bar{E}_d \bar{E}_d^D)\alpha, \\ D_{1,\infty} &= A_{1d} + \mathbb{I}_{n_1}. \end{aligned} \tag{3.86}$$

Proof. Taking into account (2.232) from (3.80), (3.82) and (3.83) for $L \to \infty$ we have

$$\begin{aligned} D_\infty &= A_{d\alpha} + \sum_{j=1}^{\infty} c_j E_d = A_d + E_d\alpha + E_d(1-\alpha) = A_d + E_d, \\ \bar{D}_\infty &= \bar{A}_{d\alpha} + \sum_{j=1}^{\infty} c_j \bar{E}_d = \bar{A}_d + \bar{E}_d\alpha + \bar{E}_d(1-\alpha) = \bar{A}_d + \bar{E}_d, \\ \bar{D}_{1,\infty} &= \bar{E}_d^D \bar{A}_{d\alpha} + \sum_{j=1}^{\infty} c_j \mathbb{I}_n = \bar{E}_d^D \bar{A}_d + \bar{E}_d \bar{E}_d^D \alpha + \mathbb{I}_n(1-\alpha), \\ D_{1,\infty} &= A_{1d\alpha} + \sum_{j=1}^{\infty} c_j \mathbb{I}_{n_1} = A_{1d} + \mathbb{I}_{n_1}\alpha + \mathbb{I}_{n_1}(1-\alpha) = A_{1d} + \mathbb{I}_{n_1}, \end{aligned} \tag{3.87}$$

which is equivalent to (3.86). □

Theorem 3.11. The fractional positive descriptor discrete-time linear system (3.73) is asymptotically stable if and only if all coefficients of the following characteristic equations

1) $p_{(E_d,D_\infty)}(z+1) = \det[E_d(z+1) - D_\infty] = \det[E_d z - A_d] = p_{(E_d,A_d)}(z) = 0,$
2) $p_{(\bar{E}_d,\bar{D}_\infty)}(z+1) = \det[\bar{E}_d(z+1) - \bar{D}_\infty] = \det[\bar{E}_d z - \bar{A}_d] = p_{(\bar{E}_d,\bar{A}_d)}(z) = 0,$
3) $p_{\bar{D}_{1,\infty}}(z+1) = \det[\mathbb{I}_n(z+1) - \bar{D}_{1,\infty}] = \frac{1}{\beta}(z+\alpha)^{n-r} p_{(\bar{E}_d,\bar{A}_d)}(z) = 0,$
4) $p_{D_{1,\infty}}(z+1) = \det[\mathbb{I}_{n_1}(z+1) - D_1] = \det[\mathbb{I}_{n_1} z - A_{1d}] = p_{A_{1d}}(z) = 0$

are positive, where D_∞, $\bar{D}_\infty$, $\bar{D}_{1,\infty}$, $D_{1,\infty}$ are given by (3.86), $r = \operatorname{rank} \bar{E}_d^D \bar{A}_d$ and $\beta \in \mathbb{R}$ is some scalar.

Proof. The proof for Eqs. (1), (2) and (4) follows immediately from Theorem 3.9 and Lemma 3.1. For the Eq. (3) we shall show that the equality

$$\begin{aligned} p_{\bar{D}_{1,\infty}}(z+1) &= \det[\mathbb{I}_n(z+1) - \bar{D}_{1,\infty}] \\ &= \det[\mathbb{I}_n z - \bar{E}_d^D \bar{A}_d + (\mathbb{I}_n - \bar{E}_d \bar{E}_d^D)\alpha] \\ &= \frac{1}{\beta}(z+\alpha)^{n-r} p_{(\bar{E}_d,\bar{A}_d)}(z) \end{aligned} \tag{3.88}$$

holds, where $\beta \in \mathbb{R}$ is some scalar. Using (1.27d) and (3.88) we obtain

$$\begin{aligned} p_{\bar{D}_{1,\infty}}(z+1) &= \det[\mathbb{I}_n(z+1) - \bar{D}_{1,\infty}] \\ &= \det[\mathbb{I}_n z - \bar{E}_d^D \bar{A}_d + (\mathbb{I}_n - \bar{E}_d \bar{E}_d^D)\alpha] \\ &= \det\left\{T \begin{bmatrix} \mathbb{I}_{n_1} z - J_d^{-1} A_{1d} & 0 \\ 0 & \mathbb{I}_{n_2}(z+\alpha) \end{bmatrix} T^{-1}\right\} \\ &= (z+\alpha)^{n_2} \det[\mathbb{I}_{n_1} z - J_d^{-1} A_{1d}] \end{aligned} \tag{3.89}$$

and

$$\begin{aligned} p_{(\bar{E}_d, \bar{A}_d)}(z) &= \det[\bar{E}_d z - \bar{A}_d] \\ &= \det\left\{T \begin{bmatrix} J_d z - A_{1d} & 0 \\ 0 & Nz - A_{2d} \end{bmatrix} T^{-1}\right\} \\ &= \det[J_d z - A_{1d}] \det[Nz - A_{2d}] \\ &= \beta \det[J_d z - A_{1d}] = \beta \det[\mathbb{I}_{n_1} z - J_d^{-1} A_{1d}] \end{aligned} \tag{3.90}$$

since $\det T \det T^{-1} = \mathbb{I}_n$ and $\det[Nz - A_{2d}] = \beta$. Taking into account that $n_2 = n - r$ from (3.89) and (3.90) we have (3.88). This completes the proof. □

Theorem 3.12. The positive descriptor discrete-time linear system (3.26) is asymptotically stable if and only if there exists a strictly positive vector $\bar{\lambda} = \begin{bmatrix} \bar{\lambda}_1 & \dots & \bar{\lambda}_n \end{bmatrix}$, $\bar{\lambda}_k > 0$, $k = 1, \dots, n$ such that

$$[\bar{E}_d^D \bar{A}_d + (G_d - \mathbb{I}_n \alpha)(\mathbb{I}_n - \bar{E}_d \bar{E}_d^D)]\bar{\lambda} < 0 \text{ for an arbitrary matrix } G_d \in \mathbb{R}^{n \times n} \tag{3.91}$$

or there exists a strictly positive vector $\lambda = \begin{bmatrix} \lambda_1 & \dots & \lambda_{n_1} \end{bmatrix}$, $\lambda_k > 0$, $k = 1, \dots, n_1$ such that

$$A_{1d}\lambda < 0, \tag{3.92}$$

Proof. The proof follows immediately from Theorem 3.10 and Lemma 3.1. □

Example 3.4. Consider the fractional descriptor discrete-time linear system (3.73) with $\alpha = 0.4$ and

$$E_d = \begin{bmatrix} 0 & -5 & 0 \\ -5 & -10 & 0 \\ 0 & -7.5 & 0 \end{bmatrix}, \quad A_d = \begin{bmatrix} 0 & 1 & 0 \\ 1 & 0 & 0 \\ 0 & 0 & 1 \end{bmatrix}. \tag{3.93}$$

The matrix pencil (E_d, A_d) of (3.93) is regular since

$$\det[Ez - A] = \begin{vmatrix} 0 & -5z-1 & 0 \\ -5z-1 & -10z & 0 \\ 0 & -7.5z & -1 \end{vmatrix} = 25z^2 + 10z + 1 \neq 0. \tag{3.94}$$

To check the asymptotic stability of the system (3.93) we can use the approach given in Sect. 2.2.5. Therefore, by Theorem 2.28 we have to compute the characteristic equation given by

$$\det\left[E_d(z-1)^{0.4}z^{0.6}-A_d\right]=[5(z-1)^{0.4}z^{0.6}+1]^2=0. \tag{3.95}$$

The system (3.93) with $\alpha = 0.4$ is asymptotically stable since roots of (3.95) are $z_1 = z_2 = -0.0656$. To check the positivity of the system we use the Drazin inverse matrix method. From (2.157b) for $c = 0$ we have

$$\bar{E}_d=[-A_d]^{-1}E_d=\begin{bmatrix}5 & 10 & 0\\ 0 & 5 & 0\\ 0 & 7.5 & 0\end{bmatrix},$$
$$\bar{A}_d=[-A_d]^{-1}A_d=\begin{bmatrix}-1 & 0 & 0\\ 0 & -1 & 0\\ 0 & 0 & -1\end{bmatrix}, \tag{3.96}$$
$$\bar{A}_{d\alpha}=\bar{A}_d+0.4\bar{E}_d=\begin{bmatrix}1 & 4 & 0\\ 0 & 1 & 0\\ 0 & 3 & -1\end{bmatrix}.$$

Next, using (B.1) and (B.2) we obtain

$$\bar{E}_d^D=\begin{bmatrix}0.2 & -0.4 & 0\\ 0 & 0.2 & 0\\ 0 & 0.3 & 0\end{bmatrix} \tag{3.97}$$

and

$$\bar{E}_d^D\bar{A}_d=\begin{bmatrix}-0.2 & 0.4 & 0\\ 0 & -0.2 & 0\\ 0 & -0.3 & 0\end{bmatrix},\quad \bar{E}_d^D\bar{A}_{d\alpha}=\begin{bmatrix}0.2 & 0.4 & 0\\ 0 & 0.2 & 0\\ 0 & 0.3 & 0\end{bmatrix}. \tag{3.98}$$

By Theorem 2.25 there exists the matrix $\bar{E}_d^D\bar{A}_{d\alpha}\in\mathbb{R}_+^{3\times 3}$ and the system (3.93) is positive. Therefore, Theorems 3.11 and 3.12 can be used to test the stability of the system. From (3.86) we have

$$D_\infty=A_d+E_d=\begin{bmatrix}0 & -4 & 0\\ -4 & -10 & 0\\ 0 & -7.5 & 1\end{bmatrix},$$
$$\bar{D}_\infty=\bar{A}_d+\bar{E}_d=\begin{bmatrix}4 & 10 & 0\\ 0 & 4 & 0\\ 0 & 7.5 & -1\end{bmatrix}, \tag{3.99}$$
$$\bar{D}_{1,\infty}=\bar{E}_d^D\bar{A}_d+\mathbb{I}_3-0.4(\mathbb{I}_3-\bar{E}_d\bar{E}_d^D)=\begin{bmatrix}0.8 & 0.4 & 0\\ 0 & 0.8 & 0\\ 0 & 0.3 & 0.6\end{bmatrix}.$$

Using (3.93), (3.66) and (3.99) we obtain

$$
\begin{aligned}
p_{(E_d,D_\infty)}(z+1) &= \det[E_d(z+1) - D_\infty] = 25z^2 + 10z + 1 = 0,\\
p_{(\bar{E}_d,\bar{D}_\infty)}(z+1) &= \det[\bar{E}_d(z+1) - \bar{D}_\infty] = 25z^2 + 10z + 1 = 0,\\
p_{\bar{D}_{1,\infty}}(z+1) &= \det[\mathbb{I}_n(z+1) - \bar{D}_{1,\infty}] = 0.04(z+0.4)(25z^2 + 10z + 1) = 0.
\end{aligned}
\tag{3.100}
$$

Note that all coefficients of the characteristic equations $p_{(E_d,D_\infty)}(z+1) = p_{(E_d,A_d)}(z)$, $p_{(\bar{E}_d,\bar{D}_\infty)}(z+1) = p_{(\bar{E}_d,\bar{A}_d)}(z)$, $p_{\bar{D}_{1,\infty}}(z+1) = \frac{1}{\beta}(z+\alpha)p_{(\bar{E}_d,\bar{A}_d)}(z)$ given by (3.100) are positive and by Theorem 3.11 the fractional positive descriptor discrete-time linear system with $\alpha = 0.4$ and (3.93) is asymptotically stable.

We obtain the same result using Theorem 3.12 since for $G_d = 0$ and $\bar{\lambda} = [\,3 \;\; 1 \;\; 1\,]^T$ we obtain

$$
\begin{aligned}
[\bar{E}_d^D \bar{A}_d + (G_d - \mathbb{I}_n\alpha)(\mathbb{I}_n - \bar{E}_d\bar{E}_d^D)]\bar{\lambda} &= \begin{bmatrix} -0.2 & 0.4 & 0 \\ 0 & -0.2 & 0 \\ 0 & 0.3 & -0.4 \end{bmatrix} \begin{bmatrix} 3 \\ 1 \\ 1 \end{bmatrix}\\
&= -\begin{bmatrix} 0.2 \\ 0.2 \\ 0.1 \end{bmatrix} < \begin{bmatrix} 0 \\ 0 \\ 0 \end{bmatrix}
\end{aligned}
\tag{3.101}
$$

and the condition (3.91) is satisfied.

3.2 Stability of Positive Interval Systems

In this section necessary and sufficient stability conditions for positive interval linear systems will be established and Kharitonov theorem [98] will be extended to positive descriptor interval linear systems. It will be shown that such systems are asymptotically stable if and only if the respective lower and upper bound systems are asymptotically stable [72–74].

3.2.1 Continuous-Time Systems

Consider the positive descriptor continuous-time linear system (3.1) with the interval matrix $A \in \mathbb{R}^{n\times n}$ defined by

$$A_L \le A \le A_U \text{ or equivalently } A \in [A_L, A_U]. \tag{3.102}$$

It is assumed that

$$\det[Es - A_L] \ne 0 \text{ and } \det[Es - A_U] \ne 0 \text{ for some } s \in \mathbb{C}. \tag{3.103}$$

Following the considerations presented in Sect. 3.1.1 the interval system (3.1) can be transformed into two equivalent forms:

1) Using Drazin inverse matrix method we obtain the interval system (3.2), where $\bar{E} \in [\bar{E}_L, \bar{E}_U]$, $\bar{A} \in [\bar{A}_L, \bar{A}_U]$ and

$$\begin{aligned} \bar{E}_L &= [Ec - A_L]^{-1}E, \quad \bar{E}_U = [Ec - A_U]^{-1}E \\ \bar{A}_L &= [Ec - A_L]^{-1}A_L, \quad \bar{A}_U = [Ec - A_U]^{-1}A_U. \end{aligned} \tag{3.104}$$

The matrix $\bar{E}^D\bar{A}$ in (3.2) will be denoted as $\bar{A}_1$. Therefore, we have $\bar{A}_1 \in [\bar{A}_{1L}, \bar{A}_{1U}]$ and $\bar{A}_{1L} = \bar{E}_L^D\bar{A}_L$, $\bar{A}_{1U} = \bar{E}_U^D\bar{A}_U$.

2) Using Weierstrass-Kronecker decomposition method we obtain the interval system (3.3), where the matrix $A_1 \in [A_{1L}, A_{1U}]$ is determined by

$$P_1 A_L Q_1 = \begin{bmatrix} A_{1L} & 0 \\ 0 & \mathbb{I}_{n_2} \end{bmatrix}, \quad P_2 A_U Q_2 = \begin{bmatrix} A_{1U} & 0 \\ 0 & \mathbb{I}_{n_2} \end{bmatrix} \tag{3.105}$$

for two pairs of nonsingular matrices (P_1, Q_1), (P_2, Q_2).

Definition 3.1. The positive descriptor interval continuous-time linear system (3.1) is called asymptotically stable if the system is asymptotically stable for all matrices E, A, where $A \in [A_L, A_U]$.

Definition 3.2. [72–74] The matrix

$$A = (1-k)A_L + kA_U, \quad 0 \le k \le 1 \tag{3.106}$$

is called the convex linear combination of matrices A_L and A_U.

According to Definition 3.2 we can also define convex linear combinations of matrices $\bar{A}_{1L}$, $\bar{A}_{1U}$ and A_{1L}, A_{1U}, i.e.

$$\bar{A}_1 = (1-k)\bar{A}_{1L} + k\bar{A}_{1U}, \quad 0 \le k \le 1, \tag{3.107a}$$

$$A_1 = (1-k)A_{1L} + kA_{1U}, \quad 0 \le k \le 1. \tag{3.107b}$$

Theorem 3.13. The convex linear combinations (3.107) are asymptotically stable if and only if:

1) the matrices $\bar{A}_{1L}$ and $\bar{A}_{1U}$ are asymptotically stable;
2) the matrices A_{1L} and A_{1U} are asymptotically stable.

Proof. If the matrices $\bar{A}_{1L}$ and $\bar{A}_{1U}$ are asymptotically stable, then by (3.4) there exist matrices $G_L \in \mathbb{R}^{n\times n}$, $G_U \in \mathbb{R}^{n\times n}$ and strictly positive vector $\bar{\lambda} \in \mathbb{R}_+^n$ such that

$$[\bar{A}_{1L} + G_L(\mathbb{I}_n - \bar{E}_L\bar{E}_L^D)]\bar{\lambda} < 0 \tag{3.108a}$$

and

$$[\bar{A}_{1U} + G_U(\mathbb{I}_n - \bar{E}_U\bar{E}_U^D)]\bar{\lambda} < 0. \tag{3.108b}$$

From (3.107a) and (3.108) we obtain

$$\begin{aligned}
[\bar{A}_1 + G(\mathbb{I}_n - \bar{E}\bar{E}^D)]\bar{\lambda} &= \{(1-k)[\bar{A}_{1L} + G_L(\mathbb{I}_n - \bar{E}_L\bar{E}_L^D)] \\
&\quad + k[\bar{A}_{1U} + G_U(\mathbb{I}_n - \bar{E}_U\bar{E}_U^D)]\}\bar{\lambda} \\
&= (1-k)[\bar{A}_{1L} + G_L(\mathbb{I}_n - \bar{E}_L\bar{E}_L^D)]\bar{\lambda} \\
&\quad + k[\bar{A}_{1U} + G_U(\mathbb{I}_n - \bar{E}_U\bar{E}_U^D)]\bar{\lambda} < 0
\end{aligned} \tag{3.109}$$

for $0 \leq k \leq 1$. Therefore, if the matrices $\bar{A}_{1L}$ and $\bar{A}_{1U}$ are asymptotically stable, then the convex linear combination (3.107a) is also asymptotically stable.

The proof for the convex linear combination (3.107b) can be performed in a similar way. If the matrices A_{1L} and A_{1U} are asymptotically stable, then by (3.5) there exist strictly positive vector $\lambda \in \mathbb{R}_+^{n_1}$ such that

$$A_{1L}\lambda < 0 \text{ and } A_{1U}\lambda < 0. \tag{3.110}$$

From (3.107b) and (3.110) we obtain

$$A_1\lambda = [(1-k)A_{1L} + kA_{1U}]\lambda = (1-k)A_{1L}\lambda + kA_{1U}\lambda < 0 \tag{3.111}$$

for $0 \leq k \leq 1$. Therefore, if the matrices A_{1L} and A_{1U} are asymptotically stable, then the convex linear combination (3.107b) is also asymptotically stable.

The necessity follows immediately from the fact that k can be equal to zero and one. □

Theorem 3.14. The positive descriptor interval continuous-time linear system (3.1) with (3.102) is asymptotically stable if and only if:

1) *(Drazin inverse matrix method case)*
the matrices $\bar{A}_{1L}$ and $\bar{A}_{1U}$ are asymptotically stable;
2) *(Weierstrass-Kronecker decomposition method case)*
the matrices A_{1L} and A_{1U} are asymptotically stable.

Proof. By (3.4) the matrices $\bar{A}_{1L}$ and $\bar{A}_{1U}$ are asymptotically stable if and only if there exist $G_L \in \mathbb{R}^{n\times n}$, $G_U \in \mathbb{R}^{n\times n}$ and strictly positive vector $\bar{\lambda} \in \mathbb{R}_+^n$ such that (3.108) holds. The convex linear combination (3.107a) satisfies the condition $[\bar{A}_1 + G(\mathbb{I}_n - \bar{E}\bar{E}^D)]\bar{\lambda} < 0$ if and only if (3.108) holds. Therefore, the positive descriptor interval continuous-time linear system (3.1) with (3.102) is asymptotically stable if and only if the matrices A_{1L} and A_{1U} are asymptotically stable.

The proof for the Weierstrass-Kronecker decomposition method case can be performed in a similar way. □

To test the stability of an interval system the Kharitonov theorem can also be used. It is recalled below and then extended to positive descriptor interval continuous-time linear systems.

Consider the set (family) of the n-th degree polynomials

$$p_n(s) = a_n s^n + a_{n-1} s^{n-1} + \ldots + a_1 s + a_0 \tag{3.112a}$$

with interval coefficients

$$\underline{a}_i \le a_i \le \overline{a}_i, \quad i = 0, 1, \ldots, n. \tag{3.112b}$$

Using (3.112a) we define the following four polynomials:

$$\begin{aligned}
p_{1n}(s) &= \underline{a}_0 + \underline{a}_1 s + \overline{a}_2 s^2 + \overline{a}_3 s^3 + \underline{a}_4 s^4 + \underline{a}_5 s^5 + \ldots, \\
p_{2n}(s) &= \underline{a}_0 + \overline{a}_1 s + \overline{a}_2 s^2 + \underline{a}_3 s^3 + \underline{a}_4 s^4 + \overline{a}_5 s^5 + \ldots, \\
p_{3n}(s) &= \overline{a}_0 + \underline{a}_1 s + \underline{a}_2 s^2 + \overline{a}_3 s^3 + \overline{a}_4 s^4 + \underline{a}_5 s^5 + \ldots, \\
p_{4n}(s) &= \overline{a}_0 + \overline{a}_1 s + \underline{a}_2 s^2 + \underline{a}_3 s^3 + \overline{a}_4 s^4 + \overline{a}_5 s^5 + \ldots .
\end{aligned} \tag{3.113}$$

Theorem 3.15 *(Kharitonov theorem).* The set of polynomials (3.112) is asymptotically stable if and only if the four polynomials (3.113) are asymptotically stable.

Theorem 3.16. The positive descriptor interval continuous-time linear system (3.1) with the characteristic equations

1) $p_{(E,A)}(s) = 0$ given by (1.100),
2) $p_{(\bar{E},\bar{A})}(s) = 0$ given by (1.114),
3) $p_{\bar{A}_1}(s) = 0$ given by (1.117),
4) $p_{A_1}(s) = 0$ given by (1.125)

with interval coefficients $\underline{a}_i \le a_i \le \overline{a}_i$ is asymptotically stable if and only if the lower bounds $\underline{a}_i$, $i = 0, 1, \ldots, r$ are positive, where $r = \operatorname{rank} \bar{A}_1 = \operatorname{rank} A_1$.

Proof. For each of the above characteristic equations we can define four characteristic equations as in (3.113). If the lower bounds $\underline{a}_i$ are positive, then the upper bounds $\overline{a}_i$ are also positive. Therefore, by Theorem 3.1 and Kharitonov theorem (Theorem 3.15) the positive descriptor interval continuous-time linear system (3.1) is asymptotically stable. □

Example 3.5. Consider the descriptor interval continuous-time linear system (3.1) with $A \in [A_L, A_U]$ and

$$E = \begin{bmatrix} 0 & 1 & 0 \\ 0.125 & 1 & 0 \\ 0 & 0 & 0 \end{bmatrix}, \quad A_L = \begin{bmatrix} 0 & -3 & 1 \\ -0.25 & -2.75 & 1 \\ 0 & 0 & 0.5 \end{bmatrix},$$
$$A_U = \begin{bmatrix} 0 & -4 & 1 \\ -0.5 & -3 & 1 \\ 0 & 0 & 1 \end{bmatrix}. \tag{3.114}$$

It is easy to check that the matrix pencils (E, A_L), (E, A_U) of (3.114) are regular. To check the stability of the interval system (3.1) with (3.114) we use two approaches as follows.

1) Drazin inverse matrix method case

From (3.104) for $c = 0$ we have

$$\bar{E}_L = \begin{bmatrix} 0.5 & 0.3333 & 0 \\ 0 & 0.3333 & 0 \\ 0 & 0 & 0 \end{bmatrix}, \quad \bar{E}_U = \begin{bmatrix} 0.25 & 0.5 & 0 \\ 0 & 0.25 & 0 \\ 0 & 0 & 0 \end{bmatrix},$$
$$\bar{A}_L = \bar{A}_U = \begin{bmatrix} -1 & 0 & 0 \\ 0 & -1 & 0 \\ 0 & 0 & -1 \end{bmatrix}. \tag{3.115}$$

Next, from (B.1) and (B.2) we have

$$\bar{E}_L^D = \begin{bmatrix} 2 & -2 & 0 \\ 0 & 3 & 0 \\ 0 & 0 & 0 \end{bmatrix}, \quad \bar{E}_U^D = \begin{bmatrix} 4 & -8 & 0 \\ 0 & 4 & 0 \\ 0 & 0 & 0 \end{bmatrix} \tag{3.116}$$

and

$$\bar{A}_{1L} = \bar{E}_L^D \bar{A}_L = \begin{bmatrix} -2 & 2 & 0 \\ 0 & -3 & 0 \\ 0 & 0 & 0 \end{bmatrix},$$
$$\bar{A}_{1U} = \bar{E}_U^D \bar{A}_U = \begin{bmatrix} -4 & 8 & 0 \\ 0 & -4 & 0 \\ 0 & 0 & 0 \end{bmatrix}. \tag{3.117}$$

By Theorem 1.10 the matrices $\bar{A}_{1L} \in \mathbb{M}_3$, $\bar{A}_{1U} \in \mathbb{M}_3$ and the interval system (3.1) with (3.114) is positive. The characteristic equations of (3.117) are given by

$$p_{\bar{A}_{1L}}(s) = s(s^2 + 5s + 6) = 0,$$
$$p_{\bar{A}_{1U}}(s) = s(s^2 + 8s + 16) = 0. \tag{3.118}$$

Therefore, by Theorem 3.1 both matrices (3.117) are asymptotically stable and by Theorem 3.14 the interval system (3.1) with (3.114) is asymptotically stable.

Also, by Theorem 3.16 the lower bounds of (3.118) are positive and the considered interval system is asymptotically stable.

2) Weierstrass-Kronecker decomposition method case

In this case

$$P_1 = \begin{bmatrix} 1 & 0 & -2 \\ -8 & 8 & 0 \\ 0 & 0 & 1 \end{bmatrix}, \quad Q_1 = \begin{bmatrix} 0 & 1 & 0 \\ 1 & 0 & 0 \\ 0 & 0 & 2 \end{bmatrix},$$
$$P_2 = \begin{bmatrix} 1 & 0 & -1 \\ -8 & 8 & 0 \\ 0 & 0 & 1 \end{bmatrix}, \quad Q_2 = \begin{bmatrix} 0 & 1 & 0 \\ 1 & 0 & 0 \\ 0 & 0 & 1 \end{bmatrix}. \tag{3.119}$$

From (3.105) and (3.115) we obtain

$$A_{1L} = \begin{bmatrix} -3 & 0 \\ 2 & -2 \end{bmatrix}, \quad A_{1U} = \begin{bmatrix} -4 & 0 \\ 8 & -4 \end{bmatrix}. \tag{3.120}$$

By Theorem 1.11 the matrices $A_{1L} \in \mathbb{M}_2$, $A_{1U} \in \mathbb{M}_2$ and $Q_1 \in \mathbb{R}_+^{3\times3}$, $Q_2 \in \mathbb{R}_+^{3\times3}$ are monomial. Therefore, the interval system (3.1) with (3.114) is positive. The characteristic equations of (3.120) are given by

$$\begin{aligned} p_{A_{1L}}(s) &= s^2 + 5s + 6 = 0, \\ p_{A_{1U}}(s) &= s^2 + 8s + 16 = 0. \end{aligned} \tag{3.121}$$

Therefore, by Theorem 3.1 both matrices (3.120) are asymptotically stable and by Theorem 3.14 the interval system (3.1) with (3.114) is asymptotically stable.

Also, by Theorem 3.16 the lower bounds of (3.121) are positive and the considered interval system is asymptotically stable.

3.2.2 Discrete-Time Systems

Consider the positive descriptor discrete-time linear system (3.26) with the interval matrix $A_d \in \mathbb{R}^{n\times n}$ defined by

$$A_{dL} \le A_d \le A_{dU} \text{ or equivalently } A_d \in [A_{dL}, A_{dU}]. \tag{3.122}$$

It is assumed that

$$\det[E_d z - A_{dL}] \ne 0 \text{ and } \det[E_d z - A_{dU}] \ne 0 \text{ for some } z \in \mathbb{C}. \tag{3.123}$$

Following the considerations presented in Sect. 3.1.2 the interval system (3.26) can be transformed into two equivalent forms:

1) Using Drazin inverse matrix method we obtain the interval system (3.27), where $\bar{E}_d \in [\bar{E}_{dL}, \bar{E}_{dU}]$, $\bar{A}_d \in [\bar{A}_{dL}, \bar{A}_{dU}]$ and

$$\begin{aligned}\bar{E}_{dL} &= [E_d c - A_{dL}]^{-1} E_d, \quad \bar{E}_{dU} = [E_d c - A_{dU}]^{-1} E_d \\ \bar{A}_{dL} &= [E_d c - A_{dL}]^{-1} A_{dL}, \quad \bar{A}_{dU} = [E_d c - A_{dU}]^{-1} A_{dU}.\end{aligned} \tag{3.124}$$

The matrix $\bar{E}_d^D \bar{A}_d$ in (3.27) will be denoted as $\bar{A}_{1d}$. Therefore, we have $\bar{A}_{1d} \in [\bar{A}_{1dL}, \bar{A}_{1dU}]$ and $\bar{A}_{1dL} = \bar{E}_{dL}^D \bar{A}_{dL}$, $\bar{A}_{1dU} = \bar{E}_{dU}^D \bar{A}_{dU}$.

2) Using Weierstrass-Kronecker decomposition method we obtain the interval system (3.28), where the matrix $A_{1d} \in [A_{1dL}, A_{1dU}]$ is determined by

$$P_{1d} A_{dL} Q_{1d} = \begin{bmatrix} A_{1dL} & 0 \\ 0 & \mathbb{I}_{n_2} \end{bmatrix}, \quad P_{2d} A_{dU} Q_{2d} = \begin{bmatrix} A_{1dU} & 0 \\ 0 & \mathbb{I}_{n_2} \end{bmatrix} \tag{3.125}$$

for two pairs of nonsingular matrices (P_{1d}, Q_{1d}), (P_{2d}, Q_{2d}).

Definition 3.3. The positive descriptor interval discrete-time linear system (3.26) is called asymptotically stable if the system is asymptotically stable for all matrices E_d, A_d, where $A_d \in [A_{dL}, A_{dU}]$.

According to Definition 3.2 we can define convex linear combinations of matrices $\bar{A}_{1dL}$, $\bar{A}_{1dU}$ and A_{1dL}, A_{1dU}, i.e.

$$\bar{A}_{1d} = (1-k)\bar{A}_{1dL} + k\bar{A}_{1dU}, \quad 0 \le k \le 1, \tag{3.126a}$$

$$A_{1d} = (1-k)A_{1dL} + kA_{1dU}, \quad 0 \le k \le 1. \tag{3.126b}$$

Theorem 3.17. The convex linear combinations (3.126) are asymptotically stable if and only if:

1) the matrices $\bar{A}_{1dL}$ and $\bar{A}_{1dU}$ are asymptotically stable;
2) the matrices A_{1dL} and A_{1dU} are asymptotically stable.

Proof. If the matrices $\bar{A}_{1dL}$ and $\bar{A}_{1dU}$ are asymptotically stable, then by (3.30) there exist matrices $G_{dL} \in \mathbb{R}^{n\times n}$, $G_{dU} \in \mathbb{R}^{n\times n}$ and strictly positive vector $\bar{\lambda} \in \mathbb{R}_+^n$ such that

$$[\bar{A}_{1dL} + G_{dL}(\mathbb{I}_n - \bar{E}_{dL}\bar{E}_{dL}^D) - \mathbb{I}_n]\bar{\lambda} < 0 \tag{3.127a}$$

and

$$[\bar{A}_{1dU} + G_{dU}(\mathbb{I}_n - \bar{E}_{dU}\bar{E}_{dU}^D) - \mathbb{I}_n]\bar{\lambda} < 0. \tag{3.127b}$$

From (3.126a) and (3.127) we obtain

$$
\begin{aligned}
&[\bar{A}_{1d} + G_d(\mathbb{I}_n - \bar{E}_d\bar{E}_d^D) - \mathbb{I}_n]\bar{\lambda} \\
&\qquad = \{(1-k)[\bar{A}_{1dL} + G_{dL}(\mathbb{I}_n - \bar{E}_{dL}\bar{E}_{dL}^D)] \\
&\qquad +k[\bar{A}_{1dU} + G_{dU}(\mathbb{I}_n - \bar{E}_{dU}\bar{E}_{dU}^D)] - \mathbb{I}_n\}\bar{\lambda} \\
&\qquad = (1-k)[\bar{A}_{1dL} + G_{dL}(\mathbb{I}_n - \bar{E}_{dL}\bar{E}_{dL}^D)]\bar{\lambda} \\
&\qquad +k[\bar{A}_{1dU} + G_{dU}(\mathbb{I}_n - \bar{E}_{dU}\bar{E}_{dU}^D)]\bar{\lambda} - \mathbb{I}_n\lambda < 0
\end{aligned}
\tag{3.128}
$$

for $0 \leq k \leq 1$. Therefore, if the matrices $\bar{A}_{1dL}$ and $\bar{A}_{1dU}$ are asymptotically stable, then the convex linear combination (3.126a) is also asymptotically stable.

The proof for the convex linear combination (3.126b) can be performed in a similar way. If the matrices A_{1dL} and A_{1dU} are asymptotically stable, then by (3.31) there exist strictly positive vector $\lambda \in \mathbb{R}_+^{n_1}$ such that

$$(A_{1dL} - \mathbb{I}_{n_1})\lambda < 0 \text{ and } (A_{1dU} - \mathbb{I}_{n_1})\lambda < 0. \tag{3.129}$$

From (3.107b) and (3.110) we obtain

$$
\begin{aligned}
(A_{1d} - \mathbb{I}_{n_1})\lambda &= [(1-k)A_{1dL} + kA_{1dU} - \mathbb{I}_{n_1}]\lambda \\
&= (1-k)A_{1dL}\lambda + kA_{1dU}\lambda - \mathbb{I}_{n_1}\lambda < 0
\end{aligned}
\tag{3.130}
$$

for $0 \leq k \leq 1$. Therefore, if the matrices A_{1dL} and A_{1dU} are asymptotically stable, then the convex linear combination (3.126b) is also asymptotically stable.

The necessity follows immediately from the fact that k can be equal to zero and one. □

Theorem 3.18. The positive descriptor interval discrete-time linear system (3.26) with (3.122) is asymptotically stable if and only if:

1) *(Drazin inverse matrix method case)*
the matrices $\bar{A}_{1dL}$ and $\bar{A}_{1dU}$ are asymptotically stable;
2) *(Weierstrass-Kronecker decomposition method case)*
the matrices A_{1dL} and A_{1dU} are asymptotically stable.

Proof. By (3.30) the matrices $\bar{A}_{1dL}$ and $\bar{A}_{1dU}$ are asymptotically stable if and only if there exist $G_{dL} \in \mathbb{R}^{n\times n}$, $G_{dU} \in \mathbb{R}^{n\times n}$ and strictly positive vector $\bar{\lambda} \in \mathbb{R}_+^n$ such that (3.127) holds. The convex linear combination (3.126a) satisfies the condition $[\bar{A}_{1d} + G_d(\mathbb{I}_n - \bar{E}_d\bar{E}_d^D) - \mathbb{I}_n]\bar{\lambda} < 0$ if and only if (3.127) holds. Therefore, the positive descriptor interval discrete-time linear system (3.26) with (3.122) is asymptotically stable if and only if the matrices A_{1dL} and A_{1dU} are asymptotically stable.

The proof for the Weierstrass-Kronecker decomposition method case can be performed in a similar way. □

To test the stability of an interval system an extension of the Kharitonov theorem (Theorem 3.15) can also be used.

Theorem 3.19. The positive descriptor interval discrete-time linear system (3.26) with the characteristic equations

1) $p_{(E_d,A_d)}(z+1) = p_{(E_d,A_d-E_d)}(z) = \det[E_d(z+1) - A_d] = 0,$
2) $p_{(\bar{E}_d,\bar{A}_d)}(z+1) = p_{(\bar{E}_d,\bar{A}_d-\bar{E}_d)}(z) = \det[\bar{E}_d(z+1) - \bar{A}_d] = 0,$
3) $p_{\bar{A}_{1d}}(z+1) = (z+1)^{n-r} p_{(\bar{E}_d,\bar{A}_d)}(z+1) = 0,$
4) $p_{A_{1d}}(z+1) = p_{A_{1d}-\mathbb{I}_{n_1}}(z) = \det[\mathbb{I}_{n_1}(z+1) - A_{1d}] = 0$

with interval coefficients $\underline{a}_{d,i} \leq a_{d,i} \leq \overline{a}_{d,i}$ is asymptotically stable if and only if the lower bounds $\underline{a}_{d,i}$, $i = 0, 1, \ldots, r$ are positive, where $r = \operatorname{rank}\bar{A}_{1d} = \operatorname{rank}A_{1d}$.

Proof. Taking into account Theorem 3.3 and Kharitonov theorem (Theorem 3.15) the proof can be performed similarly to the proof of Theorem 3.16. □

Example 3.6. Consider the descriptor interval discrete-time linear system (3.26) with $A_d \in [A_{dL}, A_{dU}]$ and

$$E_d = \begin{bmatrix} 1 & 0 & 0 \\ 2 & 2 & 0 \\ -0.8 & -0.8 & 0 \end{bmatrix}, \quad A_{dL} = \begin{bmatrix} 0.5 & 0.1 & 0 \\ 1 & 0.6 & 0 \\ -0.4 & -0.24 & 2 \end{bmatrix},$$
$$A_{dU} = \begin{bmatrix} 0.75 & 0.15 & 0 \\ 1.5 & 0.9 & 0 \\ -0.6 & -0.36 & 3 \end{bmatrix}. \tag{3.131}$$

It is easy to check that the matrix pencils (E_d, A_{dL}), (E_d, A_{dU}) of (3.131) are regular. To check the stability of the interval system (3.26) with (3.131) we use two approaches as follows.

1) Drazin inverse matrix method case

From (3.124) for $c = 0$ we have

$$\bar{E}_{dL} = \begin{bmatrix} -2 & 1 & 0 \\ 0 & -5 & 0 \\ 0 & 0 & 0 \end{bmatrix}, \quad \bar{E}_{dU} = \begin{bmatrix} -1.3333 & 0.6667 & 0 \\ 0 & -3.3333 & 0 \\ 0 & 0 & 0 \end{bmatrix},$$
$$\bar{A}_{dL} = \bar{A}_{dU} = \begin{bmatrix} -1 & 0 & 0 \\ 0 & -1 & 0 \\ 0 & 0 & -1 \end{bmatrix}. \tag{3.132}$$

Next, from (B.1) and (B.2) we have

$$\bar{E}_{dL}^D = \begin{bmatrix} -0.5 & -0.1 & 0 \\ 0 & -0.2 & 0 \\ 0 & 0 & 0 \end{bmatrix}, \quad \bar{E}_{dU}^D = \begin{bmatrix} -0.75 & -0.15 & 0 \\ 0 & -0.3 & 0 \\ 0 & 0 & 0 \end{bmatrix} \tag{3.133}$$

and

$$\bar{A}_{1dL} = \bar{E}_{dL}^{D}\bar{A}_{dL} = \begin{bmatrix} 0.5 & 0.1 & 0 \\ 0 & 0.2 & 0 \\ 0 & 0 & 0 \end{bmatrix},$$
$$\bar{A}_{1dU} = \bar{E}_{dU}^{D}\bar{A}_{dU} = \begin{bmatrix} 0.75 & 0.15 & 0 \\ 0 & 0.3 & 0 \\ 0 & 0 & 0 \end{bmatrix}. \tag{3.134}$$

By Theorem 1.29 the matrices $\bar{A}_{1dL} \in \mathbb{R}_{+}^{3\times 3}$, $\bar{A}_{1dU} \in \mathbb{R}_{+}^{3\times 3}$ and the interval system (3.26) with (3.131) is positive. The characteristic equations of (3.134) are given by

$$p_{\bar{A}_{1L}}(z+1) = (z+1)(10z^2 + 13z + 4) = 0,$$
$$p_{\bar{A}_{1U}}(z+1) = (z+1)(40z^2 + 38z + 7) = 0. \tag{3.135}$$

Therefore, by Theorem 3.3 both matrices (3.134) are asymptotically stable and by Theorem 3.18 the interval system (3.26) with (3.131) is asymptotically stable.

Also, by Theorem 3.19 the lower bounds of (3.135) are positive and the considered interval system is asymptotically stable.

2) Weierstrass-Kronecker decomposition method case

In this case

$$P_{1d} = \begin{bmatrix} 1 & 0 & 0 \\ -1 & 0.5 & 0 \\ 0 & 0.4 & 1 \end{bmatrix}, \quad Q_{1d} = \begin{bmatrix} 1 & 0 & 0 \\ 0 & 1 & 0 \\ 0 & 0 & 0.5 \end{bmatrix},$$
$$P_{2d} = \begin{bmatrix} 1 & 0 & 0 \\ -1 & 0.5 & 0 \\ 0 & 0.4 & 1 \end{bmatrix}, \quad Q_{2d} = \begin{bmatrix} 1 & 0 & 0 \\ 0 & 1 & 0 \\ 0 & 0 & 0.3333 \end{bmatrix}. \tag{3.136}$$

From (3.125) and (3.132) we obtain

$$A_{1dL} = \begin{bmatrix} 0.5 & 0.1 \\ 0 & 0.2 \end{bmatrix}, \quad A_{1dU} = \begin{bmatrix} 0.75 & 0.15 \\ 0 & 0.3 \end{bmatrix}. \tag{3.137}$$

By Theorem 1.31 the matrices $A_{1dL} \in \mathbb{R}_{+}^{2\times 2}$, $A_{1U} \in \mathbb{R}_{+}^{2\times 2}$ and $Q_{1d} \in \mathbb{R}_{+}^{2\times 2}$, $Q_{2d} \in \mathbb{R}_{+}^{2\times 2}$ are monomial. Therefore, the interval system (3.26) with (3.131) is positive. The characteristic equations of (3.137) are given by

$$p_{A_{1L}}(s) = 10z^2 + 13z + 4 = 0,$$
$$p_{A_{1U}}(s) = 40z^2 + 38z + 7 = 0. \tag{3.138}$$

Therefore, by Theorem 3.3 both matrices (3.137) are asymptotically stable and by Theorem 3.18 the interval system (3.26) with (3.131) is asymptotically stable.

Also, by Theorem 3.19 the lower bounds of (3.138) are positive and the considered interval system is asymptotically stable.

3.2.3 Fractional Continuous-Time Systems

Consider the positive fractional descriptor continuous-time linear system (3.51) with the interval matrix $A \in \mathbb{R}^{n \times n}$ defined by

$$A_L \leq A \leq A_U \text{ or equivalently } A \in [A_L, A_U]. \tag{3.139}$$

It is assumed that

$$\det[Es^\alpha - A_L] \neq 0 \text{ and } \det[Es^\alpha - A_U] \neq 0 \text{ for some } s^\alpha \in \mathbb{C}. \tag{3.140}$$

Following the considerations presented in Sect. 3.1.3 the fractional interval system (3.51) can be transformed into two equivalent forms:

1) Using Drazin inverse matrix method we obtain the fractional interval system (3.52), where $\bar{E} \in [\bar{E}_L, \bar{E}_U]$, $\bar{A} \in [\bar{A}_L, \bar{A}_U]$ and $\bar{E}_L, \bar{E}_U, \bar{A}_L, \bar{A}_U$ are determined by (3.104). The matrix $\bar{E}^D\bar{A}$ in (3.52) will be denoted as $\bar{A}_1$. Therefore, we have $\bar{A}_1 \in [\bar{A}_{1L}, \bar{A}_{1U}]$ and $\bar{A}_{1L} = \bar{E}_L^D\bar{A}_L$, $\bar{A}_{1U} = \bar{E}_U^D\bar{A}_U$.
2) Using Weierstrass-Kronecker decomposition method we obtain the fractional interval system (3.53), where the matrix $A_1 \in [A_{1L}, A_{1U}]$ is determined by (3.105) for two pairs of nonsingular matrices (P_1, Q_1), (P_2, Q_2).

Definition 3.4. The positive fractional descriptor interval continuous-time linear system (3.51) is called asymptotically stable if the system is asymptotically stable for all matrices E, A, where $A \in [A_L, A_U]$.

According to Definition 3.2 we can define convex linear combinations of matrices $\bar{A}_{1L}, \bar{A}_{1U}$ and A_{1L}, A_{1U} given by (3.107).

Theorem 3.20. The positive fractional descriptor interval continuous-time linear system (3.51) with (3.139) is asymptotically stable if and only if:

1) *(Drazin inverse matrix method case)*
the matrices $\bar{A}_{1L}$ and $\bar{A}_{1U}$ are asymptotically stable;
2) *(Weierstrass-Kronecker decomposition method case)*
the matrices A_{1L} and A_{1U} are asymptotically stable.

Proof. The proof is similar to the proof of Theorem 3.14. □

To test the stability of a fractional interval system an extension of the Kharitonov theorem (Theorem 3.15) can also be used. Consider the set (family) of the n-th degree polynomials

$$p_n(s^\alpha) = a_n s^{n\alpha} + a_{n-1}s^{(n-1)\alpha} + \ldots + a_1 s^\alpha + a_0 \tag{3.141a}$$

with interval coefficients

$$\underline{a}_i \leq a_i \leq \bar{a}_i, \quad i = 0, 1, \ldots, n. \tag{3.141b}$$

Using (3.141a) we define the following four polynomials:

$$\begin{aligned}
p_{1n}(s^\alpha) &= \underline{a}_0 + \underline{a}_1 s^\alpha + \overline{a}_2 s^{2\alpha} + \overline{a}_3 s^{3\alpha} + \underline{a}_4 s^{4\alpha} + \underline{a}_5 s^{5\alpha} + \dots,\\
p_{2n}(s^\alpha) &= \underline{a}_0 + \overline{a}_1 s^\alpha + \overline{a}_2 s^{2\alpha} + \underline{a}_3 s^{3\alpha} + \underline{a}_4 s^{4\alpha} + \overline{a}_5 s^{5\alpha} + \dots,\\
p_{3n}(s^\alpha) &= \overline{a}_0 + \underline{a}_1 s^\alpha + \underline{a}_2 s^{2\alpha} + \overline{a}_3 s^{3\alpha} + \overline{a}_4 s^{4\alpha} + \underline{a}_5 s^{5\alpha} + \dots,\\
p_{4n}(s^\alpha) &= \overline{a}_0 + \overline{a}_1 s^\alpha + \underline{a}_2 s^{2\alpha} + \underline{a}_3 s^{3\alpha} + \overline{a}_4 s^{4\alpha} + \overline{a}_5 s^{5\alpha} + \dots\,.
\end{aligned} \tag{3.142}$$

Theorem 3.21. The positive fractional descriptor interval continuous-time linear system (3.51) with the characteristic equations

1) $p_{(E,A)}(s^\alpha) = 0$ given by (2.76),
2) $p_{(\bar{E},\bar{A})}(s^\alpha) = 0$ given by (2.86),
3) $p_{\bar{A}_1}(s^\alpha) = 0$ given by (1.117),
4) $p_{A_1}(s^\alpha) = 0$ given by (2.93)

with interval coefficients $\underline{a}_i \le a_i \le \overline{a}_i$ is asymptotically stable if and only if the lower bounds $\underline{a}_i$, $i = 0, 1, \dots, r$ are positive, where $r = \operatorname{rank}\bar{A}_1 = \operatorname{rank}A_1$.

Proof. The proof is similar to the proof of Theorem 3.16. □

Example 3.7. Consider the fractional descriptor interval continuous-time linear system (3.51) with $A \in [A_L, A_U]$ and

$$E = \begin{bmatrix} -1 & -0.75 & -0.75 \\ -2 & -0.25 & -0.25 \\ 0 & -0.75 & -0.75 \end{bmatrix}, \quad A_L = \begin{bmatrix} 2 & 1 & 0 \\ 4 & 0 & 1 \\ 0 & 2 & 1 \end{bmatrix}.$$
$$A_U = \begin{bmatrix} 4 & 2 & 0 \\ 8 & 0 & 2 \\ 0 & 4 & 2 \end{bmatrix}. \tag{3.143}$$

It is easy to check that the matrix pencils (E, A_L), (E, A_U) of (3.143) are regular. To check the stability of the fractional interval system (3.51) with (3.143) we use the Drazin inverse matrix method. From (3.104) for $c = 0$ we have

$$\bar{E}_L = \begin{bmatrix} 0.5 & 0.125 & 0.125 \\ 0 & 0.5 & 0.5 \\ 0 & -0.25 & -0.25 \end{bmatrix}, \quad \bar{E}_U = \begin{bmatrix} 0.25 & 0.625 & 0.625 \\ 0 & 0.25 & 0.25 \\ 0 & -0.125 & -0.125 \end{bmatrix},$$
$$\bar{A}_L = \bar{A}_U = \begin{bmatrix} -1 & 0 & 0 \\ 0 & -1 & 0 \\ 0 & 0 & -1 \end{bmatrix}. \tag{3.144}$$

Next, using one of the methods presented in Appendix B we obtain

$$\bar{E}_L^D = \begin{bmatrix} 2 & -1 & -1 \\ 0 & 8 & 8 \\ 0 & -4 & -4 \end{bmatrix}, \quad \bar{E}_U^D = \begin{bmatrix} 4 & -2 & -2 \\ 0 & 16 & 16 \\ 0 & -8 & -8 \end{bmatrix} \tag{3.145}$$

and

$$\begin{aligned} \bar{A}_{1L} &= \bar{E}_L^D \bar{A}_L = \begin{bmatrix} -2 & 1 & 1 \\ 0 & -8 & -8 \\ 0 & 4 & 4 \end{bmatrix}, \\ \bar{A}_{1U} &= \bar{E}_U^D \bar{A}_U = \begin{bmatrix} -4 & 2 & 2 \\ 0 & -16 & -16 \\ 0 & 8 & 8 \end{bmatrix}. \end{aligned} \tag{3.146}$$

By Theorem 2.8 there exist matrices

$$G_L = \begin{bmatrix} 0 & 0 & 0 \\ 0 & -4 & 0 \\ 0 & 0 & -3 \end{bmatrix}, \quad G_U = \begin{bmatrix} 0 & 0 & 0 \\ 0 & -8 & 0 \\ 0 & 0 & -6 \end{bmatrix} \tag{3.147}$$

such that

$$\begin{aligned} \bar{A}_{1L} + G_L(\mathbb{I}_3 - \bar{E}_L \bar{E}_L^D) &= \begin{bmatrix} -2 & 1 & 1 \\ 0 & -4 & 0 \\ 0 & 1 & -2 \end{bmatrix} \in \mathbb{M}_3, \\ \bar{A}_{1U} + G_U(\mathbb{I}_3 - \bar{E}_U \bar{E}_U^D) &= \begin{bmatrix} -4 & 2 & 2 \\ 0 & -8 & 0 \\ 0 & 2 & -4 \end{bmatrix} \in \mathbb{M}_3 \end{aligned} \tag{3.148}$$

and the fractional interval system (3.1) with (3.114) is positive. The characteristic equations of (3.146) are given by

$$\begin{aligned} p_{\bar{A}_{1L}}(s^\alpha) &= s^\alpha(s^{2\alpha} + 6s^\alpha + 8) = 0, \\ p_{\bar{A}_{1U}}(s) &= s^\alpha(s^{2\alpha} + 12s^\alpha + 32) = 0. \end{aligned} \tag{3.149}$$

Therefore, by Theorem 3.5 both matrices (3.146) are asymptotically stable and by Theorem 3.20 the fractional interval system (3.51) with (3.143) is asymptotically stable.

Also, by Theorem 3.21 the lower bounds of (3.149) are positive and the considered fractional interval system is asymptotically stable.

3.2.4 Fractional Discrete-Time Systems

Consider the positive fractional descriptor discrete-time linear system (3.74) with the interval matrix $A_d \in \mathbb{R}^{n \times n}$ defined by

$$A_{dL} \le A_d \le A_{dU} \text{ or equivalently } A_d \in [A_{dL}, A_{dU}] \tag{3.150}$$

and $A_{d\alpha} \in [A_{d\alpha L}, A_{d\alpha U}]$, where $A_{d\alpha L} = A_{dL} + E_d\alpha$, $A_{d\alpha U} = A_{dU} + E_d\alpha$.
It is assumed that

$$\det[E_d z - A_{dL}] \ne 0 \text{ and } \det[E_d z - A_{dU}] \ne 0 \text{ for some } z \in \mathbb{C}. \tag{3.151}$$

Following the considerations presented in Sect. 3.1.4 the fractional interval system (3.74) can be transformed into two equivalent forms:

1) Using Drazin inverse matrix method we obtain the interval system (3.76), where $\bar{E}_d \in [\bar{E}_{dL}, \bar{E}_{dU}]$, $\bar{A}_d \in [\bar{A}_{dL}, \bar{A}_{dU}]$, $\bar{A}_{d\alpha} \in [\bar{A}_{d\alpha L}, \bar{A}_{d\alpha U}]$, the matrices $\bar{E}_{dL}$, $\bar{E}_{dU}$, $\bar{A}_{dL}$, $\bar{A}_{dU}$ are determined by (3.124) and

$$\bar{A}_{d\alpha L} = \bar{A}_{dL} + \bar{E}_{dL}\alpha, \quad \bar{A}_{d\alpha U} = \bar{A}_{dU} + \bar{E}_{dU}\alpha. \tag{3.152}$$

The matrix $\bar{E}_d^D \bar{A}_{d\alpha}$ in (3.76) will be denoted as $\bar{A}_{1d\alpha}$. Therefore, we have $\bar{A}_{1d\alpha} \in [\bar{A}_{1d\alpha L}, \bar{A}_{1d\alpha U}]$ and $\bar{A}_{1d\alpha L} = \bar{E}_{dL}^D \bar{A}_{d\alpha L}$, $\bar{A}_{1dU} = \bar{E}_{dU}^D \bar{A}_{d\alpha U}$.
2) Using Weierstrass-Kronecker decomposition method we obtain the fractional interval system (3.78), where the matrix $A_{1d} \in [A_{1dL}, A_{1dU}]$ is determined by (3.125) and

$$\begin{aligned} &A_{1d\alpha} \in [A_{1d\alpha L}, A_{1d\alpha U}], \\ &A_{1d\alpha L} = A_{1dL} + \mathbb{I}_{n_1}\alpha, \quad A_{1d\alpha U} = A_{1dU} + \mathbb{I}_{n_1}\alpha. \end{aligned} \tag{3.153}$$

Definition 3.5. The positive fractional descriptor interval discrete-time linear system (3.74) is called practically stable for given length L of practical implementation if the system is asymptotically stable for some finite L and all matrices E_d, A_d, where $A_d \in [A_{dL}, A_{dU}]$.

The positive fractional descriptor interval discrete-time linear system (3.74) is called asymptotically stable if the system is asymptotically stable for $L \to \infty$ and all matrices E_d, A_d, where $A_d \in [A_{dL}, A_{dU}]$.

Taking into account Theorems 3.7 and 3.8, to test the stability of the fractional interval system we will use the matrices $\bar{D}_1$ and D_1 defined by (3.80) and (3.82). From (3.80), (3.82) and (3.152), (3.153) we have

$$\bar{D}_1 \in [\bar{D}_{1L}, \bar{D}_{1U}], \quad D_1 \in [D_{1L}, D_{1U}], \tag{3.154a}$$

where

$$\begin{aligned} &\bar{D}_{1L} = \bar{A}_{1d\alpha L} + \sum_{j=1}^{L} c_j \mathbb{I}_n, \quad \bar{D}_{1U} = \bar{A}_{1d\alpha U} + \sum_{j=1}^{L} c_j \mathbb{I}_n \\ &D_{1L} = A_{1d\alpha L} + \sum_{j=1}^{L} c_j \mathbb{I}_n, \quad D_{1U} = A_{1d\alpha U} + \sum_{j=1}^{L} c_j \mathbb{I}_n. \end{aligned} \tag{3.154b}$$

According to Definition 3.2 we can define convex linear combinations of matrices $\bar{D}_{1L}$, $\bar{D}_{1U}$ and D_{1L}, D_{1U}, i.e.

$$\bar{D}_1 = (1-k)\bar{D}_{1L} + k\bar{D}_{1U}, \quad 0 \le k \le 1, \tag{3.155a}$$

$$D_1 = (1-k)D_{1L} + kD_{1U}, \quad 0 \le k \le 1. \tag{3.155b}$$

Theorem 3.22. The positive fractional descriptor interval discrete-time linear system (3.74) with (3.150) is practically stable for given length L of practical implementation if and only if:

1) *(Drazin inverse matrix method case)*
the matrices $\bar{D}_{1L}$ and $\bar{D}_{1U}$ are asymptotically stable;
2) *(Weierstrass-Kronecker decomposition method case)*
the matrices D_{1L} and D_{1U} are asymptotically stable.

Proof. The proof follows immediately from Theorems 3.10 and 3.18. □

Theorem 3.23. The positive fractional descriptor interval discrete-time linear system (3.74) with (3.150) is asymptotically stable if and only if:

1) *(Drazin inverse matrix method case)*
the matrices $\bar{A}_{1dL}$ and $\bar{A}_{1dU}$ are asymptotically stable;
2) *(Weierstrass-Kronecker decomposition method case)*
the matrices A_{1dL} and A_{1dU} are asymptotically stable.

Proof. The proof follows immediately from Theorems 3.12 and 3.18. □

To test the stability of a fractional interval system an extension of the Kharitonov theorem (Theorem 3.15) can also be used.

Theorem 3.24. The positive fractional descriptor interval continuous-time linear system (3.51) with the characteristic equations

1) $p_{(E_d,D)}(z+1) = p_{(E_d,D-E_d)}(z) = \det[E_d(z+1) - D] = 0$,
2) $p_{(\bar{E}_d,\bar{D})}(z+1) = p_{(\bar{E}_d,\bar{D}-\bar{E}_d)}(z) = \det[\bar{E}_d(z+1) - \bar{D}] = 0$,
3) $p_{\bar{D}_1}(z+1) = \det[\mathbb{I}_n(z+1) - \bar{D}_1] = (z+1)^{n-r} p_{(\bar{E}_d,\bar{D})}(z+1) = 0$,
4) $p_{D_1}(z+1) = p_{D_1-\mathbb{I}_{n_1}}(z) = \det[\mathbb{I}_{n_1}(z+1) - D_1] = 0$

with interval coefficients $\underline{a}_i \le a_i \le \overline{a}_i$ is practically stable for given length L of practical implementation if and only if the lower bounds $\underline{a}_i$, $i = 0, 1, \dots, r$ are positive, where $r = \operatorname{rank} \bar{A}_{1d\alpha} = \operatorname{rank} A_{1d\alpha}$ and the matrices D, $\bar{D}$ are defined by (3.83).

Proof. Taking into account Theorem 3.9 and Kharitonov theorem (Theorem 3.15) the proof can be performed similarly to the proof of Theorem 3.16. □

Theorem 3.25. The positive fractional descriptor interval continuous-time linear system (3.51) with the characteristic equations

1) $p_{(E_d,D_\infty)}(z+1) = \det[E_d z - A_d] = p_{(E_d,A_d)}(z) = 0,$
2) $p_{(\bar{E}_d,\bar{D}_\infty)}(z+1) = \det[\bar{E}_d z - \bar{A}_d] = p_{(\bar{E}_d,\bar{A}_d)}(z) = 0,$
3) $p_{\bar{D}_{1,\infty}}(z+1) = \frac{1}{\beta}(z+\alpha)^{n-r} p_{(\bar{E}_d,\bar{A}_d)}(z) = 0,$
4) $p_{D_{1,\infty}}(z+1) = \det[\mathbb{I}_{n_1} z - A_{1d}] = p_{A_{1d}}(z) = 0$

with interval coefficients $\underline{a}_i \le a_i \le \overline{a}_i$ is asymptotically stable if and only if the lower bounds $\underline{a}_i$, $i = 0, 1, \ldots, r$ are positive, where $r = \operatorname{rank} \bar{A}_{1d\alpha} = \operatorname{rank} A_{1d\alpha}$.

Proof. The proof follows immediately from Theorem 3.9, Lemma 3.1 and Kharitonov theorem (Theorem 3.15). □

Example 3.8. Consider the fractional descriptor interval discrete-time linear system (3.26) with $\alpha = 0.4$, $A_d \in [A_{dL}, A_{dU}]$ and

$$E_d = \begin{bmatrix} 0 & -5 & 0 \\ -5 & -10 & 0 \\ 0 & -7.5 & 0 \end{bmatrix}, \quad A_{dL} = \begin{bmatrix} 0 & 1 & 0 \\ 1 & 0 & 0 \\ 0 & 0 & 1 \end{bmatrix},$$
$$A_{dU} = \begin{bmatrix} 0 & 1.5 & 0 \\ 2 & 0 & 0 \\ 0 & 0 & 1.5 \end{bmatrix}. \tag{3.156}$$

It is easy to check that the matrix pencils (E_d, A_{dL}), (E_d, A_{dU}) of (3.156) are regular. To check the stability of the fractional interval system (3.26) with (3.131) we use the Drazin inverse matrix method. From (3.124) for $c = 0$ we have

$$\bar{E}_{dL} = \begin{bmatrix} 5 & 10 & 0 \\ 0 & 5 & 0 \\ 0 & 7.5 & 0 \end{bmatrix}, \quad \bar{E}_{dU} = \begin{bmatrix} 2.5 & 5 & 0 \\ 0 & 3.3333 & 0 \\ 0 & 5 & 0 \end{bmatrix},$$
$$\bar{A}_{dL} = \bar{A}_{dU} = \begin{bmatrix} -1 & 0 & 0 \\ 0 & -1 & 0 \\ 0 & 0 & -1 \end{bmatrix}, \tag{3.157}$$
$$\bar{A}_{d\alpha L} = \begin{bmatrix} 1 & 4 & 0 \\ 0 & 1 & 0 \\ 0 & 3 & -1 \end{bmatrix}, \quad \bar{A}_{d\alpha U} = \begin{bmatrix} 0 & 2 & 0 \\ 0 & 0.3333 & 0 \\ 0 & 2 & -1 \end{bmatrix},$$

Next, using (B.1) and (B.2) we obtain

$$\bar{E}_{dL}^D = \begin{bmatrix} 0.2 & -0.4 & 0 \\ 0 & 0.2 & 0 \\ 0 & 0.3 & 0 \end{bmatrix}, \quad \bar{E}_{dU}^D = \begin{bmatrix} 0.4 & -0.6 & 0 \\ 0 & 0.3 & 0 \\ 0 & 0.45 & 0 \end{bmatrix} \tag{3.158}$$

and

$$\begin{aligned}\bar{A}_{1dL} = \bar{E}_{dL}^D \bar{A}_{dL} &= \begin{bmatrix} -0.2 & 0.4 & 0 \\ 0 & -0.2 & 0 \\ 0 & -0.3 & 0 \end{bmatrix}, \\ \bar{A}_{1dU} = \bar{E}_{dU}^D \bar{A}_{dU} &= \begin{bmatrix} -0.4 & 0.6 & 0 \\ 0 & -0.3 & 0 \\ 0 & -0.45 & 0 \end{bmatrix},\end{aligned} \tag{3.159}$$

$$\begin{aligned}\bar{A}_{1d\alpha L} = \bar{E}_{dL}^D \bar{A}_{d\alpha L} &= \begin{bmatrix} 0.2 & 0.4 & 0 \\ 0 & 0.2 & 0 \\ 0 & 0.3 & 0 \end{bmatrix}, \\ \bar{A}_{1d\alpha U} = \bar{E}_{dU}^D \bar{A}_{d\alpha U} &= \begin{bmatrix} 0 & 0.6 & 0 \\ 0 & 0.1 & 0 \\ 0 & 0.15 & 0 \end{bmatrix}.\end{aligned} \tag{3.160}$$

By Theorem 1.29 the matrices $\bar{A}_{1d\alpha L} \in \mathbb{R}_+^{3\times 3}$, $\bar{A}_{1d\alpha U} \in \mathbb{R}_+^{3\times 3}$ and the fractional interval system (3.74) with (3.156) is positive. The characteristic equations of (3.159) are given by

$$\begin{aligned} p_{\bar{A}_{1dL}}(z) &= z(25z^2 + 10z + 1) = 0, \\ p_{\bar{A}_{1dU}}(z) &= z(50z^2 + 35z + 6) = 0. \end{aligned} \tag{3.161}$$

Therefore, by Theorem 3.9 both matrices (3.159) are asymptotically stable and by Theorem 3.22 the interval system (3.74) with (3.156) is asymptotically stable.

Also, by Theorem 3.24 the lower bounds of (3.161) are positive and the considered interval system is asymptotically stable.

3.3 Stability of Nonlinear Systems with Positive Linear Parts

In this section the stability of nonlinear feedback systems with positive linear parts will be analyzed [75–78, 80, 81, 91, 93]. Sufficient conditions for the global stability will be established using Lyapunov method (Sect. 3.3.1–3.3.4) and Kudrewicz theorem [104] will be extended to positive descriptor continuous-time nonlinear systems (Sect. 3.3.5).

3.3.1 Continuous-Time Systems

Consider the continuous-time nonlinear feedback system shown in Fig. 3.1, which consists of a positive descriptor linear part, a nonlinear element with characteristic

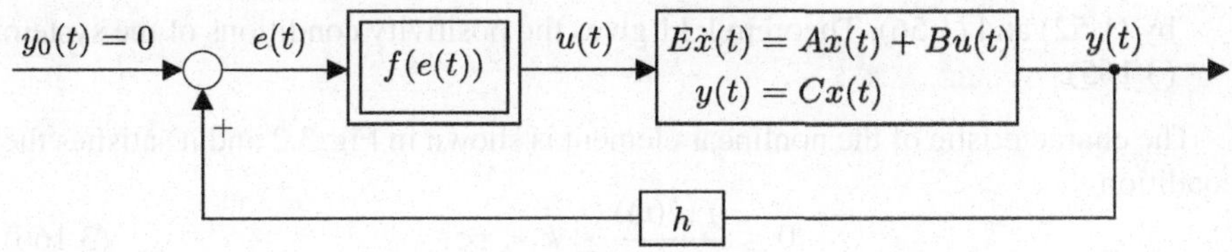

Fig. 3.1 Continuous-time nonlinear feedback system

$u(t) = f(e(t))$ and a positive scalar gain feedback h. The linear part is described by the equations

$$\begin{aligned} E\dot{x}(t) &= Ax(t) + Bu(t), \\ y(t) &= Cx(t), \end{aligned} \tag{3.162}$$

where $x(t) \in \mathbb{R}_+^n$, $u(t) \in \mathbb{R}_+$, $y(t) \in \mathbb{R}_+$ are state, input and output vectors and $E, A \in \mathbb{R}^{n\times n}, B \in \mathbb{R}^{n\times 1}, C \in \mathbb{R}^{1\times n}$. It is assumed that $\det E = 0$ and $\det[Es - A] \neq 0$ for some $s \in \mathbb{C}$.

Following the considerations presented in Sect. 1.1 the system (3.162) can be transformed into the following equivalent forms.

1) Using the Drazin inverse matrix method we obtain

$$\begin{aligned} \dot{x}_1(t) &= \bar{A}_1 x_1(t) + \bar{B}_1 u(t), \\ \bar{N}\dot{x}_2(t) &= x_2(t) + \bar{B}_2 u(t), \\ y(t) &= Cx(t) = C[x_1(t) + x_2(t)], \end{aligned} \tag{3.163}$$

where $x_1(t), x_2(t) \in \mathbb{R}_+^n$ are given by (1.28) and

$$\begin{aligned} \bar{A}_1 = \bar{E}^D \bar{A}, \quad \bar{B}_1 = \bar{E}^D \bar{B}, \\ \bar{B}_2 = (\mathbb{I}_n - \bar{E}\bar{E}^D)\bar{A}^D \bar{B}, \quad \bar{N} = (\mathbb{I}_n - \bar{E}\bar{E}^D)\bar{A}^D \bar{E}. \end{aligned} \tag{3.164}$$

The matrices $\bar{E}, \bar{A} \in \mathbb{R}^{n\times n}$ and $\bar{B} \in \mathbb{R}^{n\times 1}$ have the form (1.24b). Theorem 1.10 gives the positivity conditions of the system (3.163).

2) Using the Weierstrass-Kronecker decomposition method we have

$$\begin{aligned} \dot{\bar{x}}_1(t) &= A_1 \bar{x}_1(t) + B_1 u(t), \\ N\dot{\bar{x}}_2(t) &= \bar{x}_2(t) + B_2 u(t), \\ y(t) &= C_1 \bar{x}_1(t) + C_2 \bar{x}_2(t), \end{aligned} \tag{3.165}$$

where $\bar{x}_1(t) \in \mathbb{R}_+^{n_1}, \bar{x}_2(t) \in \mathbb{R}_+^{n_2}$ are given by (1.53) and the matrices $A_1 \in \mathbb{R}^{n_1\times n_1}$, $B_1 \in \mathbb{R}^{n_1\times 1}$, $B_2 \in \mathbb{R}^{n_2\times 1}$, $N \in \mathbb{R}^{n_2\times n_2}$, $C_1 \in \mathbb{R}^{1\times n_1}$, $C_2 \in \mathbb{R}^{1\times n_2}$ are determined by (1.52) and (1.56). Theorem 1.11 gives the positivity conditions of the system (3.165).

The characteristic of the nonlinear element is shown in Fig. 3.2 and it satisfies the condition

$$0 \le \frac{f(e(t))}{e(t)} \le k < \infty. \tag{3.166}$$

It is assumed that the positive descriptor linear part is asymptotically stable.

Definition 3.6. The positive descriptor continuous-time nonlinear system is called globally stable if it is asymptotically stable for all admissible nonnegative initial conditions $x_0 \in \mathbb{R}_+^n$.

The following theorem gives sufficient conditions for the global stability of the positive descriptor continuous-time nonlinear system.

Theorem 3.26. The continuous-time nonlinear system consisting of the positive and asymptotically stable descriptor linear part, the nonlinear element satisfying the condition (3.166) and the gain feedback h is globally stable if:

1) *(Drazin inverse matrix method case)*

   a) $\bar{B}_2 = 0$ and there exists a matrix $G \in \mathbb{R}^{n\times n}$ such that

$$\bar{A}_{C1} + G(\mathbb{I}_n - \bar{E}\bar{E}^D) \in \mathbb{M}_n \tag{3.167}$$

   is asymptotically stable, where

$$\bar{A}_{C1} = \bar{A}_1 + kh\bar{B}_1 C; \tag{3.168}$$

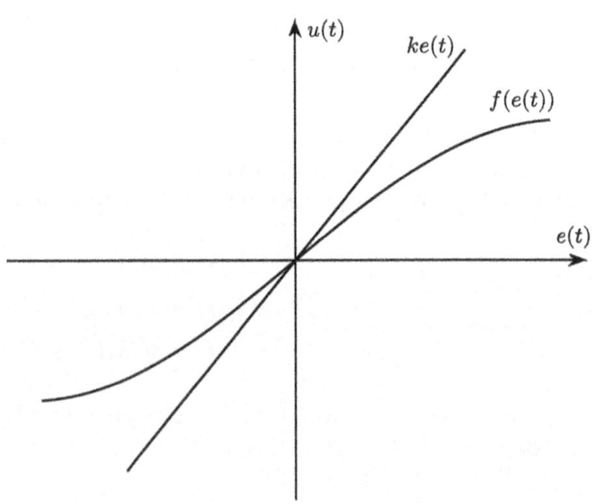

Fig. 3.2 Characteristic of the nonlinear element

b) $C(\mathbb{I}_n - \bar{E}\bar{E}^D) = 0$ and there exists a matrix $G \in \mathbb{R}^{n\times n}$ such that (3.167) is asymptotically stable and

$$-\bar{N}^j \bar{A}_{C2}\bar{A}_{C1}^j \in \mathbb{R}_+^{n\times n},\ \ j = 1,\ldots,q-1, \tag{3.169}$$

where $\bar{A}_{C1}$ is given by (3.168),

$$\bar{A}_{C2} = kh\bar{B}_2 C \tag{3.170}$$

and q is the index of $\bar{E}$.

2) *(Weierstrass-Kronecker decomposition method case)*

a) $B_2 = 0$ and

$$A_{C1} \in \mathbb{M}_{n_1} \tag{3.171}$$

is asymptotically stable, where

$$A_{C1} = A_1 + khB_1C_1; \tag{3.172}$$

b) $C_2 = 0$, (3.171) is asymptotically stable and

$$-N^j A_{C2}A_{C1}^j \in \mathbb{R}_+^{n_2\times n_1},\ \ j = 1,\ldots,\mu-1, \tag{3.173}$$

where A_{C1} is given by (3.172),

$$A_{C2} = khB_2C_1 \tag{3.174}$$

and μ is the nilpotency index of N.

Proof. The proof will be accomplished for the Weierstrass-Kronecker decomposition method. Taking into account (3.165) for $B_2 = 0$ we obtain

$$\begin{aligned}\dot{\bar{x}}_1(t) &= A_1\bar{x}_1(t) + B_1u(t),\\ N\dot{\bar{x}}_2(t) &= \bar{x}_2(t),\\ y(t) &= C_1\bar{x}_1(t) + C_2\bar{x}_2(t).\end{aligned} \tag{3.175}$$

From the second equation of (3.175) it follows that $\bar{x}_2(t) = 0$ since N is a nilpotent matrix. Assuming $u(t) = ke(t)$ and taking into account that $e(t) = hy(t)$ from (3.175) we obtain

$$\begin{aligned}\dot{\bar{x}}_1(t) &= (A_1 + khB_1C_1)\bar{x}_1(t),\\ \bar{x}_2(t) &= 0\end{aligned} \tag{3.176}$$

As the Lyapunov function $V(\bar{x}_1(t))$ we choose

$$V(\bar{x}_1(t)) = \lambda^T \bar{x}_1(t) \geq 0 \text{ for } \bar{x}_1(t) \in \mathbb{R}_+^{n_1}, \tag{3.177}$$

where $\lambda = \begin{bmatrix} \lambda_1 & \dots & \lambda_{n_1} \end{bmatrix}$ is a strictly positive vector, i.e. $\lambda_j > 0,\ j = 1, \dots, n_1$. Using (3.175)–(3.177) we obtain

$$\begin{aligned} \dot{V}(\bar{x}_1(t)) &= \lambda^T \dot{\bar{x}}_1(t) = \lambda^T [A_1 \bar{x}_1(t) + B_1 u(t)] \\ &= \lambda^T [A_1 \bar{x}_1(t) + B_1 h f(e(t))] \leq \lambda^T [(A_1 + khB_1C_1)\bar{x}_1(t)] \end{aligned} \tag{3.178}$$

since $u(t) = f(e(t)) \leq ke(t) = kC_1\bar{x}_1(t)$. From (3.178) it follows that $\dot{V}(\bar{x}_1(t)) < 0$ if the condition (3.171) is satisfied with $B_2 = 0$ and the nonlinear system is globally stable.

Now let us consider (3.165) for $C_2 = 0$. In this case we have

$$\begin{aligned} \dot{\bar{x}}_1(t) &= A_1 \bar{x}_1(t) + B_1 u(t), \\ N\dot{\bar{x}}_2(t) &= \bar{x}_2(t) + B_2 u(t), \\ y(t) &= C_1 \bar{x}_1(t). \end{aligned} \tag{3.179}$$

Assuming $u(t) = ke(t)$ and taking into account that $e(t) = hy(t)$ from (3.179) we obtain

$$\begin{aligned} \dot{\bar{x}}_1(t) &= (A_1 + khB_1C_1)\bar{x}_1(t), \\ N\dot{\bar{x}}_2(t) &= khB_2C_1\bar{x}_1(t) + \bar{x}_2(t). \end{aligned} \tag{3.180}$$

The second equation of (3.180) can be solved in a similar way to the Eq. (1.55b). From (3.180) and (1.59) we have

$$\bar{x}_2(t) = -\sum_{j=0}^{\mu-1} khN^j B_2 C_1 \bar{x}_1^{(j)}(t), \tag{3.181}$$

where μ is the nilpotency index of the matrix N and $\bar{x}_1^{(j)}(t)$ is the j-th time derivative of the vector $\bar{x}_1(t)$. Let $A_{C1} = A_1 + khB_1C_1$ and $A_{C2} = khB_2C_1$. Taking into account that

$$\begin{aligned} \dot{\bar{x}}_1(t) &= A_{C1}\bar{x}_1(t), \\ \ddot{\bar{x}}_1(t) &= A_{C1}\dot{\bar{x}}_1(t) = A_{C1}^2\bar{x}_1(t), \\ &\vdots \\ \bar{x}_1^{(j)}(t) &= A_{C1}\bar{x}_1^{(j-1)}(t) = A_{C1}^j\bar{x}_1(t) \end{aligned} \tag{3.182}$$

from (3.181) we obtain

$$\bar{x}_2(t) = -\sum_{j=0}^{\mu-1} N^j A_{C2} A_{C1}^j \bar{x}_1(t). \tag{3.183}$$

Choosing the Lyapunov function in the form (3.177) we get (3.178) and $\dot{V}(\bar{x}_1(t)) < 0$ if the condition (3.171) is satisfied with $C_2 = 0$. From (3.183) it follows that if $\lim_{t\to\infty} \bar{x}_1(t) = 0$ then also $\lim_{t\to\infty} \bar{x}_2(t) = 0$. The condition (3.173) ensures $\bar{x}_2(t) \in \mathbb{R}_+^{n_2}$. Therefore, the nonlinear system is globally stable.

The proof for the Drazin inverse matrix method can be accomplished similarly. Note that if $C(\mathbb{I}_n - \bar{E}\bar{E}^D) = 0$, then $Cx_2(t) = 0$ since $x_2(t) = (\mathbb{I}_n - \bar{E}\bar{E}^D)x(t)$. The matrix $G \in \mathbb{R}^{n\times n}$ and the term $G(\mathbb{I}_n - \bar{E}\bar{E}^D)$ eliminates from the matrix $\bar{A}_1 = \bar{E}^D\bar{A}$ unimportant entires that are further canceled through multiplication by $x_1(t) \in \mathrm{Im}\bar{E}\bar{E}^D$. □

Remark 3.2. Theorems 3.1 and 3.2 can be used to test the stability of (3.167) and (3.171).

Example 3.9. Consider the nonlinear system consisting of the descriptor continuous-time linear part (3.162) with

$$E = \begin{bmatrix} 0.5 & 0 & 0 \\ 0 & 0.25 & 0 \\ 0 & 0.25 & 0 \end{bmatrix}, \quad A = \begin{bmatrix} -1 & 0 & 0 \\ 0.5 & -0.5 & 0 \\ 0.5 & -0.5 & 1 \end{bmatrix}, \quad B = \begin{bmatrix} 1 \\ 2 \\ -1 \end{bmatrix}, \quad C = \begin{bmatrix} 2 & 1 & 0 \end{bmatrix}, \tag{3.184}$$

the nonlinear element satisfying the condition (3.166) and the gain feedback $h = 0.5$. Find k satisfying (3.166) for which the nonlinear system is globally stable.

The matrix pencil (E, A) of (3.184) is regular since

$$\det[Es - A] = \begin{vmatrix} 0.5s + 1 & 0 & 0 \\ -0.5 & 0.25s + 0.5 & 0 \\ -0.5 & 0.25s + 0.5 & -1 \end{vmatrix} = -0.125(s + 2)^2 \neq 0. \tag{3.185}$$

The characteristic equation of (3.184) is given by

$$p_{(E,A)}(s) = \det[Es - A] = -0.125(s + 2)^2 = 0. \tag{3.186}$$

From (3.186) it follows that the pair (E, A) has two eigenvalues $s_1 = s_2 = -2$ and by condition (1.101) the considered linear part is asymptotically stable. To check the positivity and global stability of the system we use two approaches as follows.

1) Drazin inverse matrix method case

We choose $c = 0$ and using (1.24b) we obtain

$$\bar{E} = [-A]^{-1}E = \begin{bmatrix} 0.5 & 0 & 0 \\ 0.5 & 0.5 & 0 \\ 0 & 0 & 0 \end{bmatrix},$$
$$\bar{A} = [-A]^{-1}A = \begin{bmatrix} -1 & 0 & 0 \\ 0 & -1 & 0 \\ 0 & 0 & -1 \end{bmatrix}, \quad \bar{B} = [-A]^{-1}B = \begin{bmatrix} 1 \\ 5 \\ 3 \end{bmatrix} \tag{3.187}$$

and $q = 1$ since $\operatorname{rank}\bar{E} = \operatorname{rank}\bar{E}^2$. Next, from (B.1) and (B.2) we have

$$\bar{E}^D = \begin{bmatrix} 2 & 0 & 0 \\ -2 & 2 & 0 \\ 0 & 0 & 0 \end{bmatrix}, \quad \bar{A}^D = \bar{A}^{-1} = \bar{A} = \begin{bmatrix} -1 & 0 & 0 \\ 0 & -1 & 0 \\ 0 & 0 & -1 \end{bmatrix} \tag{3.188}$$

and

$$\bar{A}_1 = \bar{E}^D\bar{A} = \begin{bmatrix} -2 & 0 & 0 \\ 2 & -2 & 0 \\ 0 & 0 & 0 \end{bmatrix}, \quad \bar{B}_1 = \bar{E}^D\bar{B} = \begin{bmatrix} 2 \\ 8 \\ 0 \end{bmatrix},$$
$$\bar{B}_2 = (\mathbb{I}_3 - \bar{E}\bar{E}^D)\bar{A}^D\bar{B} = \begin{bmatrix} 0 \\ 0 \\ -3 \end{bmatrix}, \quad \bar{N} = (\mathbb{I}_3 - \bar{E}\bar{E}^D)\bar{A}^D\bar{E} = \begin{bmatrix} 0 & 0 & 0 \\ 0 & 0 & 0 \\ 0 & 0 & 0 \end{bmatrix},$$
$$C(\mathbb{I}_3 - \bar{E}\bar{E}^D) = \begin{bmatrix} 0 & 0 & 0 \end{bmatrix}. \tag{3.189}$$

By Theorem 1.10 the conditions (1.88) for the matrices (3.189) are satisfied and the descriptor linear part described by (3.184) is positive. Also, from (3.189) it follows that $\bar{B}_2 \neq 0$ and $C(\mathbb{I}_3 - \bar{E}\bar{E}^D) = 0$. Therefore, the case b) of Theorem 3.26 can be used to test the global stability of the nonlinear system. Taking into account (3.168), (3.170), (3.184), (3.189) and $h = 0.5$ we obtain

$$\bar{A}_{C1} = \bar{A}_1 + 0.5k\bar{B}_1C = \begin{bmatrix} 2k-2 & k & 0 \\ 8k+2 & 4k-2 & 0 \\ 0 & 0 & 0 \end{bmatrix},$$
$$\bar{A}_{C2} = 0.5k\bar{B}_2C = \begin{bmatrix} 0 & 0 & 0 \\ 0 & 0 & 0 \\ -3k & -1.5k & 0 \end{bmatrix}. \tag{3.190}$$

The matrix $\bar{A}_{C1} \in \mathbb{M}_3$ for any $k > 0$ and by Theorem 3.1 it is asymptotically stable if and only if all coefficients of the characteristic equation

$$p_{\bar{A}_{C1}}(s) = \det[\mathbb{I}_3 s - \bar{A}_{C1}] = s[s^2 + (4-6k)s + 4 - 14k] = 0 \tag{3.191}$$

are positive, i.e. if and only if $4 - 6k > 0$ and $4 - 14k > 0$, and this implies $k < 0.2857$. Taking into account (3.169) the matrix

$$-\bar{N}^0\bar{A}_{C2}\bar{A}_{C1}^0 = -\bar{A}_{C2} = \begin{bmatrix} 0 & 0 & 0 \\ 0 & 0 & 0 \\ 3k & 1.5k & 0 \end{bmatrix} \in \mathbb{R}_+^{3\times 3} \tag{3.192}$$

for any $k > 0$. Therefore, the positive nonlinear system is globally stable if the characteristic $u(t) = f(e(t))$ of the nonlinear element satisfies the condition (3.166) for $k < 0.2857$.

2) Weierstrass-Kronecker decomposition method case

In this case

$$P = \begin{bmatrix} 0 & 1 & 0 \\ 1 & 0 & 0 \\ 0 & -1 & 1 \end{bmatrix}, \quad Q = \begin{bmatrix} 0 & 2 & 0 \\ 4 & 0 & 0 \\ 0 & 0 & 1 \end{bmatrix} \tag{3.193}$$

and

$$\begin{bmatrix} \mathbb{I}_{n_1} & 0 \\ 0 & N \end{bmatrix} = PEQ = \begin{bmatrix} 1 & 0 & 0 \\ 0 & 1 & 0 \\ 0 & 0 & 0 \end{bmatrix},$$

$$\begin{bmatrix} A_1 & 0 \\ 0 & \mathbb{I}_{n_2} \end{bmatrix} = PAQ = \begin{bmatrix} -2 & 1 & 0 \\ 0 & -2 & 0 \\ 0 & 0 & 1 \end{bmatrix}, \tag{3.194}$$

$$\begin{bmatrix} B_1 \\ B_2 \end{bmatrix} = PB = \begin{bmatrix} 2 \\ 1 \\ -3 \end{bmatrix}, \quad \begin{bmatrix} C_1 & C_2 \end{bmatrix} = CQ = \begin{bmatrix} 4 & 4 & 0 \end{bmatrix}$$

where $n_1 = 2, n_2 = 1$. By Theorem 1.11 the conditions (1.94) for the matrices (3.194) are satisfied and $Q \in \mathbb{R}_+^{3\times 3}$ is monomial. Thus, the descriptor linear part described by (3.184) is positive. Also, from (3.194) it follows that $B_2 \neq 0$ and $C_2 = 0$. Therefore, the case b) of Theorem 3.26 can be used to test the global stability of the nonlinear system. Taking into account (3.172), (3.174), (3.194) and $h = 0.5$ we obtain

$$\begin{aligned} A_{C1} &= A_1 + 0.5kB_1C_1 = \begin{bmatrix} 4k-2 & 4k+1 \\ 2k & 2k-2 \end{bmatrix}, \\ A_{C2} &= 0.5kB_2C_1 = \begin{bmatrix} -6k & -6k \end{bmatrix}. \end{aligned} \tag{3.195}$$

The matrix $A_{C1} \in \mathbb{M}_2$ for any $k > 0$ and by Theorem 3.1 it is asymptotically stable if and only if all coefficients of the characteristic equation

$$p_{A_{C1}}(s) = \det[\mathbb{I}_2 s - A_{C1}] = s^2 + (4-6k)s + 4 - 14k = 0 \tag{3.196}$$

are positive, i.e. if and only if $4 - 6k > 0$ and $4 - 14k > 0$, and this implies $k < 0.2857$. Taking into account (3.173) the matrix

$$-N^0 A_{C2} A_{C1}^0 = -A_{C2} = \begin{bmatrix} 6k & 6k \end{bmatrix} \in \mathbb{R}_+^{1\times 2} \tag{3.197}$$

for any $k > 0$. Therefore, the positive nonlinear system is globally stable if the characteristic $u(t) = f(e(t))$ of the nonlinear element satisfies the condition (3.166) for $k < 0.2857$.

3.3.2 Discrete-Time Systems

Consider the discrete-time nonlinear feedback system shown in Fig. 3.3, which consists of a positive descriptor linear part, a nonlinear element with characteristic $u_i = f(e_i)$ and a positive scalar gain feedback h. The linear part is described by the equations

$$\begin{aligned} E_d x_{i+1} &= A_d x_i + B_d u_i, \quad i \in \mathbb{Z}_+, \\ y_i &= C_d x_i, \end{aligned} \tag{3.198}$$

where $x_i \in \mathbb{R}_+^n$, $u_i \in \mathbb{R}_+$, $y_i \in \mathbb{R}_+$ are state, input and output vectors and $E_d, A_d \in \mathbb{R}^{n\times n}$, $B_d \in \mathbb{R}^{n\times 1}$, $C_d \in \mathbb{R}^{1\times n}$. It is assumed that $\det E_d = 0$ and $\det[E_d z - A_d] \neq 0$ for some $z \in \mathbb{C}$.

Following the considerations presented in Sect. 1.2 the system (3.198) can be transformed into the following equivalent forms.

1) Using the Drazin inverse matrix method we obtain

$$\begin{aligned} x_{1,i+1} &= \bar{A}_{1d} x_{1,i} + \bar{B}_{1d} u_i, \\ \bar{N}_d x_{2,i+1} &= x_{2,i} + \bar{B}_{2d} u_i, \\ y_i &= C_d x_i = C_d[x_{1,i} + x_{2,i}], \end{aligned} \tag{3.199}$$

where $x_{1,i}, x_{2,i} \in \mathbb{R}_+^n$ are given by (2.160c) and

$$\begin{aligned} \bar{A}_{1d} &= \bar{E}_d^D \bar{A}_d, \quad \bar{B}_{1d} = \bar{E}_d^D \bar{B}_d, \\ \bar{B}_{2d} &= (\mathbb{I}_n - \bar{E}_d \bar{E}_d^D) \bar{A}_d^D \bar{B}_d, \quad \bar{N}_d = (\mathbb{I}_n - \bar{E}_d \bar{E}_d^D) \bar{A}_d^D \bar{E}_d. \end{aligned} \tag{3.200}$$

The matrices $\bar{E}_d, \bar{A}_d \in \mathbb{R}^{n\times n}$ and $\bar{B}_d \in \mathbb{R}^{n\times 1}$ have the form (1.182b). Theorem 1.29 gives the positivity conditions of the system (3.199).

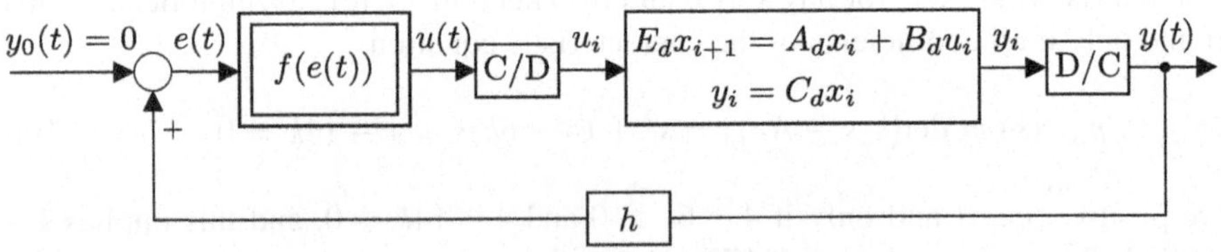

C/D - continuous-time to discrete-time converter
D/C - discrete-time to continuous-time converter

Fig. 3.3 Discrete-time nonlinear feedback system

2) Using the Weierstrass-Kronecker decomposition method we have

$$\begin{aligned} \bar{x}_{1,i+1} &= A_{1d}\bar{x}_{1,i} + B_{1d}u_i, \\ N_d\bar{x}_{2,i+1} &= x_{2,i} + B_{2d}u_i, \\ y_i &= C_{1d}\bar{x}_{1,i} + C_{2d}\bar{x}_{2,i}, \end{aligned} \tag{3.201}$$

where $\bar{x}_{1,i} \in \mathbb{R}_+^{n_1}, \bar{x}_{2,i} \in \mathbb{R}_+^{n_2}$ are given by (1.193) and the matrices $A_{1d} \in \mathbb{R}^{n_1 \times n_1}$, $B_{1d} \in \mathbb{R}^{n_1 \times 1}$, $B_{2d} \in \mathbb{R}^{n_2 \times 1}$, $N_d \in \mathbb{R}^{n_2 \times n_2}$, $C_{1d} \in \mathbb{R}^{1 \times n_1}$, $C_{2d} \in \mathbb{R}^{1 \times n_2}$ are determined by (1.192) and (1.196). Theorem 1.31 gives the positivity conditions of the system (3.201).

The characteristic of the nonlinear element is shown in Fig. 3.2 and it satisfies the condition (3.166). It is assumed that the positive descriptor linear part is asymptotically stable.

Definition 3.7. The positive descriptor discrete-time nonlinear system is called globally stable if it is asymptotically stable for all admissible nonnegative initial conditions $x_0 \in \mathbb{R}_+^n$.

The following theorem gives sufficient conditions for the global stability of the positive descriptor discrete-time nonlinear system.

Theorem 3.27. The discrete-time nonlinear system consisting of the positive and asymptotically stable descriptor linear part, the nonlinear element satisfying the condition (3.166) and the gain feedback h is globally stable if:

1) *(Drazin inverse matrix method case)*

a) $\bar{B}_{2d} = 0$ and there exists a matrix $G_d \in \mathbb{R}^{n \times n}$ such that

$$\bar{A}_{C1d} + G_d(\mathbb{I}_n - \bar{E}_d\bar{E}_d^D) \in \mathbb{R}_+^{n \times n} \tag{3.202}$$

is asymptotically stable, where

$$\bar{A}_{C1d} = \bar{A}_{1d} + kh\bar{B}_{1d}C_d; \tag{3.203}$$

b) $C_d(\mathbb{I}_n - \bar{E}_d\bar{E}_d^D) = 0$ and there exists a matrix $G_d \in \mathbb{R}^{n \times n}$ such that (3.202) is asymptotically stable and

$$-\bar{N}_d^j\bar{A}_{C2d}\bar{A}_{C1d}^j \in \mathbb{R}_+^{n \times n}, \quad j = 1, \dots, q-1, \tag{3.204}$$

where $\bar{A}_{C1d}$ is given by (3.203),

$$\bar{A}_{C2d} = kh\bar{B}_{2d}C_d \tag{3.205}$$

and q is the index of $\bar{E}_d$.

2) *(Weierstrass-Kronecker decomposition method case)*

a) $B_{2d} = 0$ and

$$A_{C1d} \in \mathbb{R}_+^{n_1 \times n_1} \tag{3.206}$$

is asymptotically stable, where

$$A_{C1d} = A_{1d} + khB_{1d}C_{1d}; \tag{3.207}$$

b) $C_{2d} = 0$, (3.206) is asymptotically stable and

$$-N_d^j A_{C2d} A_{C1d}^j \in \mathbb{R}_+^{n_2 \times n_1}, \quad j = 1, \ldots, \mu - 1, \tag{3.208}$$

where A_{C1d} is given by (3.207),

$$A_{C2d} = khB_{2d}C_{1d} \tag{3.209}$$

and μ is the nilpotency index of N_d.

Proof. The proof will be accomplished for the Weierstrass-Kronecker decomposition method. Taking into account (3.201) for $B_{2d} = 0$ we obtain

$$\begin{aligned} \bar{x}_{1,i+1} &= A_{1d}\bar{x}_{1,i} + B_{1d}u_i, \\ N_d\bar{x}_{2,i+1} &= \bar{x}_{2,i}, \\ y_i &= C_{1d}\bar{x}_{1,i} + C_{2d}\bar{x}_{2,i}. \end{aligned} \tag{3.210}$$

From the second equation of (3.210) it follows that $\bar{x}_{2,i} = 0$ since N_d is a nilpotent matrix. Assuming $u_i = ke_i$ and taking into account that $e_i = hy_i$ from (3.210) we obtain

$$\begin{aligned} \bar{x}_{1,i+1} &= (A_{1d} + khB_{1d}C_{1d})\bar{x}_{1,i}, \\ \bar{x}_{2,i} &= 0 \end{aligned} \tag{3.211}$$

As the Lyapunov function $V(\bar{x}_{1,i})$ we choose

$$V(\bar{x}_{1,i}) = \lambda^T \bar{x}_{1,i} \geq 0 \text{ for } \bar{x}_{1,i} \in \mathbb{R}_+^{n_1}, \tag{3.212}$$

where $\lambda = \begin{bmatrix} \lambda_1 & \ldots & \lambda_{n_1} \end{bmatrix}$ is a strictly positive vector, i.e. $\lambda_j > 0$, $j = 1, \ldots, n_1$. Using (3.210)–(3.212) we obtain

$$\begin{aligned} \Delta V(\bar{x}_{1,i}) &= V(\bar{x}_{1,i+1}) - V(\bar{x}_{1,i}) = \lambda^T \bar{x}_{1,i+1} - \lambda^T \bar{x}_{1,i} \\ &= \lambda^T [(A_{1d} - \mathbb{I}_{n_1})\bar{x}_{1,i} + B_{1d}u_i] \\ &= \lambda^T [(A_{1d} - \mathbb{I}_{n_1})\bar{x}_{1,i} + B_{1d}hf(e_i)] \\ &\leq \lambda^T [(A_{1d} + khB_{1d}C_{1d} - \mathbb{I}_{n_1})\bar{x}_{1,i}] \end{aligned} \tag{3.213}$$

since $u_i = f(e_i) \leq ke_i = kC_{1d}\bar{x}_{1,i}$. It is well-known [49] that elements on the main diagonal of the state matrix of a positive and asymptotically stable discrete-time system are less than 1. Therefore, from (3.213) it follows that $\Delta V(\bar{x}_{1,i}) < 0$ if the condition (3.206) is satisfied with $B_{2d} = 0$ and the nonlinear system is globally stable.

Now let us consider (3.201) for $C_{2d} = 0$. In this case we have

$$\begin{aligned} \bar{x}_{1,i+1} &= A_{1d}\bar{x}_{1,i} + B_{1d}u_i, \\ N_d\bar{x}_{2,i+1} &= \bar{x}_{2,i} + B_{2d}u_i, \\ y_i &= C_{1d}\bar{x}_{1,i}. \end{aligned} \tag{3.214}$$

Assuming $u_i = ke_i$ and taking into account that $e_i = hy_i$ from (3.214) we obtain

$$\begin{aligned} \bar{x}_{1,i+1} &= (A_{1d} + khB_{1d}C_{1d})\bar{x}_{1,i}, \\ N_d\bar{x}_{2,i+1} &= khB_{2d}C_{1d}\bar{x}_{1,i} + \bar{x}_{2,i}. \end{aligned} \tag{3.215}$$

The second equation of (3.215) can be solved in a similar way to the Eq. (1.195b). From (3.215) and (1.198) we have

$$\bar{x}_{2,i} = -\sum_{j=0}^{\mu-1} khN^j B_{2d}C_{1d}\bar{x}_{1,i+j}, \tag{3.216}$$

where μ is the nilpotency index of the matrix N_d. Let $A_{C1d} = A_{1d} + khB_{1d}C_{1d}$ and $A_{C2d} = khB_{2d}C_{1d}$. Taking into account that

$$\begin{aligned} \bar{x}_{1,i+1} &= A_{C1d}\bar{x}_{1,i}, \\ \bar{x}_{1,i+2} &= A_{C1d}\bar{x}_{1,i+1} = A_{C1d}^2\bar{x}_{1,i}, \\ &\vdots \\ \bar{x}_{1,i+j} &= A_{C1d}\bar{x}_{1,i+j-1} = A_{C1}^j\bar{x}_{1,i} \end{aligned} \tag{3.217}$$

from (3.216) we obtain

$$\bar{x}_{2,i} = -\sum_{j=0}^{\mu-1} N_d^j A_{C2d}A_{C1d}^j\bar{x}_{1,i}. \tag{3.218}$$

Choosing the Lyapunov function in the form (3.212) we get (3.213) and $\Delta V(\bar{x}_{1,i}) < 0$ if the condition (3.206) is satisfied with $C_{2d} = 0$. From (3.218) it follows that if $\lim_{i\to\infty} \bar{x}_{1,i} = 0$ then also $\lim_{i\to\infty} \bar{x}_{2,i} = 0$. The condition (3.208) ensures $\bar{x}_{2,i} \in \mathbb{R}_+^{n_2}$.

Therefore, the nonlinear system is globally stable.

The proof for the Drazin inverse matrix method can be accomplished similarly. Note that if $C_d(\mathbb{I}_n - \bar{E}_d\bar{E}_d^D) = 0$, then $C_d x_{2,i} = 0$ since $x_{2,i} = (\mathbb{I}_n - \bar{E}_d\bar{E}_d^D)x_i$.

The matrix $G_d \in \mathbb{R}^{n\times n}$ and the term $G_d(\mathbb{I}_n - \bar{E}_d\bar{E}_d^D)$ eliminates from the matrix $\bar{A}_{1d} = \bar{E}_d^D\bar{A}_d$ unimportant entires that are further canceled through multiplication by $x_{1,i} \in \operatorname{Im}\bar{E}_d\bar{E}_d^D$. □

Remark 3.3. Theorems 3.3 and 3.4 can be used to test the stability of (3.202) and (3.206).

Example 3.10. Consider the nonlinear system consisting of the descriptor discrete-time linear part (3.162) with

$$E_d = \begin{bmatrix} 1 & 2 & 0 \\ 0 & -1 & 0 \\ 0.5 & 1 & 0 \end{bmatrix},\quad A_d = \begin{bmatrix} 0.4 & 1.2 & 0 \\ 0 & -0.5 & 0 \\ 0.2 & 0.6 & -0.5 \end{bmatrix},\quad B_d = \begin{bmatrix} 3 \\ -1 \\ 1.5 \end{bmatrix},\quad C_d = \begin{bmatrix} 1 & 2 & 1 \end{bmatrix}, \tag{3.219}$$

the nonlinear element satisfying the condition (3.166) and the gain feedback $h = 0.5$. Find k satisfying (3.166) for which the nonlinear system is globally stable.

The matrix pencil (E_d, A_d) of (3.219) is regular since

$$\det[E_d z - A_d] = \begin{vmatrix} z-0.4 & 2z-1.2 & 0 \\ 0 & -z+0.5 & 0 \\ 0.5z-0.2 & z-0.6 & 0.5 \end{vmatrix} = -0.05(2z-1)(5z-2) \neq 0. \tag{3.220}$$

The characteristic equation of (3.219) is given by

$$p_{(E_d,A_d)}(z) = \det[E_d z - A_d] = -0.05(2z-1)(5z-2) = 0. \tag{3.221}$$

From (3.221) it follows that the pair (E_d, A_d) has two eigenvalues $z_1 = 0.5, z_2 = 0.4$ and by condition (1.239) the considered linear part is asymptotically stable. To check the positivity and global stability of the system we use two approaches as follows.

1) Drazin inverse matrix method case

We choose $c = 0$ and using (1.182b) we obtain

$$\bar{E}_d = [-A_d]^{-1}E_d = \begin{bmatrix} -2.5 & 1 & 0 \\ 0 & -2 & 0 \\ 0 & 0 & 0 \end{bmatrix},$$
$$\bar{A}_d = [-A_d]^{-1}A_d = \begin{bmatrix} -1 & 0 & 0 \\ 0 & -1 & 0 \\ 0 & 0 & -1 \end{bmatrix},\quad \bar{B}_d = [-A_d]^{-1}B_d = \begin{bmatrix} -1.5 \\ -2 \\ 0 \end{bmatrix} \tag{3.222}$$

and $q = 1$ since $\operatorname{rank}\bar{E} = \operatorname{rank}\bar{E}^2$. Next, from (B.1) and (B.2) we have

$$\bar{E}_d^D = \begin{bmatrix} -0.4 & -0.2 & 0 \\ 0 & -0.5 & 0 \\ 0 & 0 & 0 \end{bmatrix}, \quad \bar{A}_d^D = \bar{A}_d^{-1} = \bar{A}_d = \begin{bmatrix} -1 & 0 & 0 \\ 0 & -1 & 0 \\ 0 & 0 & -1 \end{bmatrix} \tag{3.223}$$

and

$$\bar{A}_{1d} = \bar{E}_d^D \bar{A}_d = \begin{bmatrix} 0.4 & 0.2 & 0 \\ 0 & 0.5 & 0 \\ 0 & 0 & 0 \end{bmatrix}, \quad \bar{B}_{1d} = \bar{E}_d^D \bar{B}_d = \begin{bmatrix} 1 \\ 1 \\ 0 \end{bmatrix},$$

$$\bar{B}_{2d} = (\mathbb{I}_3 - \bar{E}_d \bar{E}_d^D)\bar{A}_d^D \bar{B}_d = \begin{bmatrix} 0 \\ 0 \\ 0 \end{bmatrix}, \quad C_d(\mathbb{I}_3 - \bar{E}\bar{E}^D) = \begin{bmatrix} 0 & 0 & 1 \end{bmatrix} \tag{3.224}$$

$$\bar{N}_d = (\mathbb{I}_3 - \bar{E}_d \bar{E}_d^D)\bar{A}_d^D \bar{E}_d = \begin{bmatrix} 0 & 0 & 0 \\ 0 & 0 & 0 \\ 0 & 0 & 0 \end{bmatrix}.$$

By Theorem 1.29 the conditions (1.223) for the matrices (3.224) are satisfied and the descriptor linear part described by (3.219) is positive. Also, from (3.224) it follows that $\bar{B}_{2d} = 0$ and $C_d(\mathbb{I}_3 - \bar{E}\bar{E}^D) \neq 0$. Therefore, the case a) of Theorem 3.27 can be used to test the global stability of the nonlinear system. Taking into account (3.203), (3.219), (3.224) and $h = 0.5$ we obtain

$$\bar{A}_{C1d} = \bar{A}_{1d} + 0.5k\bar{B}_{1d}C_d = \begin{bmatrix} 0.5k + 0.4 & k + 0.2 & 0.5k \\ 0.5k & k + 0.5 & 0.5k \\ 0 & 0 & 0 \end{bmatrix}. \tag{3.225}$$

The matrix $\bar{A}_{C1d} \in \mathbb{R}_+^{3\times 3}$ for any $k > 0$ and by Theorem 3.3 it is asymptotically stable if and only if all coefficients of the characteristic equation

$$\begin{aligned} p_{\bar{A}_{C1d}}(z+1) &= \det[\mathbb{I}_3(z+1) - \bar{A}_{C1d}] \\ &= (z+1)[z^2 + (1.1 - 1.5k)z + 0.3 - 0.95k] = 0 \end{aligned} \tag{3.226}$$

are positive, i.e. if and only if $1.1 - 1.5k > 0$ and $0.3 - 0.95k > 0$, and this implies $k < 0.3158$. Therefore, the positive nonlinear system is globally stable if the characteristic $u(t) = f(e(t))$ of the nonlinear element satisfies the condition (3.166) for $k < 0.3158$.

2) Weierstrass-Kronecker decomposition method case

In this case

$$P_d = \begin{bmatrix} 1 & 2 & 0 \\ 0 & -1 & 0 \\ 1 & 0 & -2 \end{bmatrix}, \quad Q_d = \begin{bmatrix} 1 & 0 & 0 \\ 0 & 1 & 0 \\ 0 & 0 & 1 \end{bmatrix} \tag{3.227}$$

and

$$\begin{bmatrix} \mathbb{I}_{n_1} & 0 \\ 0 & N_d \end{bmatrix} = P_d E_d Q_d = \begin{bmatrix} 1 & 0 & 0 \\ 0 & 1 & 0 \\ 0 & 0 & 0 \end{bmatrix},$$

$$\begin{bmatrix} A_{1d} & 0 \\ 0 & \mathbb{I}_{n_2} \end{bmatrix} = P_d A_d Q_d = \begin{bmatrix} 0.4 & 0.2 & 0 \\ 0 & 0.5 & 0 \\ 0 & 0 & 1 \end{bmatrix}, \tag{3.228}$$

$$\begin{bmatrix} B_{1d} \\ B_{2d} \end{bmatrix} = P_d B_d = \begin{bmatrix} 1 \\ 1 \\ 0 \end{bmatrix}, \quad \begin{bmatrix} C_{1d} & C_{2d} \end{bmatrix} = C_d Q_d = \begin{bmatrix} 1 & 2 & 1 \end{bmatrix}$$

where $n_1 = 2$, $n_2 = 1$. By Theorem 1.31 the conditions (1.231) for the matrices (3.228) are satisfied and $Q_d \in \mathbb{R}_+^{3\times 3}$ is monomial. Thus, the descriptor linear part described by (3.219) is positive. Also, from (3.228) it follows that $B_2 = 0$ and $C_2 \neq 0$. Therefore, the case a) of Theorem 3.27 can be used to test the global stability of the nonlinear system. Taking into account (3.207), (3.228) and $h = 0.5$ we obtain

$$A_{C1d} = A_{1d} + 0.5kB_{1d}C_{1d} = \begin{bmatrix} 0.5k + 0.4 & k + 0.2 \\ 0.5k & k + 0.5 \end{bmatrix}. \tag{3.229}$$

The matrix $A_{C1d} \in \mathbb{R}_+^{2\times 2}$ for any $k > 0$ and by Theorem 3.3 it is asymptotically stable if and only if all coefficients of the characteristic equation

$$p_{A_{C1d}}(z+1) = \det[\mathbb{I}_2(z+1) - A_{C1}] = z^2 + (1.1 - 1.5k)z + 0.3 - 0.95k = 0 \tag{3.230}$$

are positive, i.e. if and only if $1.1 - 1.5k > 0$ and $0.3 - 0.95k > 0$, and this implies $k < 0.3158$. Therefore, the positive nonlinear system is globally stable if the characteristic $u(t) = f(e(t))$ of the nonlinear element satisfies the condition (3.166) for $k < 0.3158$.

3.3.3 Fractional Continuous-Time Systems

Consider the fractional continuous-time nonlinear feedback system shown in Fig. 3.4, which consists of a fractional positive descriptor linear part, a nonlinear element with characteristic $u(t) = f(e(t))$ and a positive scalar gain feedback h. The linear part is described by the equations

$$\begin{aligned} E\,{}_0D_t^\alpha x(t) &= Ax(t) + Bu(t), \quad 0 < \alpha < 1, \\ y(t) &= Cx(t), \end{aligned} \tag{3.231}$$

where $x(t) \in \mathbb{R}_+^n$, $u(t) \in \mathbb{R}_+$, $y(t) \in \mathbb{R}_+$ are state, input and output vectors, ${}_0D_t^\alpha$ is the Caputo derivative-integral operator defined by (2.8) and $E, A \in \mathbb{R}^{n\times n}$, $B \in \mathbb{R}^{n\times 1}$, $C \in \mathbb{R}^{1\times n}$. It is assumed that $\det E = 0$ and $\det[Es^\alpha - A] \neq 0$ for some $s^\alpha \in \mathbb{C}$.

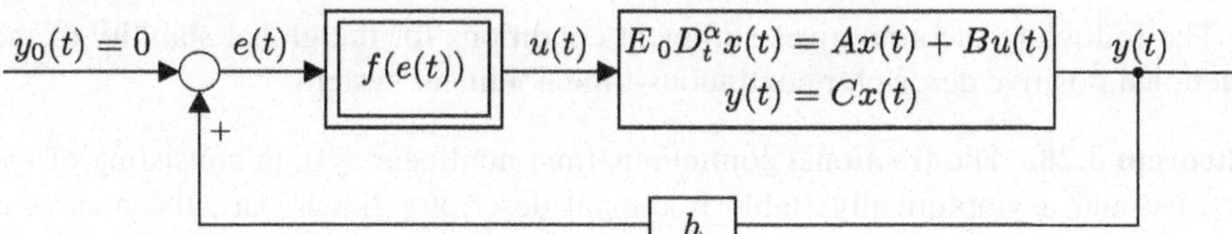

Fig. 3.4 Fractional continuous-time nonlinear feedback system

Following the considerations presented in Sect. 2.1 the system (3.231) can be transformed into the following equivalent forms.

1) Using the Drazin inverse matrix method we obtain

$$\begin{aligned} {}_0D_t^\alpha x_1(t) &= \bar{A}_1 x_1(t) + \bar{B}_1 u(t), \\ \bar{N}\,{}_0D_t^\alpha x_2(t) &= x_2(t) + \bar{B}_2 u(t), \\ y(t) &= Cx(t) = C[x_1(t) + x_2(t)], \end{aligned} \tag{3.232}$$

where $x_1(t), x_2(t) \in \mathbb{R}_+^n$ are given by (2.29c) and

$$\begin{aligned} \bar{A}_1 &= \bar{E}^D \bar{A}, \;\; \bar{B}_1 = \bar{E}^D \bar{B}, \\ \bar{B}_2 &= (\mathbb{I}_n - \bar{E}\bar{E}^D)\bar{A}^D \bar{B}, \;\; \bar{N} = (\mathbb{I}_n - \bar{E}\bar{E}^D)\bar{A}^D \bar{E}. \end{aligned} \tag{3.233}$$

The matrices $\bar{E}, \bar{A} \in \mathbb{R}^{n\times n}$ and $\bar{B} \in \mathbb{R}^{n\times 1}$ have the form (2.28b). Theorem 2.8 gives the positivity conditions of the system (3.232).

2) Using the Weierstrass-Kronecker decomposition method we have

$$\begin{aligned} {}_0D_t^\alpha \bar{x}_1(t) &= A_1 \bar{x}_1(t) + B_1 u(t), \\ N\,{}_0D_t^\alpha \bar{x}_2(t) &= \bar{x}_2(t) + B_2 u(t), \\ y(t) &= C_1 \bar{x}_1(t) + C_2 \bar{x}_2(t), \end{aligned} \tag{3.234}$$

where $\bar{x}_1(t) \in \mathbb{R}_+^{n_1}, \bar{x}_2(t) \in \mathbb{R}_+^{n_2}$ are given by (2.38) and the matrices $A_1 \in \mathbb{R}^{n_1\times n_1}$, $B_1 \in \mathbb{R}^{n_1\times 1}$, $B_2 \in \mathbb{R}^{n_2\times 1}$, $N \in \mathbb{R}^{n_2\times n_2}$, $C_1 \in \mathbb{R}^{1\times n_1}$, $C_2 \in \mathbb{R}^{1\times n_2}$ are determined by (2.37) and (2.41). Theorem 2.9 gives the positivity conditions of the system (3.234).

The characteristic of the nonlinear element is shown in Fig. 3.2 and it satisfies the condition (3.166). It is assumed that the fractional positive descriptor linear part is asymptotically stable.

Definition 3.8. The fractional positive descriptor continuous-time nonlinear system is called globally stable if it is asymptotically stable for all admissible nonnegative initial conditions $x_0 \in \mathbb{R}_+^n$.

The following theorem gives sufficient conditions for the global stability of the fractional positive descriptor continuous-time nonlinear system.

Theorem 3.28. The fractional continuous-time nonlinear system consisting of the positive and asymptotically stable fractional descriptor linear part, the nonlinear element satisfying the condition (3.166) and the gain feedback h is globally stable if:

1) *(Drazin inverse matrix method case)*

   a) $\bar{B}_2 = 0$ and there exists a matrix $G \in \mathbb{R}^{n\times n}$ such that

   $$\bar{A}_{C1} + G(\mathbb{I}_n - \bar{E}\bar{E}^D) \in \mathbb{M}_n \tag{3.235}$$

   is asymptotically stable, where

   $$\bar{A}_{C1} = \bar{A}_1 + kh\bar{B}_1 C; \tag{3.236}$$

   b) $C(\mathbb{I}_n - \bar{E}\bar{E}^D) = 0$ and there exists a matrix $G \in \mathbb{R}^{n\times n}$ such that (3.235) is asymptotically stable and

   $$-\bar{N}^j \bar{A}_{C2} \bar{A}_{C1}^j \in \mathbb{R}_+^{n\times n},\quad j = 1,\ldots,q-1, \tag{3.237}$$

   where $\bar{A}_{C1}$ is given by (3.236),

   $$\bar{A}_{C2} = kh\bar{B}_2 C \tag{3.238}$$

   and q is the index of $\bar{E}$.

2) *(Weierstrass-Kronecker decomposition method case)*

   a) $B_2 = 0$ and

   $$A_{C1} \in \mathbb{M}_{n_1} \tag{3.239}$$

   is asymptotically stable, where

   $$A_{C1} = A_1 + khB_1 C_1; \tag{3.240}$$

   b) $C_2 = 0$, (3.239) is asymptotically stable and

   $$-N^j A_{C2} A_{C1}^j \in \mathbb{R}_+^{n_2\times n_1},\quad j = 1,\ldots,\mu-1, \tag{3.241}$$

   where A_{C1} is given by (3.172),

   $$A_{C2} = khB_2 C_1 \tag{3.242}$$

   and μ is the nilpotency index of N.

Proof. The proof can be accomplished in a similar way as for Theorem 3.26. □

Remark 3.4. Theorems 3.5 and 3.6 can be used to test the stability of (3.235) and (3.239).

Example 3.11. Consider the nonlinear system consisting of the fractional descriptor continuous-time linear part (3.231) with

$$E = \begin{bmatrix} -1 & -0.5 & -0.5 \\ -2 & -0.5 & -0.5 \\ 0 & -0.5 & -0.5 \end{bmatrix}, \quad A = \begin{bmatrix} 2 & 1 & 0 \\ 4 & 0 & 1 \\ 0 & 2 & 1 \end{bmatrix}, \quad B = \begin{bmatrix} -4 \\ -4 \\ -4 \end{bmatrix}, \quad C = \begin{bmatrix} 3 & 1 & 2 \end{bmatrix}, \tag{3.243}$$

the nonlinear element satisfying the condition (3.166) and the gain feedback $h = 0.5$. Find k satisfying (3.166) for which the fractional nonlinear system is globally stable.

The matrix pencil (E, A) of (3.243) is regular since

$$\det[Es^\alpha - A] = \begin{vmatrix} -s^\alpha - 2 & -0.5s^\alpha - 1 & -0.5s^\alpha \\ -2s^\alpha - 4 & -0.5s^\alpha & -0.5s^\alpha - 1 \\ 0 & -0.5s^\alpha - 2 & -0.5s^\alpha - 1 \end{vmatrix} = s^{2\alpha} + 6s^\alpha + 8 \neq 0. \tag{3.244}$$

The characteristic equation of (3.243) is given by

$$p_{(E,A)}(s^\alpha) = \det[Es^\alpha - A] = s^{2\alpha} + 6s^\alpha + 8 = 0. \tag{3.245}$$

From (3.245) it follows that the pair (E, A) has two eigenvalues $s_1^\alpha = -4$, $s_2^\alpha = -2$ and by condition (2.77) the considered fractional linear part is asymptotically stable for any $\alpha \in (0; 1)$. To check the positivity and global stability of the system we use the Drazin inverse matrix method. From (2.28b) for $c = 0$ we have

$$\bar{E} = [-A]^{-1}E = \begin{bmatrix} 0.5 & 0.125 & 0.125 \\ 0 & 0.25 & 0.25 \\ 0 & 0 & 0 \end{bmatrix}, \quad \bar{A} = [-A]^{-1}A = \begin{bmatrix} -1 & 0 & 0 \\ 0 & -1 & 0 \\ 0 & 0 & -1 \end{bmatrix}, \quad \bar{B} = [-A]^{-1}B = \begin{bmatrix} 1 \\ 2 \\ 0 \end{bmatrix} \tag{3.246}$$

and $q = 1$ since $\operatorname{rank}\bar{E} = \operatorname{rank}\bar{E}^2$. Next, from (B.1) and (B.2) we have

$$\bar{E}^D = \begin{bmatrix} 2 & -1 & -1 \\ 0 & 4 & 4 \\ 0 & 0 & 0 \end{bmatrix}, \quad \bar{A}^D = \bar{A}^{-1} = \bar{A} = \begin{bmatrix} -1 & 0 & 0 \\ 0 & -1 & 0 \\ 0 & 0 & -1 \end{bmatrix} \tag{3.247}$$

and

$$\bar{A}_1 = \bar{E}^D\bar{A} = \begin{bmatrix} -2 & 1 & 1 \\ 0 & -4 & -4 \\ 0 & 0 & 0 \end{bmatrix}, \quad \bar{B}_1 = \bar{E}^D\bar{B} = \begin{bmatrix} 0 \\ 8 \\ 0 \end{bmatrix},$$
$$\bar{B}_2 = (\mathbb{I}_3 - \bar{E}\bar{E}^D)\bar{A}^D\bar{B} = \begin{bmatrix} 0 \\ 0 \\ 0 \end{bmatrix}, \quad \bar{N} = (\mathbb{I}_3 - \bar{E}\bar{E}^D)\bar{A}^D\bar{E} = \begin{bmatrix} 0 & 0 & 0 \\ 0 & 0 & 0 \\ 0 & 0 & 0 \end{bmatrix}, \tag{3.248}$$
$$C(\mathbb{I}_3 - \bar{E}\bar{E}^D) = \begin{bmatrix} 0 & 0 & 1 \end{bmatrix}.$$

By Theorem 2.8 there exists a matrix

$$G = \begin{bmatrix} 0 & 0 & -1 \\ 0 & -4 & 0 \\ 0 & 0 & -1 \end{bmatrix} \tag{3.249}$$

such that

$$\bar{A}_1 + G(\mathbb{I}_3 - \bar{E}\bar{E}^D) = \begin{bmatrix} -2 & 1 & 0 \\ 0 & -4 & 0 \\ 0 & 0 & -1 \end{bmatrix} \tag{3.250}$$

and the rest of the conditions (2.64) for the matrices (3.248) are satisfied. Thus, the fractional descriptor linear part described by (3.243) is positive. Also, from (3.248) it follows that $\bar{B}_2 = 0$ and $C(\mathbb{I}_3 - \bar{E}\bar{E}^D) \neq 0$. Therefore, the case a) of Theorem 3.28 can be used to test the global stability of the fractional nonlinear system. Taking into account (3.236), (3.243), (3.248) and $h = 0.5$ we obtain

$$\bar{A}_{C1} = \bar{A}_1 + 0.5k\bar{B}_1 C = \begin{bmatrix} -2 & 1 & 1 \\ 12k & 4k-4 & 8k-4 \\ 0 & 0 & 0 \end{bmatrix}. \tag{3.251}$$

From (3.235), (3.249) and (3.251) we have

$$\bar{A}_{C1} + G(\mathbb{I}_3 - \bar{E}\bar{E}^D) = \begin{bmatrix} -2 & 1 & 0 \\ 12k & 4k-4 & 8k \\ 0 & 0 & -1 \end{bmatrix} \in \mathbb{M}_3 \tag{3.252}$$

for any $k > 0$. By Theorem 3.5 the matrix $\bar{A}_{C1}$ is asymptotically stable if and only if all coefficients of the characteristic equation

$$p_{\bar{A}_{C1}}(s^\alpha) = \det[\mathbb{I}_3 s^\alpha - \bar{A}_{C1}] = s^\alpha[s^{2\alpha} + (6-4k)s^\alpha + 8 - 20k] = 0 \tag{3.253}$$

are positive, i.e. if and only if $6 - 4k > 0$ and $8 - 20k > 0$, and this implies $k < 0.4$. Therefore, the positive fractional nonlinear system is globally stable if the characteristic $u(t) = f(e(t))$ of the nonlinear element satisfies the condition (3.166) for $k < 0.4$.

3.3.4 Fractional Discrete-Time Systems

Consider the fractional discrete-time nonlinear feedback system shown in Fig. 3.5, which consists of a fractional positive descriptor linear part, a nonlinear element with characteristic $u_i = f(e_i)$ and a positive scalar gain feedback h. The linear part is described by the equations

$$\begin{aligned} E_d \Delta^\alpha x_{i+1} &= A_d x_i + B_d u_i, \quad i \in \mathbb{Z}_+, \quad 0 < \alpha < 1, \\ y_i &= C_d x_i, \end{aligned} \tag{3.254}$$

where $x_i \in \mathbb{R}_+^n$, $u_i \in \mathbb{R}_+$, $y_i \in \mathbb{R}_+$ are state, input and output vectors, Δ^α is the Grünwald-Letnikov fractional difference operator defined by (2.136) and $E_d, A_d \in \mathbb{R}^{n\times n}$, $B_d \in \mathbb{R}^{n\times 1}$, $C_d \in \mathbb{R}^{1\times n}$. It is assumed that $\det E_d = 0$ and $\det[E_d z - A_d] \neq 0$ for some $z \in \mathbb{C}$.

Substituting (2.136) into (3.254) we get

$$\begin{aligned} E_d x_{i+1} &= A_{d\alpha} x_i + \sum_{j=1}^{i} E_d c_j x_{i-j} + B_d u_i, \\ y_i &= C_d x_i, \end{aligned} \tag{3.255}$$

where $A_{d\alpha}$ is given by (2.140b) and the coefficients c_j are determined by (2.140c). Following the considerations presented in Sect. 2.2 the system (3.255) can be transformed into the following equivalent forms.

1) Using the Drazin inverse matrix method we obtain

$$\begin{aligned} \Delta^\alpha x_{1,i+1} &= \bar{A}_{1d} x_{1,i} + \bar{B}_{1d} u_i, \\ \bar{N}_d \Delta^\alpha x_{2,i+1} &= x_{2,i} + \bar{B}_{2d} u_i, \\ y_i &= C_d x_i = C_d[x_{1,i} + x_{2,i}], \end{aligned} \tag{3.256}$$

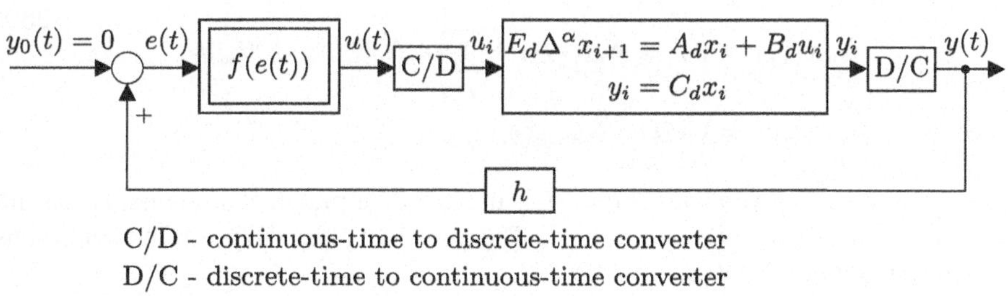

Fig. 3.5 Fractional discrete-time nonlinear feedback system

where $x_{1,i}, x_{2,i} \in \mathbb{R}_+^n$ are given by (2.160c) and

$$\begin{aligned} \bar{A}_{1d} &= \bar{E}_d^D \bar{A}_d, \quad \bar{B}_{1d} = \bar{E}_d^D \bar{B}_d, \\ \bar{B}_{2d} &= (\mathbb{I}_n - \bar{E}_d \bar{E}_d^D) \bar{A}_d^D \bar{B}_d, \quad \bar{N}_d = (\mathbb{I}_n - \bar{E}_d \bar{E}_d^D) \bar{A}_d^D \bar{E}_d. \end{aligned} \tag{3.257}$$

The matrices $\bar{E}_d, \bar{A}_d \in \mathbb{R}^{n \times n}$ and $\bar{B}_d \in \mathbb{R}^{n \times 1}$ have the form (2.157b). Substituting (2.136) into (3.256) we get

$$\begin{aligned} x_{1,i+1} &= \bar{E}_d^D \bar{A}_{d\alpha} x_{1,i} + \sum_{j=1}^{L} c_j x_{1,i-j} + \bar{B}_{1d} u_i, \\ \bar{N}_d x_{2,i+1} &= (\mathbb{I}_{n_2} + \bar{N}_d \alpha) x_{2,i} + \sum_{j=1}^{L} c_j \bar{N}_d x_{2,i-j} + \bar{B}_{2d} u_i, \\ y_i &= C_d x_i = C_d[x_{1,i} + x_{2,i}], \end{aligned} \tag{3.258}$$

where $\bar{A}_{d\alpha}$ is given by (2.159), L is the length of practical implementation and $x_{1,-k} = 0, x_{2,-k} = 0, k = 1, 2, \ldots$. Theorem 2.25 gives the positivity conditions of the system (3.258).

2) Using the Weierstrass-Kronecker decomposition method we have

$$\begin{aligned} \Delta^\alpha \bar{x}_{1,i+1} &= A_{1d} \bar{x}_{1,i} + B_{1d} u_i, \\ N_d \Delta^\alpha \bar{x}_{2,i+1} &= x_{2,i} + B_{2d} u_i, \\ y_i &= C_{1d} \bar{x}_{1,i} + C_{2d} \bar{x}_{2,i}, \end{aligned} \tag{3.259}$$

where $\bar{x}_{1,i} \in \mathbb{R}_+^{n_1}, \bar{x}_{2,i} \in \mathbb{R}_+^{n_2}$ are given by (1.193) and the matrices $A_{1d} \in \mathbb{R}^{n_1 \times n_1}$, $B_{1d} \in \mathbb{R}^{n_1 \times 1}$, $B_{2d} \in \mathbb{R}^{n_2 \times 1}$, $N_d \in \mathbb{R}^{n_2 \times n_2}$, $C_{1d} \in \mathbb{R}^{1 \times n_1}$, $C_{2d} \in \mathbb{R}^{1 \times n_2}$ are determined by (1.192) and (1.196). Substituting (2.136) into (3.259) we get

$$\begin{aligned} \bar{x}_{1,i+1} &= A_{1d\alpha} \bar{x}_{1,i} + \sum_{j=1}^{L} c_j \bar{x}_{1,i-j} + B_{1d} u_i, \\ N_d \bar{x}_{2,i+1} &= (\mathbb{I}_{n_2} + N_d \alpha) \bar{x}_{2,i} + \sum_{j=1}^{L} c_j N_d \bar{x}_{2,i-j} + B_{2d} u_i, \\ y_i &= C_{1d} \bar{x}_{1,i} + C_{2d} \bar{x}_{2,i}, \end{aligned} \tag{3.260}$$

where $A_{1d\alpha}$ is given by (2.179c), L is the length of practical implementation and $\bar{x}_{1,-k} = 0, \bar{x}_{2,-k} = 0, k = 1, 2, \ldots$. Theorem 2.26 gives the positivity conditions of the system (3.260).

The characteristic of the nonlinear element is shown in Fig. 3.2 and it satisfies the condition (3.166). It is assumed that the fractional positive descriptor linear part is asymptotically stable.

Definition 3.9. The fractional positive descriptor discrete-time nonlinear system is called:

1) globally stable for given length of practical implementation L if it is asymptotically stable for all admissible nonnegative initial conditions $x_0 \in \mathbb{R}_+^n$ and some finite L;
2) globally stable independent of the length of practical implementation L if it is asymptotically stable for all admissible nonnegative initial conditions $x_0 \in \mathbb{R}_+^n$ and $L \to \infty$.

The following theorems give sufficient conditions for the global stability of the fractional positive descriptor discrete-time nonlinear system.

Theorem 3.29. The fractional discrete-time nonlinear system consisting of the positive and asymptotically stable fractional descriptor linear part, the nonlinear element satisfying the condition (3.166) and the gain feedback h is globally stable for given length of practical implementation L if:

1) *(Drazin inverse matrix method case)*

   a) $\bar{B}_{2d} = 0$ and there exists a matrix $G_d \in \mathbb{R}^{n \times n}$ such that

$$\bar{A}_{C1d} + \bar{E}\bar{E}^D\alpha + \sum_{j=1}^{L} c_j \mathbb{I}_n + G_d(\mathbb{I}_n - \bar{E}_d\bar{E}_d^D) \in \mathbb{R}_+^{n \times n} \tag{3.261}$$

   is asymptotically stable, where

$$\bar{A}_{C1d} = \bar{A}_{1d} + kh\bar{B}_{1d}C_d; \tag{3.262}$$

   b) $C_d(\mathbb{I}_n - \bar{E}_d\bar{E}_d^D) = 0$ and there exists a matrix $G_d \in \mathbb{R}^{n \times n}$ such that (3.261) is asymptotically stable and

$$-\bar{N}_d^j \bar{A}_{C2d} \bar{A}_{C1d}^j \in \mathbb{R}_+^{n \times n}, \quad j = 1, \ldots, q-1, \tag{3.263}$$

   where $\bar{A}_{C1d}$ is given by (3.262),

$$\bar{A}_{C2d} = kh\bar{B}_{2d}C_d \tag{3.264}$$

   and q is the index of $\bar{E}_d$.

2) *(Weierstrass-Kronecker decomposition method case)*

   a) $B_{2d} = 0$ and

$$A_{C1d} + \mathbb{I}_{n_1}\left(\alpha + \sum_{j=1}^{L} c_j\right) \in \mathbb{R}_+^{n_1 \times n_1} \tag{3.265}$$

is asymptotically stable, where

$$A_{C1d} = A_{1d} + khB_{1d}C_{1d}; \tag{3.266}$$

b) $C_{2d} = 0$, (3.265) is asymptotically stable and

$$-N_d^j A_{C2d} A_{C1d}^j \in \mathbb{R}_+^{n_2 \times n_1}, \quad j = 1, \ldots, \mu - 1, \tag{3.267}$$

where A_{C1d} is given by (3.266),

$$A_{C2d} = khB_{2d}C_{1d} \tag{3.268}$$

and μ is the nilpotency index of N_d.

Proof. The proof will be accomplished for the Weierstrass-Kronecker decomposition method. Taking into account (3.260) for $B_{2d} = 0$ we obtain

$$\begin{aligned}
\bar{x}_{1,i+1} &= A_{1d\alpha}\bar{x}_{1,i} + \sum_{j=1}^{L} c_j \bar{x}_{1,i-j} + B_{1d}u_i, \\
N_d\bar{x}_{2,i+1} &= (\mathbb{I}_{n_2} + N_d\alpha)\bar{x}_{2,i} + \sum_{j=1}^{L} c_j N_d \bar{x}_{2,i-j}, \\
y_i &= C_{1d}\bar{x}_{1,i} + C_{2d}\bar{x}_{2,i},
\end{aligned} \tag{3.269}$$

From the second equation of (3.210) it follows that $\bar{x}_{2,i} = 0$ since N_d is a nilpotent matrix. Assuming $u_i = ke_i$ and taking into account that $e_i = hy_i$ from (3.269) we obtain

$$\begin{aligned}
\bar{x}_{1,i+1} &= (A_{1d\alpha} + khB_1C_1)\bar{x}_{1,i} + \sum_{j=1}^{L} c_j \bar{x}_{1,i-j}, \\
\bar{x}_{2,i} &= 0
\end{aligned} \tag{3.270}$$

By Remark 3.1 the positive fractional system (3.270) is asymptotically stable if and only if the positive system

$$\begin{aligned}
\bar{x}_{1,i+1} &= \left[A_{C1d} + \mathbb{I}_{n_1}\left(\alpha + \sum_{j=1}^{L} c_j\right)\right]\bar{x}_{1,i}, \\
\bar{x}_{2,i} &= 0
\end{aligned} \tag{3.271}$$

is asymptotically stable, where A_{C1d} is given by (3.266). As the Lyapunov function $V(\bar{x}_{1,i})$ we choose

$$V(\bar{x}_{1,i}) = \lambda^T \bar{x}_{1,i} \geq 0 \text{ for } \bar{x}_{1,i} \in \mathbb{R}_+^{n_1}, \tag{3.272}$$

where $\lambda = \begin{bmatrix} \lambda_1 & \dots & \lambda_{n_1} \end{bmatrix}$ is a strictly positive vector, i.e. $\lambda_j > 0,\ j = 1, \dots, n_1$. Using (3.269)–(3.272) we obtain

$$\begin{aligned}
\Delta V(\bar{x}_{1,i}) &= V(\bar{x}_{1,i+1}) - V(\bar{x}_{1,i}) = \lambda^T \bar{x}_{1,i+1} - \lambda^T \bar{x}_{1,i} \\
&= \lambda^T \left[\left(A_{1d\alpha} + \sum_{j=1}^{L} c_j \mathbb{I}_{n_1} - \mathbb{I}_{n_1} \right) \bar{x}_{1,i} + B_{1d} u_i \right] \\
&= \lambda^T \left[\left(A_{1d\alpha} + \sum_{j=1}^{L} c_j \mathbb{I}_{n_1} - \mathbb{I}_{n_1} \right) \bar{x}_{1,i} + B_{1d} h f(e_i) \right] \\
&\le \lambda^T \left\{ \left[A_{C1d} + \mathbb{I}_{n_1} \left(\alpha + \sum_{j=1}^{L} c_j \right) - \mathbb{I}_{n_1} \right] \bar{x}_{1,i} \right\}
\end{aligned} \tag{3.273}$$

since $u_i = f(e_i) \le k e_i = k C_{1d} \bar{x}_{1,i}$. It is well-known [49] that elements on the main diagonal of the state matrix of a positive and asymptotically stable discrete-time system are less than 1. Therefore, from (3.273) it follows that $\Delta V(\bar{x}_{1,i}) < 0$ if the condition (3.265) is satisfied with $B_{2d} = 0$ and the nonlinear system is globally stable.

Now let us consider (3.260) for $C_{2d} = 0$. In this case we have

$$\begin{aligned}
\bar{x}_{1,i+1} &= A_{1d\alpha} \bar{x}_{1,i} + \sum_{j=1}^{L} c_j \bar{x}_{1,i-j} + B_{1d} u_i, \\
N_d \bar{x}_{2,i+1} &= (\mathbb{I}_{n_2} + N_d \alpha) \bar{x}_{2,i} + \sum_{j=1}^{L} c_j N_d \bar{x}_{2,i-j} + B_{2d} u_i, \\
y_i &= C_{1d} \bar{x}_{1,i},
\end{aligned} \tag{3.274}$$

Assuming $u_i = k e_i$ and taking into account that $e_i = h y_i$ from (3.274) we obtain

$$\begin{aligned}
\bar{x}_{1,i+1} &= (A_{C1d} + \mathbb{I}_{n_1} \alpha) \bar{x}_{1,i} + \sum_{j=1}^{L} c_j \bar{x}_{1,i-j}, \\
N_d \bar{x}_{2,i} &= k h B_{2d} C_{1d} \bar{x}_{1,i} + (\mathbb{I}_{n_2} + N_d \alpha) \bar{x}_{2,i} + \sum_{j=1}^{L} c_j N_d \bar{x}_{2,i-j},
\end{aligned} \tag{3.275}$$

where A_{C1d} is given by (3.266). The second equation of (3.275) can be solved in a similar way to the Eq. (2.179b). From (3.275) and (2.181) we have

$$\bar{x}_{2,i} = -\sum_{j=0}^{\mu-1} N^j A_{C2d} \theta_{i,j}, \tag{3.276}$$

where μ is the nilpotency index of the matrix N_d, A_{C2d} is given by (3.268) and

$$\begin{aligned}
\theta_{i,0} &= \bar{x}_{1,i}, \\
\theta_{i,1} &= \theta_{i+1,0} - \alpha\theta_{i,0} - \sum_{j=1}^{L} c_j\theta_{i-j,0} = \bar{x}_{1,i+1} - \alpha\bar{x}_{1,i} - \sum_{j=1}^{L} c_j\bar{x}_{1,i-j}, \\
\theta_{i,2} &= \theta_{i+1,1} - \alpha\theta_{i,1} - \sum_{j=1}^{L} c_j\theta_{i-j,1} \\
&= \bar{x}_{1,i+2} - 2\alpha\bar{x}_{1,i+1} + \alpha^2\bar{x}_{1,i} + 2\alpha\sum_{j=1}^{L} c_j\bar{x}_{1,i-j} - \sum_{j=1}^{L+1} c_j\bar{x}_{1,i-j+1} \\
&\quad - \sum_{j=1}^{L} c_j\bar{x}_{1,i-j+1} + \sum_{j=1}^{L} c_j\sum_{l=1}^{L-j} c_l\bar{x}_{1,i-j-l} \\
&\ \vdots \\
\theta_{i,\mu-1} &= \theta_{i+1,\mu-2} - \alpha\theta_{i,\mu-2} - \sum_{j=1}^{L} c_j\theta_{i-j,\mu-2}.
\end{aligned} \tag{3.277}$$

Taking into account that

$$\begin{aligned}
\bar{x}_{1,i+1} &= (A_{C1d} + \mathbb{I}_{n_1}\alpha)\bar{x}_{1,i} + \sum_{j=1}^{L} c_j\bar{x}_{1,i-j}, \\
\bar{x}_{1,i+2} &= (A_{C1d} + \mathbb{I}_{n_1}\alpha)\bar{x}_{1,i+1} + \sum_{j=1}^{L+1} c_j\bar{x}_{1,i-j+1}, \\
&= (A_{C1d} + \mathbb{I}_{n_1}\alpha)^2\bar{x}_{1,i} + (A_{C1d} + \mathbb{I}_{n_1}\alpha)\sum_{j=1}^{L} c_j\bar{x}_{1,i-j} + \sum_{j=1}^{L+1} c_j\bar{x}_{1,i-j+1} \\
&\ \vdots \\
\bar{x}_{1,i+l} &= (A_{C1d} + \mathbb{I}_{n_1}\alpha)\bar{x}_{1,i+l-1} + \sum_{j=1}^{L+l-1} c_j\bar{x}_{1,i-j+l-1}
\end{aligned} \tag{3.278}$$

from (3.277) and (3.278) we obtain

$$\begin{aligned}\theta_{i,0} &= \bar{x}_{1,i},\\ \theta_{i,1} &= A_{C1d}\bar{x}_{1,i},\\ \theta_{i,2} &= A_{C1d}^2\bar{x}_{1,i},\\ &\vdots\\ \theta_{i,\mu-1} &= A_{C1d}^{\mu-1}\bar{x}_{1,i}.\end{aligned} \tag{3.279}$$

Using (3.276) and (3.279) we have

$$\bar{x}_{2,i} = -\sum_{j=0}^{\mu-1} N_d^j A_{C2d} A_{C1d}^j \bar{x}_{1,i}. \tag{3.280}$$

Choosing the Lyapunov function in the form (3.272) we get (3.273) and $\Delta V(\bar{x}_{1,i}) < 0$ if the condition (3.265) is satisfied with $C_{2d} = 0$. From (3.280) it follows that if $\lim_{i\to\infty} \bar{x}_{1,i} = 0$ then also $\lim_{i\to\infty} \bar{x}_{2,i} = 0$. The condition (3.267) ensures $\bar{x}_{2,i} \in \mathbb{R}_+^{n_2}$. Therefore, the nonlinear system is globally stable.

The proof for the Drazin inverse matrix method can be accomplished similarly. Note that if $C_d(\mathbb{I}_n - \bar{E}_d\bar{E}_d^D) = 0$, then $C_d x_{2,i} = 0$ since $x_{2,i} = (\mathbb{I}_n - \bar{E}_d\bar{E}_d^D)x_i$. The matrix $G_d \in \mathbb{R}^{n\times n}$ and the term $G_d(\mathbb{I}_n - \bar{E}_d\bar{E}_d^D)$ eliminates from the matrix $\bar{A}_{1d} = \bar{E}_d^D\bar{A}_d$ unimportant entires that are further canceled through multiplication by $x_{1,i} \in \mathrm{Im}\,\bar{E}_d\bar{E}_d^D$. □

Theorem 3.30. The fractional discrete-time nonlinear system consisting of the positive and asymptotically stable fractional descriptor linear part, the nonlinear element satisfying the condition (3.166) and the gain feedback h is globally stable independent of the length of practical implementation L if:

1) *(Drazin inverse matrix method case)*

   a) $\bar{B}_{2d} = 0$ and there exists a matrix $G_d \in \mathbb{R}^{n\times n}$ such that

$$\bar{A}_{C1d} + \mathbb{I}_n + (G_d - \mathbb{I}_n\alpha)(\mathbb{I}_n - \bar{E}_d\bar{E}_d^D) \in \mathbb{R}_+^{n\times n} \tag{3.281}$$

   is asymptotically stable, where $\bar{A}_{C1d}$ is given by (3.262);

   b) $C_d(\mathbb{I}_n - \bar{E}_d\bar{E}_d^D) = 0$ and there exists a matrix $G_d \in \mathbb{R}^{n\times n}$ such that (3.281) is asymptotically stable and

$$-\bar{N}_d^j\bar{A}_{C2d}\bar{A}_{C1d}^j \in \mathbb{R}_+^{n\times n},\ j = 1, \ldots, q-1, \tag{3.282}$$

   where $\bar{A}_{C1d}$ and $\bar{A}_{C2d}$ are given by (3.262) and (3.264), respectively and q is the index of $\bar{E}_d$.

2) *(Weierstrass-Kronecker decomposition method case)*

a) $B_{2d} = 0$ and

$$A_{C1d} + \mathbb{I}_{n_1} \in \mathbb{R}_+^{n_1 \times n_1} \tag{3.283}$$

is asymptotically stable, where $\bar{A}_{C1d}$ is given by (3.262);

b) $C_{2d} = 0$, (3.283) is asymptotically stable and

$$-N_d^j A_{C2d} A_{C1d}^j \in \mathbb{R}_+^{n_2 \times n_1}, \quad j = 1, \ldots, \mu - 1, \tag{3.284}$$

where $\bar{A}_{C1d}$ and $\bar{A}_{C2d}$ are given by (3.262) and (3.264), respectively and μ is the nilpotency index of N_d.

Proof. The proof follows immediately from Theorem 3.29 and (2.232). □

Remark 3.5. Theorems 3.9 and 3.10 can be used to test the stability of (3.261) and (3.265). Theorems 3.11 and 3.12 can be used to test the stability of (3.281) and (3.283).

Example 3.12. Consider the nonlinear system consisting of the fractional descriptor discrete-time linear part (3.254) with $\alpha = 0.3$ and

$$E_d = \begin{bmatrix} -10 & -30 & 0 \\ 0 & -10 & 0 \\ 0 & -20 & 0 \end{bmatrix}, \quad A_d = \begin{bmatrix} 1 & 1 & 0 \\ 0 & 1 & 0 \\ 0 & 0 & 1 \end{bmatrix}, \quad B_d = \begin{bmatrix} -1 \\ 0 \\ -1 \end{bmatrix}, \tag{3.285}$$
$$C_d = \begin{bmatrix} 2 & 2 & 0 \end{bmatrix},$$

the nonlinear element satisfying the condition (3.166) and the gain feedback $h = 0.5$. Find k satisfying (3.166) for which the fractional nonlinear system is globally stable independent of the length of practical implementation L.

The matrix pencil (E_d, A_d) of (3.285) is regular since

$$\det[E_d z - A_d] = \begin{vmatrix} -10z - 1 & -30z - 1 & 0 \\ 0 & -10z - 1 & 0 \\ 0 & -20z & -1 \end{vmatrix} = (-10z + 1)^2 \neq 0. \tag{3.286}$$

To check the asymptotic stability of the fractional linear part (3.285) we can use the approach given in Sect. 2.2.5. Therefore, by Theorem 2.28 we have to compute the characteristic equation given by

$$\det[E_d(z-1)^{0.3} z^{0.7} - A_d] = -[10(z-1)^{0.3} z^{0.7} + 1]^2 = 0. \tag{3.287}$$

The fractional linear part (3.285) with $\alpha = 0.3$ is asymptotically stable since roots of (3.287) are $z_1 = z_2 = -0.0367$. To check the positivity and global stability of the system we use the Drazin inverse matrix method. From (2.157b) for $c = 0$ we have

$$\bar{E}_d = [-A_d]^{-1}E_d = \begin{bmatrix} 10 & 20 & 0 \\ 0 & 10 & 0 \\ 0 & 20 & 0 \end{bmatrix}, \quad \bar{A}_d = [-A_d]^{-1}A_d = \begin{bmatrix} -1 & 0 & 0 \\ 0 & -1 & 0 \\ 0 & 0 & -1 \end{bmatrix},$$

$$\bar{A}_{d\alpha} = \bar{A}_d + 0.4\bar{E}_d = \begin{bmatrix} 2 & 6 & 0 \\ 0 & 2 & 0 \\ 0 & 6 & -1 \end{bmatrix} \quad \bar{B}_d = [-A_d]^{-1}B_d = \begin{bmatrix} 1 \\ 0 \\ 1 \end{bmatrix} \tag{3.288}$$

and $q = 1$ since $\operatorname{rank}\bar{E}_d = \operatorname{rank}\bar{E}_d^2$. Next, from (B.1) and (B.2) we have

$$\bar{E}_d^D = \begin{bmatrix} 0.1 & -0.2 & 0 \\ 0 & 0.1 & 0 \\ 0 & 0.2 & 0 \end{bmatrix}, \quad \bar{A}_d^D = \bar{A}_d^{-1} = \bar{A}_d = \begin{bmatrix} -1 & 0 & 0 \\ 0 & -1 & 0 \\ 0 & 0 & -1 \end{bmatrix} \tag{3.289}$$

and

$$\bar{A}_{1d} = \bar{E}_d^D\bar{A}_d = \begin{bmatrix} -0.1 & 0.2 & 0 \\ 0 & -0.1 & 0 \\ 0 & -0.2 & 0 \end{bmatrix}, \quad \bar{E}_d^D\bar{A}_{d\alpha} = \begin{bmatrix} 0.2 & 0.2 & 0 \\ 0 & 0.2 & 0 \\ 0 & 0.4 & 0 \end{bmatrix}$$

$$\bar{B}_{1d} = \bar{E}_d^D\bar{B}_d = \begin{bmatrix} 0.1 \\ 0 \\ 0 \end{bmatrix}, \quad \bar{B}_{2d} = (\mathbb{I}_3 - \bar{E}_d\bar{E}_d^D)\bar{A}_d^D\bar{B}_d = \begin{bmatrix} 0 \\ 0 \\ -1 \end{bmatrix}, \tag{3.290}$$

$$\bar{N}_d = (\mathbb{I}_3 - \bar{E}_d\bar{E}_d^D)\bar{A}_d^D\bar{E}_d = \begin{bmatrix} 0 & 0 & 0 \\ 0 & 0 & 0 \\ 0 & 0 & 0 \end{bmatrix}, \quad C_d(\mathbb{I}_3 - \bar{E}\bar{E}^D) = \begin{bmatrix} 0 & 0 & 0 \end{bmatrix}$$

By Theorem 2.25 the conditions (2.197) for the matrices (3.290) are satisfied and the fractional descriptor linear part described by (3.285) is positive. Also, from (3.290) it follows that $\bar{B}_{2d} \neq 0$ and $C_d(\mathbb{I}_3 - \bar{E}\bar{E}^D) = 0$. Therefore, the case b) of Theorem 3.30 can be used to test the global stability independent of the length of practical implementation L for the considered system. Taking into account (3.262), (3.264), (3.285), (3.290) and $h = 0.5$ we obtain

$$\begin{aligned} \bar{A}_{C1d} &= \bar{A}_{1d} + 0.5k\bar{B}_{1d}C_d = \begin{bmatrix} 2k-0.1 & 2k+0.2 & 0 \\ 0 & -0.1 & 0 \\ 0 & -0.2 & 0 \end{bmatrix}, \\ \bar{A}_{C2d} &= \bar{A}_{1d} + 0.5k\bar{B}_{1d}C_d = \begin{bmatrix} 0 & 0 & 0 \\ 0 & 0 & 0 \\ -10k & -10k & 0 \end{bmatrix}. \end{aligned} \tag{3.291}$$

From (3.281) for $G_d = 0$ and (3.291) we have

$$\bar{D}_{1,\infty} = \bar{A}_{C1d} + \mathbb{I}_3 - \alpha(\mathbb{I}_3 - \bar{E}\bar{E}^D) = \begin{bmatrix} 2k+0.9 & 2k+0.2 & 0 \\ 0 & 0.9 & 0 \\ 0 & 0.4 & 0.7 \end{bmatrix} \in \mathbb{R}_+^{3\times 3} \tag{3.292}$$

for any $k > 0$. By Theorem 3.11 the matrix $\bar{D}_{1,\infty}$ is asymptotically stable if and only if all coefficients of the characteristic equation

$$\begin{aligned} p_{\bar{D}_{1,\infty}}(z+1) &= \det[\mathbb{I}_3(z+1) - \bar{D}_{1,\infty}] \\ &= 0.1(z+0.3)[10z^2 + (2-20k)z + 0.1 - 2k] = 0 \end{aligned} \tag{3.293}$$

are positive, i.e. if and only if $2 - 20k > 0$ and $0.1 - 2k > 0$, and this implies $k < 0.05$. Taking into account (3.282) the matrix

$$-\bar{N}_d^0 \bar{A}_{C2d} \bar{A}_{C1d}^0 = -\bar{A}_{C2d} = \begin{bmatrix} 0 & 0 & 0 \\ 0 & 0 & 0 \\ 10k & 10k & 0 \end{bmatrix} \in \mathbb{R}_+^{3\times 3} \tag{3.294}$$

for any $k > 0$. Therefore, the positive fractional nonlinear system is globally stable independent of the length of practical implementation L if the characteristic $u(t) = f(e(t))$ of the nonlinear element satisfies the condition (3.166) for $k < 0.05$.

3.3.5 *Analysis of Global Stability of Descriptor Continuous-Time Nonlinear Feedback Systems by the Use of Nyquist Plots*

The following analysis presents an extension of Kudrewicz theorem [104] to positive descriptor continuous-time nonlinear systems (see also [78, 80]).

Consider the continuous-time nonlinear feedback system shown in Fig. 3.6, which consists of a positive descriptor linear part and a nonlinear element with characteristic $u(t) = f(e(t))$. The linear part is described by (3.162).

It is assumed that the linear part with transfer function $T(s)$ is positive, but not necessarily asymptotically stable. The characteristic of the nonlinear element is shown

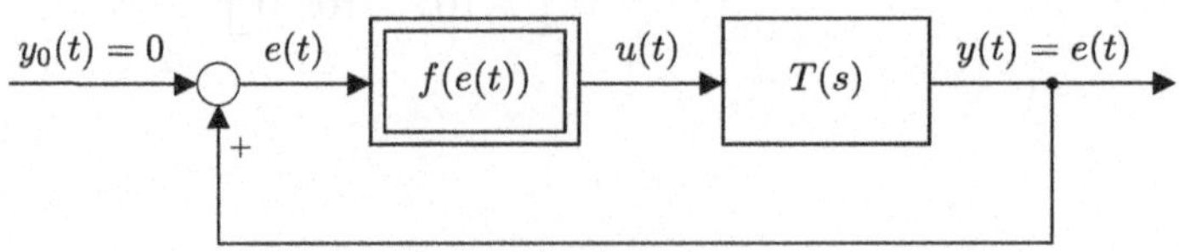

Fig. 3.6 Continuous-time nonlinear feedback system

in Fig. 3.2 and it satisfies the condition (3.166). It is well-known [47] that the transfer function of a descriptor system consists of two parts, i.e.

$$T(s) = C[Es - A]^{-1}B = T_{sp}(s) + T_p(s), \tag{3.295}$$

where $T_{sp}(s)$ is a strictly proper part and $T_p(s)$ is a polynomial part. The following equivalent forms of transfer matrices can be obtained using Drazin inverse matrix method and Weierstrass-Kronecker decomposition method.

Theorem 3.31. The transfer matrix of the descriptor continuous-time linear system (3.162) has the form:

1) for the Drazin inverse matrix method

$$T(s) = C[\mathbb{I}_n s - \bar{A}_1]^{-1}\bar{B}_1 + C[\bar{N}s - \mathbb{I}_n]^{-1}\bar{B}_2 = T_{sp}(s) + T_p(s); \tag{3.296}$$

2) for the Weierstrass-Kronecker decomposition method

$$T(s) = C_1[\mathbb{I}_{n_1} s - A_1]^{-1}B_1 + C_2[Ns - \mathbb{I}_{n_2}]^{-1}B_2 = T_{sp}(s) + T_p(s), \tag{3.297}$$

where the matrices $\bar{A}_1$, $\bar{B}_1$, $\bar{B}_2$, $\bar{N}$ are given by (3.164) and the matrices A_1, B_1, B_2, C_1, C_2, N are determined by (1.52) and (1.56).

Proof. Applying the Laplace transform to (3.163) with zero initial conditions we obtain

$$\begin{aligned} sX_1(s) &= \bar{A}_1X_1(s) + \bar{B}_1U(s), \\ \bar{N}sX_2(s) &= X_2(s) + \bar{B}_2U(s), \\ Y(s) &= C[X_1(s) + X_2(s)], \end{aligned} \tag{3.298}$$

and

$$\begin{aligned} X_1(s) &= [\mathbb{I}_n s - \bar{A}_1]^{-1}\bar{B}_1U(s), \\ X_2(s) &= [\bar{N}s - \mathbb{I}_n]^{-1}\bar{B}_2U(s), \\ Y(s) &= C[X_1(s) + X_2(s)]. \end{aligned} \tag{3.299}$$

From (3.299) we have

$$T(s) = \frac{Y(s)}{U(s)} = C[\mathbb{I}_n s - \bar{A}_1]^{-1}\bar{B}_1 + C[\bar{N}s - \mathbb{I}_n]^{-1}\bar{B}_2, \tag{3.300}$$

which is equal to (3.296). The proof for the transfer matrix (3.297) can be performed similarly applying the Laplace transform to (3.165) with zero initial conditions. □

In many cases, if the linear part is unstable, then by a suitable choice of the gain k_1 we may obtain (Fig. 3.7) an asymptotically stable positive linear part with the transfer function

$$T_1(s) = \frac{T(s)}{1 + k_1T(s)} \tag{3.301}$$

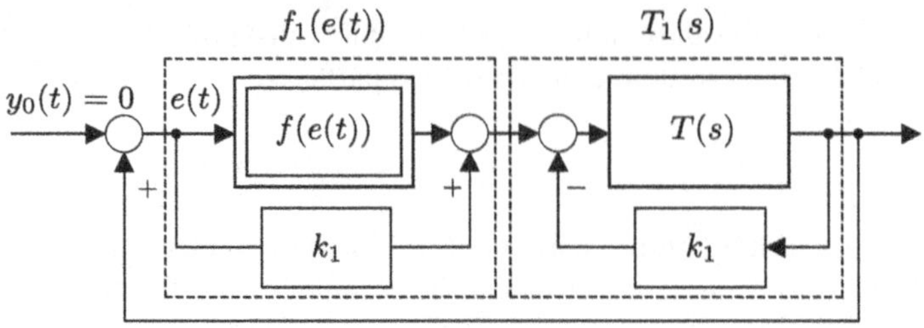

Fig. 3.7 Continuous-time nonlinear feedback system with the gain k_1

Fig. 3.8 Characteristic of the nonlinear element with the gain k_1

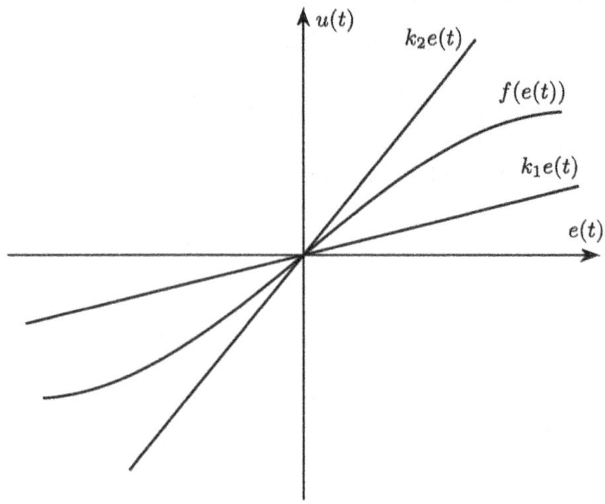

and a nonlinear element with the characteristic $f_1(e(t)) = f(e(t)) - k_1e(t)$ (Fig. 3.8) satisfying the condition

$$k_1 \leq \frac{f(e(t))}{e(t)} \leq k_2, \ \ k_2 < \infty. \tag{3.302}$$

According to Definition 3.6 the nonlinear continuous-time system is called globally stable if it is asymptotically stable for all admissible nonnegative initial conditions $x_0 \in \mathbb{R}^n_+$.

Definition 3.10. The circle in the plane $(P(\omega), Q(\omega))$ with center in the point $\left(-\frac{k_1+k_2}{2k_1k_2}, 0\right)$ and radius $\frac{k_2-k_1}{2k_1k_2}$ is called the $\left(-\frac{1}{k_1}, -\frac{1}{k_2}\right)$ circle (see Fig. 3.9).

Theorem 3.32. The continuous-time nonlinear system consisting of the positive and asymptotically stable descriptor linear part with the transfer function $T_1(s)$ and the nonlinear element satisfying the condition (3.302) is globally stable if

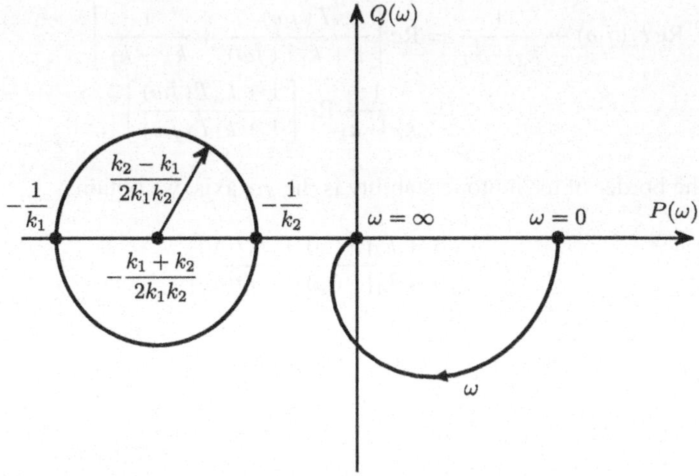

Fig. 3.9 Illustration of Definition 3.10 and Theorem 3.32

1) *(Drazin inverse matrix method case)*

$$\bar{B}_2 = 0 \text{ or } C(\mathbb{I}_n - \bar{E}\bar{E}^D) = 0; \tag{3.303}$$

2) *(Weierstrass-Kronecker decomposition method case)*

$$B_2 = 0 \text{ or } C_2 = 0 \tag{3.304}$$

and the Nyquist plot of $T_1(j\omega) = P(\omega) + jQ(\omega)$ of the linear part is located on the right-hand side of the circle $\left(-\frac{1}{k_1}, -\frac{1}{k_2}\right)$.

Proof. The proof will be accomplished for the Weierstrass-Kronecker decomposition method. From (3.295) and (3.301) we have

$$T_1(s) = \frac{T(s)}{1 + k_1 T(s)} = \frac{T_{sp}(s) + T_p(s)}{1 + k_1[T_{sp}(s) + T_p(s)]}, \tag{3.305}$$

where $T_{sp}(s) = C_1[\mathbb{I}_{n_1}s - A_1]^{-1}B_1$ and $T_p(s) = C_2[Ns - \mathbb{I}_{n_2}]^{-1}B_2$. If the condition (3.304) is satisfied, then from (3.305) we obtain

$$T_1(s) = \frac{T_{sp}(s)}{1 + k_1 T_{sp}(s)}. \tag{3.306}$$

From (3.306) we have

$$\text{Re}T_1(j\omega) + \frac{1}{k} > 0 \tag{3.307}$$

for $\omega \geq 0$ and $k = k_2 - k_1 > 0$. Taking into account that

$$\begin{aligned}\mathrm{Re}T_1(j\omega) + \frac{1}{k_2 - k_1} &= \mathrm{Re}\left[\frac{T(j\omega)}{1 + k_1 T(j\omega)} + \frac{1}{k_2 - k_1}\right] \\ &= \frac{1}{k_2 - k_1}\mathrm{Re}\left[\frac{1 + k_2 T(j\omega)}{1 + k_1 T(j\omega)}\right]\end{aligned} \tag{3.308}$$

and that the border of asymptotic stability is the $j\omega$ axis, we obtain

$$j\omega = \frac{1 + k_2[P(\omega) + jQ(\omega)]}{1 + k_1[P(\omega) + jQ(\omega)]} \tag{3.309}$$

or

$$j\omega\{1 + k_1[P(\omega) + jQ(\omega)]\} = 1 + k_2[P(\omega) + jQ(\omega)]. \tag{3.310}$$

From (3.310) we have

$$\begin{aligned}-\omega k_1 Q(\omega) &= 1 + k_2 P(\omega), \\ \omega[1 + k_1 P(\omega)] &= k_2 Q(\omega)\end{aligned} \tag{3.311}$$

and after elimination of ω we obtain

$$[1 + k_1 P(\omega)][1 + k_2 P(\omega)] + k_1 k_2 Q^2(\omega) = 0 \tag{3.312}$$

or

$$\frac{1}{k_1 k_2} + \frac{k_1 + k_2}{k_1 k_2} P(\omega) + P^2(\omega) + Q^2(\omega) = 0. \tag{3.313}$$

Note that (3.313) can be rewritten in the form of the equation

$$\left[P(\omega) + \frac{k_1 + k_2}{2k_1 k_2}\right]^2 + Q^2(\omega) = \left(\frac{k_2 - k_1}{2k_1 k_2}\right)^2, \tag{3.314}$$

which describes the circle $\left(-\frac{1}{k_1}, -\frac{1}{k_2}\right)$ (Fig. 3.9). This completes the proof. □

Example 3.13. (Continuation of Example 3.9) Consider the nonlinear system consisting of the positive descriptor continuous-time linear part (3.162) with (3.184) and the nonlinear element with the characteristic $u(t) = f(e(t))$ satisfying the condition (3.302). Using (3.184) and (3.295) we obtain the transfer function

$$T(s) = C[Es - A]^{-1}B = \frac{12s + 28}{s^2 + 4s + 4}, \tag{3.315a}$$

where

$$T_{sp}(s) = T(s) = \frac{12s + 28}{s^2 + 4s + 4}, \quad T_p(s) = 0. \tag{3.315b}$$

To check the global stability of the system we use two approaches as follows.

1) Drazin inverse matrix method case

From (3.189) it follows that $C(\mathbb{I}_3 - \bar{E}\bar{E}^D) = 0$. Using (3.184), (3.189) and (3.296) we have

$$T(s) = T_{sp}(s) = C[\mathbb{I}_3 s - \bar{A}_1]^{-1}\bar{B}_1 = \frac{12s + 28}{s^2 + 4s + 4} \tag{3.316}$$

and $T_p(s) = 0$, which is equal to (3.315). The Nyquist plot of

$$T(j\omega) = \frac{12j\omega + 28}{(j\omega)^2 + 4(j\omega) + 4} \tag{3.317}$$

is shown in Fig. 3.10 and it is located in the fourth quarter of the plane $(P(\omega), Q(\omega))$. Therefore, by Theorem 3.32 the nonlinear system is globally stable for all nonlinear elements with characteristic $u(t) = f(e(t))$ located in the first and third quarter (Fig. 3.8) since for any positive $k_2 > k_1 \geq 0$ the Nyquist plot is located on the right-hand side of the circle $\left(-\frac{1}{k_1}, -\frac{1}{k_2}\right)$.

2) Weierstrass-Kronecker decomposition method case

From (3.194) it follows that $C_2 = 0$. Using (3.194) and (3.297) we have

$$T(s) = T_{sp}(s) = C_1[\mathbb{I}_2 s - A_1]^{-1}B_1 = \frac{12s + 28}{s^2 + 4s + 4} \tag{3.318}$$

and $T_p(s) = 0$, which is equal to (3.315) and (3.316). The conclusions are the same as for the Drazin inverse matrix method case.

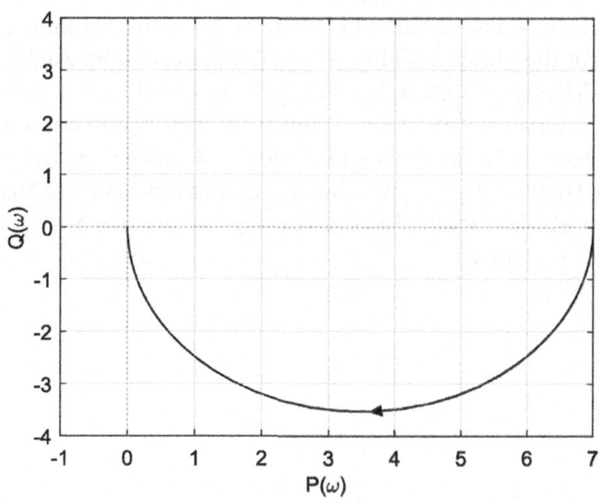

Fig. 3.10 Nyquist plot of (3.317)

Remark 3.6. Similar analysis can be performed for fractional positive descriptor continuous-time nonlinear systems with the linear part described by the transfer function

$$T(s^\alpha) = C[Es^\alpha - A]^{-1}B = T_{sp}(s^\alpha) + T_p(s^\alpha), \tag{3.319}$$

where $T_{sp}(s^\alpha)$ is a strictly proper part and $T_p(s^\alpha)$ is a polynomial part.

3.4 Concluding Remarks

In this chapter stability of positive descriptor systems have been investigated. In the subsequent sections stability tests for positive linear systems, stability of positive interval systems and stability of nonlinear systems with positive linear parts have been analyzed.

In positive descriptor systems some other methods of stability analysis can be applied, in contrast to nonpositive descriptor systems, where an approach based on the location of the roots of the characteristic equation is mostly used. To make use of the first of the presented methods we have to check whether the coefficients of the corresponding characteristic equations are positive (Theorems 3.1, 3.3, 3.5, 3.9 and 3.11). The second of the methods is based on the search for some strictly positive vector and it results from the Lyapunov method (Theorems 3.2, 3.4, 3.6, 3.10 and 3.12).

To investigate the stability of positive descriptor interval systems the approach based on the convex linear combination of matrices can be applied (Theorems 3.14, 3.18, 3.20, 3.22 and 3.23). Also, an extension of the Kharitonov theorem (Theorem 3.15) can be used, from which it follows that positive descriptor interval systems are asymptotically stable if and only if the lower bounds of corresponding characteristic equations are positive (Theorems 3.16, 3.19, 3.21, 3.24 and 3.25).

The positive descriptor nonlinear system is called globally stable if it is asymptotically stable for all admissible nonnegative initial conditions. Sufficient conditions for the global stability of such systems can be established using Lyapunov method (Theorems 3.26, 3.27, 3.28, 3.29 and 3.30).

Analysis of global stability of descriptor nonlinear feedback systems can be performed using Nyquist plots, i.e. using an extension of Kudrewicz theorem (Theorem 3.32). The results presented in Sect. 3.3.5 can be easily extended to fractional and interval systems using the approaches given in the previous sections of this chapter.

Appendix A
Extensions of the Cayley-Hamilton Theorem for Descriptor Linear Systems

A.1 Cayley-Hamilton Theorem for Descriptor Linear Systems with Commuting Matrices

Consider the descriptor linear system (1.1a) with the commuting matrices

$$EA = AE. \tag{A.1}$$

Theorem A.1. Let the condition (1.2) be satisfied and

$$\det[E\lambda - A] = \sum_{k=0}^{r} a_k \lambda^k. \tag{A.2}$$

Then the matrices E, A satisfy the equation

$$\sum_{k=0}^{r} a_k A^k E^{n-k} = 0. \tag{A.3}$$

Proof. Let

$$[E\lambda - A]_{ad} = B_{n-1}\lambda^{n-1} + \ldots + B_1\lambda + B_0 \tag{A.4}$$

be the adjoint matrix of the matrix $[E\lambda - A]$. From the definition of the inverse matrix and (A.2), (A.4) we have

$$[E\lambda - A][B_{n-1}\lambda^{n-1} + \ldots + B_1\lambda + B_0] = \mathbb{I}_n(a_r\lambda^r + a_{r-1}\lambda^{r-1} + \ldots + a_1\lambda + a_0). \tag{A.5}$$

T. Kaczorek and K. Borawski, *Descriptor Systems of Integer and Fractional Orders*, Studies in Systems, Decision and Control 367,
https://doi.org/10.1007/978-3-030-72480-1

Comparing the coefficients at the same powers of λ in (A.5) we obtain

$$\begin{bmatrix} E & 0 & 0 & \dots & 0 & 0 & 0 \\ -A & E & 0 & \dots & 0 & 0 & 0 \\ 0 & -A & E & \dots & 0 & 0 & 0 \\ \vdots & \vdots & \vdots & \ddots & \vdots & \vdots & \vdots \\ 0 & 0 & 0 & \dots & -A & E & 0 \\ 0 & 0 & 0 & \dots & 0 & -A & E \\ 0 & 0 & 0 & \dots & 0 & 0 & -A \end{bmatrix} \begin{bmatrix} B_{n-1} \\ B_{n-2} \\ B_{n-3} \\ \vdots \\ B_2 \\ B_1 \\ B_0 \end{bmatrix} = \begin{bmatrix} 0 \\ \vdots \\ 0 \\ a_r \mathbb{I}_n \\ \vdots \\ a_1 \mathbb{I}_n \\ a_0 \mathbb{I}_n \end{bmatrix}. \tag{A.6}$$

Premultiplying (A.6) by the matrix

$$\left[\, A^n \; A^{n-1}E \; A^{n-2}E \; \dots \; AE^{n-1} \; E^n \,\right] \tag{A.7}$$

and using (A.1) we obtain the Eq. (A.3). □

Example A.1. Consider the pair of matrices

$$E = \begin{bmatrix} 0 & 1 & 0 \\ 0 & 0 & 1 \\ 0 & 1 & 0 \end{bmatrix}, \quad A = \begin{bmatrix} a & 0 & 0 \\ 0 & a & 0 \\ 0 & 0 & a \end{bmatrix}, \quad a \neq 0, \tag{A.8}$$

which satisfies the condition (A.1) for any value of a. The characteristic polynomial of the pair (A.8) has the form

$$\det[E\lambda - A] = \begin{vmatrix} -a & \lambda & 0 \\ 0 & -a & \lambda \\ 0 & \lambda & -a \end{vmatrix} = a(\lambda^2 - a^2) \tag{A.9}$$

and its characteristic equation is

$$\lambda^2 - a^2 = 0. \tag{A.10}$$

In this case we have

$$r = 2 \text{ and } a_2 = 1, \ a_1 = 0, \ a_0 = -a^2. \tag{A.11}$$

Taking into account that

$$E^3 = \begin{bmatrix} 0 & 1 & 0 \\ 0 & 0 & 1 \\ 0 & 1 & 0 \end{bmatrix}, \quad A^2E = \begin{bmatrix} 0 & a^2 & 0 \\ 0 & 0 & a^2 \\ 0 & a^2 & 0 \end{bmatrix} \tag{A.12}$$

and using (A.11) we obtain

$$\sum_{k=0}^{2} a_k A^k E^{n-k} = -a^2 E^3 + A^2 E$$
$$= -a^2 \begin{bmatrix} 0 & 1 & 0 \\ 0 & 0 & 1 \\ 0 & 1 & 0 \end{bmatrix} + \begin{bmatrix} 0 & a^2 & 0 \\ 0 & 0 & a^2 \\ 0 & a^2 & 0 \end{bmatrix} \tag{A.13}$$
$$= \begin{bmatrix} 0 & 0 & 0 \\ 0 & 0 & 0 \\ 0 & 0 & 0 \end{bmatrix}.$$

Therefore, the matrices (A.8) satisfy their characteristic Eq. (A.10).

A.2 Cayley-Hamilton Theorem for Descriptor Linear Systems with Noncommuting Matrices

Consider the descriptor linear system (1.1a) with the noncommuting matrices

$$EA \neq AE. \tag{A.14}$$

Let the condition (1.2) hold and a constant c be chosen such that $\det[Ec - A] \neq 0$. Following Lemma 1.1 the matrices

$$\bar{E} = [Ec - A]^{-1} E, \ \ \bar{A} = [Ec - A]^{-1} A \tag{A.15}$$

satisfy the condition

$$\bar{E}\bar{A} = \bar{A}\bar{E}. \tag{A.16}$$

Theorem A.2. Let (A.2) be the characteristic polynomial of the pair satisfying the condition (A.16). Then

$$\sum_{k=0}^{r} b_k \bar{E}^{n-k} = 0 \tag{A.17}$$

or

$$\sum_{k=0}^{n} c_k \bar{A}^k = 0, \tag{A.18}$$

where

$$b_k = (-1)^k \sum_{j=k}^{r} \binom{j}{k} c^{j-k} a_j \tag{A.19}$$

and

$$c_k = \sum_{j=n-k}^{n} \binom{j}{n-k} \frac{a_{n-j}}{c^j}, \quad a_k = 0 \text{ for } k > r. \tag{A.20}$$

Proof. By Theorem A.1 for the pair $(\bar{E}, \bar{A})$ we have

$$\sum_{k=0}^{r} a_k \bar{A}^k \bar{E}^{n-k} = 0. \tag{A.21}$$

From (A.15) we get

$$\bar{E}c - \bar{A} = [Ec - A]^{-1}[Ec - A] = \mathbb{I}_n, \tag{A.22}$$

and

$$\bar{A} = \bar{E}c - \mathbb{I}_n. \tag{A.23}$$

Substitution of (A.23) into (A.21) yields

$$\sum_{k=0}^{r} a_k (\bar{E}c - \mathbb{I}_n)^k \bar{E}^{n-k} = \sum_{k=0}^{r} b_k \bar{E}^{n-k} = 0, \tag{A.24}$$

where the coefficients b_k are defined by (A.19). Similarly, substituting

$$\bar{E} = \frac{1}{c}[\bar{A} + \mathbb{I}_n] \tag{A.25}$$

(obtained from (A.23)) into (A.21) we obtain

$$\sum_{k=0}^{r} a_k \bar{A}^k \left[\frac{1}{c}(\bar{A} + \mathbb{I}_n)\right]^{n-k} = \sum_{k=0}^{n} c_k \bar{A}^k = 0, \tag{A.26}$$

where the coefficients c_k are defined by (A.20). □

Example A.2. Consider the pair of matrices

$$E = \begin{bmatrix} 1 & 0 & 0 \\ 0 & 1 & 0 \\ 0 & 0 & 0 \end{bmatrix}, \quad A = \begin{bmatrix} 0 & -1 & 0 \\ 0 & -1 & -2 \\ 0 & 0 & -1 \end{bmatrix}, \tag{A.27}$$

which satisfies the condition (A.14) since

$$EA = \begin{bmatrix} 0 & -1 & 0 \\ 0 & -1 & -2 \\ 0 & 0 & 0 \end{bmatrix}, \quad AE = \begin{bmatrix} 0 & -1 & 0 \\ 0 & -1 & 0 \\ 0 & 0 & 0 \end{bmatrix}. \tag{A.28}$$

Choosing in (A.15) $c = 1$ we obtain

$$\bar{E} = [E - A]^{-1}E = \begin{bmatrix} 1 & -0.5 & 0 \\ 0 & 0.5 & 0 \\ 0 & 0 & 0 \end{bmatrix} \tag{A.29}$$

and

$$\bar{A} = [E - A]^{-1}A = \begin{bmatrix} 0 & -0.5 & 0 \\ 0 & -0.5 & 0 \\ 0 & 0 & -1 \end{bmatrix}. \tag{A.30}$$

The characteristic polynomial of the pair $(\bar{E}, \bar{A})$ has the form

$$\det[\bar{E}\lambda - \bar{A}] = \begin{vmatrix} \lambda & -0.5\lambda + 0.5 & 0 \\ 0 & 0.5\lambda + 0.5 & 0 \\ 0 & 0 & 1 \end{vmatrix} = 0.5\lambda^2 + 0.5\lambda \tag{A.31}$$

and

$$a_2 = 0.5, \quad a_1 = 0.5, \quad a_0 = 0. \tag{A.32}$$

Using (A.21), (A.32) and (A.29), (A.30) we obtain

$$\begin{aligned} \sum_{k=0}^{2} a_k \bar{A}^k \bar{E}^{3-k} &= a_1 \bar{A}\bar{E}^2 + a_2 \bar{A}^2 \bar{E} \\ &= 0.5 \begin{bmatrix} 0 & -0.5 & 0 \\ 0 & -0.5 & 0 \\ 0 & 0 & -1 \end{bmatrix} \begin{bmatrix} 1 & -0.75 & 0 \\ 0 & 0.25 & 0 \\ 0 & 0 & 0 \end{bmatrix} \\ &+ 0.5 \begin{bmatrix} 0 & 0.25 & 0 \\ 0 & 0.25 & 0 \\ 0 & 0 & -1 \end{bmatrix} \begin{bmatrix} 1 & -0.5 & 0 \\ 0 & 0.5 & 0 \\ 0 & 0 & 0 \end{bmatrix} \\ &= \begin{bmatrix} 0 & 0 & 0 \\ 0 & 0 & 0 \\ 0 & 0 & 0 \end{bmatrix}. \end{aligned} \tag{A.33}$$

Using (A.32) and (A.19) we obtain

$$b_0 = a_1 + a_2 = 1, \quad b_1 = -(a_1 + 2a_2) = -1.5, \quad b_2 = a_2 = 0.5 \tag{A.34}$$

and from (A.17), (A.34) we have

$$\sum_{k=0}^{2} b_k \bar{E}^{3-k} = b_0\bar{E}^3 + b_1\bar{E}^2 + b_2\bar{E}$$
$$= \begin{bmatrix} 1 & -0.5 & 0 \\ 0 & 0.5 & 0 \\ 0 & 0 & 0 \end{bmatrix}^3 - 1.5\begin{bmatrix} 1 & -0.5 & 0 \\ 0 & 0.5 & 0 \\ 0 & 0 & 0 \end{bmatrix}^2 \tag{A.35}$$
$$+ 0.5\begin{bmatrix} 1 & -0.5 & 0 \\ 0 & 0.5 & 0 \\ 0 & 0 & 0 \end{bmatrix} = \begin{bmatrix} 0 & 0 & 0 \\ 0 & 0 & 0 \\ 0 & 0 & 0 \end{bmatrix}.$$

Similarly, using (A.20) and (A.32) we obtain

$$c_0 = 0,\ \ c_1 = a_1 = 0.5,\ \ c_2 = 2a_1 + a_2 = 1.5,\ \ c_3 = a_1 + a_2 = 1 \tag{A.36}$$

and from (A.18), (A.36) we have

$$\sum_{k=0}^{3} c_k \bar{A}^k = c_1\bar{A} + c_2\bar{A}^2 + c_3\bar{A}^3$$
$$= 0.5\begin{bmatrix} 0 & -0.5 & 0 \\ 0 & -0.5 & 0 \\ 0 & 0 & -1 \end{bmatrix} + 1.5\begin{bmatrix} 0 & -0.5 & 0 \\ 0 & -0.5 & 0 \\ 0 & 0 & -1 \end{bmatrix}^2 \tag{A.37}$$
$$+ \begin{bmatrix} 0 & -0.5 & 0 \\ 0 & -0.5 & 0 \\ 0 & 0 & -1 \end{bmatrix}^3 = \begin{bmatrix} 0 & 0 & 0 \\ 0 & 0 & 0 \\ 0 & 0 & 0 \end{bmatrix}.$$

Therefore, the matrices (A.29) and (A.30) satisfy their characteristic equations.

A.3 Cayley-Hamilton Theorem for Drazin Inverse Matrices

Theorem A.3. If

$$\det[\mathbb{I}_n\lambda - E] = \lambda^n + a_{n-1}\lambda^{n-1} + \ldots + a_1\lambda, \tag{A.38}$$

then

$$a_1 E^D + a_2\left(E^D\right)^2 + \ldots + a_{n-1}\left(E^D\right)^{n-1} + \left(E^D\right)^n = 0, \tag{A.39}$$

where $E^D \in \mathbb{R}^{n\times n}$ is the Drazin inverse of the matrix E.

Proof. Using (A.38) and the classical Cayley-Hamilton theorem we obtain

$$E^n + a_{n-1}E^{n-1} + \ldots + a_2E^2 + a_1E = 0. \tag{A.40}$$

Premultiplying and postmultiplying (A.40) by the Drazin inverse matrix E^D we obtain

$$E^DE^nE^D + a_{n-1}E^DE^{n-1}E^D + \ldots + a_2E^DE^2E^D + a_1E^DEE^D = 0 \tag{A.41}$$

and using (1.26a), (1.26b) we have

$$E^DE^{n-1} + a_{n-1}E^DE^{n-2} + \ldots + a_2E^DE + a_1E^D = 0 \tag{A.42}$$

since

$$E^DE^kE^D = E^DEE^DE^{k-1} = E^DE^{k-1} \text{ for } k = 1, 2, \ldots, n. \tag{A.43}$$

Postmultiplying (A.42) by E^D and using (A.43) we obtain

$$E^DE^{n-2} + a_{n-1}E^DE^{n-3} + \ldots + a_2E^D + a_1\left(E^D\right)^2 = 0. \tag{A.44}$$

Repeating $n-2$ times this procedure we obtain (A.39). □

Example A.3. The Drazin inverse of the singular matrix

$$E = \begin{bmatrix} 1 & 0 & -1 \\ 0 & 1 & 0 \\ 0 & -1 & 0 \end{bmatrix} \tag{A.45}$$

has the form

$$E^D = \begin{bmatrix} 1 & -2 & -1 \\ 0 & 1 & 0 \\ 0 & -1 & 0 \end{bmatrix}. \tag{A.46}$$

The characteristic polynomial of (A.45) is given by

$$\det[\mathbb{I}_3\lambda - E] = \begin{vmatrix} \lambda - 1 & 0 & 1 \\ 0 & \lambda - 1 & 0 \\ 0 & 1 & s \end{vmatrix} = \lambda^3 - 2\lambda^2 + \lambda. \tag{A.47}$$

From the classical Cayley-Hamilton theorem we have

$$E^3 - 2E^2 + E = \begin{bmatrix} 1 & 0 & -1 \\ 0 & 1 & 0 \\ 0 & -1 & 0 \end{bmatrix}^3 - 2\begin{bmatrix} 1 & 0 & -1 \\ 0 & 1 & 0 \\ 0 & -1 & 0 \end{bmatrix}^2 + \begin{bmatrix} 1 & 0 & -1 \\ 0 & 1 & 0 \\ 0 & -1 & 0 \end{bmatrix} = \begin{bmatrix} 0 & 0 & 0 \\ 0 & 0 & 0 \\ 0 & 0 & 0 \end{bmatrix}. \tag{A.48}$$

Applying Theorem A.3 to (A.46) we obtain

$$E^D - 2\left(E^D\right)^2 + \left(E^D\right)^3 = \begin{bmatrix} 1 & -2 & -1 \\ 0 & 1 & 0 \\ 0 & -1 & 0 \end{bmatrix} - 2\begin{bmatrix} 1 & -2 & -1 \\ 0 & 1 & 0 \\ 0 & -1 & 0 \end{bmatrix}^2 + \begin{bmatrix} 1 & -2 & -1 \\ 0 & 1 & 0 \\ 0 & -1 & 0 \end{bmatrix} = \begin{bmatrix} 0 & 0 & 0 \\ 0 & 0 & 0 \\ 0 & 0 & 0 \end{bmatrix}. \tag{A.49}$$

Postmultiplying (A.39) by $\left(E^D\right)^k$, $k = 1 = 2, \ldots$ we obtain the following remark.

Remark A.1. If (A.38) is the characteristic polynomial of E, then

$$a_1\left(E^D\right)^{k+1} + a_2\left(E^D\right)^{k+2} + \ldots + a_{n-1}\left(E^D\right)^{n+k-1} + \left(E^D\right)^{n+k} = 0 \tag{A.50}$$

for $k = 1, 2, \ldots$.

Appendix B
Computation of the Drazin Inverse

B.1 Method 1

Theorem B.1. If the matrix $E \in \mathbb{R}^{n\times n}$ has one of the following forms

$$\begin{aligned} E_1 &= \begin{bmatrix} E_{11} & E_{12} \\ 0 & 0 \end{bmatrix}, \quad E_2 = \begin{bmatrix} E_{21} & 0 \\ E_{22} & 0 \end{bmatrix}, \\ E_3 &= \begin{bmatrix} 0 & 0 \\ E_{31} & E_{32} \end{bmatrix}, \quad E_4 = \begin{bmatrix} 0 & E_{41} \\ 0 & E_{42} \end{bmatrix}, \end{aligned} \tag{B.1}$$

where $E_{11}, E_{21} \in \mathbb{R}^{n_1\times n_1}$ and $E_{32}, E_{42} \in \mathbb{R}^{n_2\times n_2}$ are nonsingular and $E_{12}, E_{41} \in \mathbb{R}^{n_1\times n_2}$, $E_{22}, E_{31} \in \mathbb{R}^{n_2\times n_1}$, $n = n_1 + n_2$, then its Drazin inverse $E^D \in \mathbb{R}^{n\times n}$ has one of the forms, respectively:

$$\begin{aligned} E_1^D &= \begin{bmatrix} E_{11}^{-1} & E_{11}^{-2}E_{12} \\ 0 & 0 \end{bmatrix}, \quad E_2^D = \begin{bmatrix} E_{21}^{-1} & 0 \\ E_{22}E_{21}^{-2} & 0 \end{bmatrix}, \\ E_3^D &= \begin{bmatrix} 0 & 0 \\ E_{32}^{-2}E_{31} & E_{32}^{-1} \end{bmatrix}, \quad E_4^D = \begin{bmatrix} 0 & E_{41}E_{42}^{-2} \\ 0 & E_{42}^{-1} \end{bmatrix}. \end{aligned} \tag{B.2}$$

Proof. Using Definition 1.2 we will show that the matrices (B.1) and (B.2) satisfy the conditions (1.26). Consider the matrix E_1 and its Drazin inverse E_1^D. From (1.26a) we have

$$E_1E_1^D = \begin{bmatrix} E_{11} & E_{12} \\ 0 & 0 \end{bmatrix}\begin{bmatrix} E_{11}^{-1} & E_{11}^{-2}E_{12} \\ 0 & 0 \end{bmatrix} = \begin{bmatrix} \mathbb{I}_{n_1} & E_{11}^{-1}E_{12} \\ 0 & 0 \end{bmatrix} \tag{B.3a}$$

and

$$E_1^DE_1 = \begin{bmatrix} E_{11}^{-1} & E_{11}^{-2}E_{12} \\ 0 & 0 \end{bmatrix}\begin{bmatrix} E_{11} & E_{12} \\ 0 & 0 \end{bmatrix} = \begin{bmatrix} \mathbb{I}_{n_1} & E_{11}^{-1}E_{12} \\ 0 & 0 \end{bmatrix}. \tag{B.3b}$$

T. Kaczorek and K. Borawski, *Descriptor Systems of Integer and Fractional Orders*, Studies in Systems, Decision and Control 367,
https://doi.org/10.1007/978-3-030-72480-1

Therefore, $E_1^D E_1 = E_1 E_1^D$. Next, from (1.26b) we have

$$\begin{aligned} E_1^D E_1 E_1^D &= \begin{bmatrix} \mathbb{I}_{n_1} & E_{11}^{-1}E_{12} \\ 0 & 0 \end{bmatrix} \begin{bmatrix} E_{11}^{-1} & E_{11}^{-2}E_{12} \\ 0 & 0 \end{bmatrix} \\ &= \begin{bmatrix} E_{11}^{-1} & E_{11}^{-2}E_{12} \\ 0 & 0 \end{bmatrix} = E_1^D. \end{aligned} \tag{B.4}$$

We will show that

$$E_1^j = \begin{bmatrix} E_{11}^j & E_{11}^{j-1}E_{12} \\ 0 & 0 \end{bmatrix} \tag{B.5}$$

for $j = 1, 2, \ldots$. For example, consider the matrix (B.5) for $j = 3$. Then

$$\begin{aligned} E_1^3 = E_1 E_1 E_1 &= \begin{bmatrix} E_{11} & E_{12} \\ 0 & 0 \end{bmatrix} \begin{bmatrix} E_{11} & E_{12} \\ 0 & 0 \end{bmatrix} \begin{bmatrix} E_{11} & E_{12} \\ 0 & 0 \end{bmatrix} \\ &= \begin{bmatrix} E_{11}^2 & E_{11}E_{12} \\ 0 & 0 \end{bmatrix} \begin{bmatrix} E_{11} & E_{12} \\ 0 & 0 \end{bmatrix} = \begin{bmatrix} E_{11}^3 & E_{11}^2E_{12} \\ 0 & 0 \end{bmatrix}. \end{aligned} \tag{B.6}$$

It is easy to check that the matrix E_1^j, $j = 1, 2, \ldots$ have the form (B.5). Finally, from (1.26c) we have

$$\begin{aligned} E_1^D E_1^{q+1} &= \begin{bmatrix} E_{11}^{-1} & E_{11}^{-2}E_{12} \\ 0 & 0 \end{bmatrix} \begin{bmatrix} E_{11}^{q+1} & E_{11}^{q}E_{12} \\ 0 & 0 \end{bmatrix} \\ &= \begin{bmatrix} E_{11}^{q} & E_{11}^{q-1}E_{12} \\ 0 & 0 \end{bmatrix} = E_1^q. \end{aligned} \tag{B.7}$$

Therefore, E_1^D is the Drazin inverse of E_1. In a similar way we can show that E_k^D, $k = 2, 3, 4$ are the Drazin inverses of E_k, $k = 2, 3, 4$, respectively. □

Now consider the singular matrix E in a different form than (B.1). Let $P_k \in \mathbb{R}^{n\times n}$ be a nonsingular matrix such that

$$E_k = P_k E P_k^{-1}, \quad k \in \{1, 2, 3, 4\}, \tag{B.8}$$

i.e. transforms E to the one of the forms (B.1). If for a given matrix E there exists the similarity transformation (B.8), then we can use the following approach.

Theorem B.2. The Drazin inverse of (B.8) has the form

$$E_k^D = P_k E^D P_k^{-1}, \quad k \in \{1, 2, 3, 4\}, \tag{B.9}$$

where E^D is the Drazin inverse of E.

Proof. Using Definition 1.2 we will show that the matrices (B.8) and (B.9) satisfy the conditions (1.26). From (1.26a) we have

$$E_k E_k^D = P_k E P_k^{-1} P_k E^D P_k^{-1} = P_k E E^D P_k^{-1} \tag{B.10a}$$

and

$$E_k^D E_k = P_k E^D P_k^{-1} P_k E P_k^{-1} = P_k E^D E P_k^{-1} = P_k E E^D P_k^{-1}. \tag{B.10b}$$

Therefore, $E_k E_k^D = E_k^D E_k$. Next, from (1.26b) we have

$$\begin{aligned} E_k^D E_k E_k^D &= P_k E^D P_k^{-1} P_k E P_k^{-1} P_k E^D P_k^{-1} = P_k E^D E E^D P_k^{-1} \\ &= P_k E^D P_k^{-1} = E_k^D. \end{aligned} \tag{B.11}$$

Finally, from (1.26c) we get

$$\begin{aligned} E_k^D E_k^{q+1} &= P_k E^D P_k^{-1} P_k E^{q+1} P_k^{-1} = P_k E^D E^{q+1} P_k^{-1} \\ &= P E^q P_k^{-1} = E_k^q, \end{aligned} \tag{B.12}$$

since $E_k^j = P_k E^j P_k^{-1}$, $j = 1, 2, \ldots$. □

From the considerations above we have the following theorem.

Theorem B.3. The Drazin inverse of E has the form

$$E^D = P_k^{-1} E_k^D P_k, \ \ k \in \{1, 2, 3, 4\}, \tag{B.13}$$

where E_k^D is one of the matrices (B.2) obtained from the similarity transformation.

Proof. The proof follows immediately from the equation (B.9). □

To compute the Drazin inverse E^D of the matrix $E \in \mathbb{R}^{n\times n}$ the following procedure can be used.

Procedure B.1.

Step 1. According to (B.8) find the nonsingular matrix P_k, $k \in \{1, 2, 3, 4\}$ such that it transforms the matrix E to one of the forms (B.1).

Step 2. Compute the Drazin inverse E_k^D of the transformed matrix E_k, $k \in \{1, 2, 3, 4\}$ using (B.2).

Step 3. Compute the Drazin inverse E^D of the matrix E using (B.13).

Example B.1. Consider the matrix E in the form

$$E = \begin{bmatrix} 1 & 2 & 1 \\ 0 & 2 & 1 \\ 1 & 0 & 0 \end{bmatrix}. \tag{B.14}$$

It is easy to check that $\det E = 0$.

Step 1. According to (B.8) we find the matrix

$$P_1 = \begin{bmatrix} 1 & 0 & 0 \\ 0 & 1 & 0 \\ -1 & 1 & 1 \end{bmatrix}, \tag{B.15}$$

which transforms (B.14) to the form E_1 from (B.1). Thus

$$\begin{aligned} E_1 = P_1 E P_1^{-1} &= \begin{bmatrix} 1 & 0 & 0 \\ 0 & 1 & 0 \\ -1 & 1 & 1 \end{bmatrix} \begin{bmatrix} 1 & 2 & 1 \\ 0 & 2 & 1 \\ 1 & 0 & 0 \end{bmatrix} \begin{bmatrix} 1 & 0 & 0 \\ 0 & 1 & 0 \\ 1 & -1 & 1 \end{bmatrix} \\ &= \begin{bmatrix} 2 & 1 & 1 \\ 1 & 1 & 1 \\ 0 & 0 & 0 \end{bmatrix}. \end{aligned} \tag{B.16}$$

Step 2. From (B.16) we have $E_{11} = \begin{bmatrix} 2 & 1 \\ 1 & 1 \end{bmatrix}$, $\det E_{11} \neq 0$ and $E_{12} = \begin{bmatrix} 1 \\ 1 \end{bmatrix}$. Using (B.2) we obtain

$$E_1^D = \begin{bmatrix} E_{11}^{-1} & E_{11}^{-2} E_{12} \\ 0 & 0 \end{bmatrix} = \begin{bmatrix} 1 & -1 & -1 \\ -1 & 2 & 2 \\ 0 & 0 & 0 \end{bmatrix}. \tag{B.17}$$

Step 3. The Drazin inverse of (B.14) can be computed using (B.13):

$$\begin{aligned} E^D = P_1^{-1} E_1^D P_1 &= \begin{bmatrix} 1 & 0 & 0 \\ 0 & 1 & 0 \\ 1 & -1 & 1 \end{bmatrix} \begin{bmatrix} 1 & -1 & -1 \\ -1 & 2 & 2 \\ 0 & 0 & 0 \end{bmatrix} \begin{bmatrix} 1 & 0 & 0 \\ 0 & 1 & 0 \\ -1 & 1 & 1 \end{bmatrix} \\ &= \begin{bmatrix} 2 & -2 & -1 \\ -3 & 4 & 2 \\ 5 & -6 & -3 \end{bmatrix}. \end{aligned} \tag{B.18}$$

To avoid the calculation of negative matrix powers in (B.2) we can use the Cayley-Hamilton theorem as follows. Consider the matrix E in the form E_1. From (A.38), (A.40) and (B.1) we have

$$\det[\mathbb{I}_{n_1}\lambda - E_{11}] = s^{n_1} + a_{n_1-1}s^{n_1-1} + \ldots + a_1 s + a_0 \tag{B.19}$$

and

$$E_{11}^{n_1} + a_{n_1-1}E_{11}^{n_1-1} + \ldots + a_1 E_{11} + a_0\mathbb{I}_{n_1} = 0. \tag{B.20}$$

Multiplying the equation (B.20) by E_{11}^{-1} we obtain

$$E_{11}^{n_1-1} + a_{n_1-1}E_{11}^{n_1-2} + \ldots + a_1\mathbb{I}_{n_1} + a_0 E_{11}^{-1} = 0 \tag{B.21}$$

and

$$E_{11}^{-1} = -\frac{1}{a_0}\left(E_{11}^{n_1-1} + a_{n_1-1}E_{11}^{n_1-2} + \ldots + a_1\mathbb{I}_{n_1}\right) \tag{B.22}$$

since $a_0 = \det E_{11} \neq 0$. Similarly, multiplying the Eq. (B.21) by E_{11}^{-1} we obtain

$$E_{11}^{n_1-2} + a_{n_1-1}E_{11}^{n_1-3} + \ldots + a_1E_{11}^{-1} + a_0E_{11}^{-2} = 0 \tag{B.23}$$

and

$$E_{11}^{-2} = -\frac{1}{a_0}\left(E_{11}^{n_1-2} + a_{n_1-1}E_{11}^{n_1-3} + \ldots + a_1E_{11}^{-1}\right). \tag{B.24}$$

Substitution of (B.22) into (B.24) yields

$$E_{11}^{-2} = \frac{1}{a_0^2}\left(b_0E_{11}^{n_1-1} + b_1E_{11}^{n_1-2} + \ldots + b_{n_1-1}\mathbb{I}_{n_1}\right), \tag{B.25a}$$

where

$$\begin{gathered} b_k = a_1a_{n_1-k} - a_0a_{n_1-k+1},\ \ k = 0, \ldots, n_1 - 1, \\ a_{n_1} = 1,\ \ a_{n_1+1} = 0. \end{gathered} \tag{B.25b}$$

Similar analysis can be performed for the matrix E in the form E_k, $k = 2, 3, 4$ from (B.1).

Example B.2. (Continuation of Example B.1) Consider the matrix E_1 in the form (B.16), where

$$E_{11} = \begin{bmatrix} 2 & 1 \\ 1 & 1 \end{bmatrix},\ \ E_{12} = \begin{bmatrix} 1 \\ 1 \end{bmatrix}. \tag{B.26}$$

From (B.19), (B.25b) and (B.26) we have

$$\det[\mathbb{I}_2\lambda - E_{11}] = \lambda^2 - 3\lambda + 1 \tag{B.27}$$

and

$$a_1 = -3,\ \ a_0 = 1,\ \ b_1 = 8,\ \ b_0 = -3. \tag{B.28}$$

Using (B.22), (B.25), (B.26) and (B.28) we have

$$\begin{aligned} E_{11}^{-1} &= -E_{11} + 3\mathbb{I}_2 = \begin{bmatrix} 1 & -1 \\ -1 & 2 \end{bmatrix}, \\ E_{11}^{-2} &= -3E_{11} + 8\mathbb{I}_2 = \begin{bmatrix} 2 & -3 \\ -3 & 5 \end{bmatrix}. \end{aligned} \tag{B.29}$$

Therefore, the matrix E_1^D has the form

$$E_1^D = \begin{bmatrix} E_{11}^{-1} & E_{11}^{-2}E_{12} \\ 0 & 0 \end{bmatrix} = \begin{bmatrix} 1 & -1 & -1 \\ -1 & 2 & 2 \\ 0 & 0 & 0 \end{bmatrix}, \tag{B.30}$$

which is equivalent to (B.17).

B.2 Method 2

Consider a singular matrix $E \in \mathbb{R}^{n \times n}$ with the index q satisfying the condition

$$q \leq n - 1 \tag{B.31}$$

and the characteristic polynomial (A.38). From (1.26a), (1.26c) and (A.42) we have

$$\begin{aligned} &E^D E^{n-1} + a_{n-1} E^D E^{n-2} + \ldots + a_2 E^D E + a_1 E^D \\ &\quad = E^{n-2} + a_{n-1} E^{n-3} + \ldots + a_{q+2} E^q + (a_{q+1} E^q + \ldots + a_2 E + a_1 \mathbb{I}_n) E^D \\ &\quad = -A + BE^D = 0, \end{aligned} \tag{B.32}$$

where

$$A = -E^{n-2} - a_{n-1} E^{n-3} - \ldots - a_{q+2} E^q, \tag{B.33}$$

$$B = a_{q+1} E^q + \ldots + a_2 E + a_1 \mathbb{I}_n. \tag{B.34}$$

If

$$\det B \neq 0, \tag{B.35}$$

then from (B.32) we have

$$E^D = B^{-1} A. \tag{B.36}$$

Therefore, the following theorem has been proved.

Theorem B.4. If the conditions (B.31) and (B.35) are satisfied, then the Drazin inverse E^D of a singular matrix E is given by (B.36).

According to Theorem B.4 the Drazin inverse can be computed using the following procedure.

Procedure B.2.

Step 1. Using (1.25) compute the index q of E and check the condition (B.31).

Step 2. Compute the coefficients a_k, $k = 1, \ldots, n - 1$ of the characteristic polynomial (A.38).

Step 3. Compute the matrix B given by (B.34) and check the condition (B.35).

Step 4. Compute the matrix B^{-1} and using (B.33) compute the matrix A.

Step 5. Using (B.36) compute the Drazin inverse matrix E^D.

Example B.3. Compute the Drazin inverse of the matrix

$$E = \begin{bmatrix} 1 & 2 & 1 \\ 0 & 2 & 1 \\ 1 & 0 & 0 \end{bmatrix}. \tag{B.37}$$

Step 1. The index of the matrix (B.37) is $q = 1$ since $\operatorname{rank} E = \operatorname{rank} E^2 = 2$.
Step 2. The characteristic polynomial of (B.37) is

$$\det[\mathbb{I}_3\lambda - E] = \lambda^3 - 3\lambda^2 + \lambda \tag{B.38}$$

and

$$a_2 = -3, \ \ a_1 = 1. \tag{B.39}$$

Step 3. In this case we have

$$B = -3E + \mathbb{I}_3 = \begin{bmatrix} -2 & -6 & -3 \\ 0 & -5 & -3 \\ -3 & 0 & 1 \end{bmatrix} \tag{B.40}$$

and the condition (B.35) is satisfied since $\det B = 1$.
Step 4. Using (B.33) we compute

$$A = -E = \begin{bmatrix} -1 & -2 & -1 \\ 0 & -2 & -1 \\ -1 & 0 & 0 \end{bmatrix} \tag{B.41}$$

and from (B.40) we have

$$B^{-1} = \begin{bmatrix} -5 & 6 & 3 \\ 9 & -11 & -6 \\ -15 & 18 & 10 \end{bmatrix}. \tag{B.42}$$

Step 5. Using (B.36), (B.41) and (B.42) we obtain

$$E^D = B^{-1}A = \begin{bmatrix} 2 & -2 & -1 \\ -3 & 4 & 2 \\ 5 & -6 & -3 \end{bmatrix}. \tag{B.43}$$

Similarly to Method 1, to avoid calculating the inverse of the matrix B we can compute it as follows. Let

$$\det[\mathbb{I}_n\lambda - B] = \lambda^n + b_{n-1}\lambda^{n-1} + \ldots + b_1\lambda + b_0, \tag{B.44}$$

then by Cayley-Hamilton theorem we have

$$B^n + b_{n-1}B^{n-1} + \ldots + b_1 B + b_0 = 0 \tag{B.45}$$

and

$$B^{-1} = -\frac{1}{b_0}\left(B^{n-1} + b_{n-1}B^{n-2} + \ldots + b_2 B + b_1\mathbb{I}_n\right) \tag{B.46}$$

since $b_0 = \det B \neq 0$.
Knowing the matrix B and the coefficients b_k, $k = 0, 1, \ldots, n-1$ we may compute the inverse matrix B^{-1}.

Example B.4. (Continuation of Example B.3) The characteristic polynomial of the matrix (B.40) has the form

$$\det[\mathbb{I}_3\lambda - B] = \lambda^3 + 6\lambda^2 - 6\lambda - 1 \tag{B.47}$$

and

$$b_2 = 6,\ \ b_1 = -6,\ \ b_0 = -1. \tag{B.48}$$

From (B.40), (B.46) and (B.48) we have

$$B^{-1} = B^2 + 6B - 6\mathbb{I}_3 = \begin{bmatrix} -5 & 6 & 3 \\ 9 & -11 & -6 \\ -15 & 18 & 10 \end{bmatrix}, \tag{B.49}$$

which is equivalent to (B.42).

B.3 Method 3

To compute the Drazin inverse E^D of a singular matrix $E \in \mathbb{R}^{n\times n}$ the following procedure can be used.

Procedure B.3.

Step 1. Find the pair of matrices $V \in \mathbb{R}^{n\times r}$, $W \in \mathbb{R}^{r\times n}$ such that

$$E = VW, \quad \text{rank}V = \text{rank}W = \text{rank}E = r. \tag{B.50}$$

As the r columns (rows) of the matrix V (W) the r linearly independent columns (rows) of the matrix E can be chosen.

Step 2. Compute the nonsingular matrix

$$WEV \in \mathbb{R}^{r\times r}. \tag{B.51}$$

Step 3. The desired Drazin inverse matrix is given by

$$E^D = V[WEV]^{-1}W. \tag{B.52}$$

Proof. It will be shown that the matrix (B.52) satisfies the conditions (1.26). Taking into account that $\det WV \neq 0$ and (B.50) we obtain

$$[WEV]^{-1} = [WVWV]^{-1} = [WV]^{-1}[WV]^{-1}. \tag{B.53}$$

Using (1.26a), (B.50) and (B.53) we obtain

$$EE^D = VWV[WEV]^{-1}W = VWV[WV]^{-1}[WV]^{-1}W = V[WV]^{-1}W \tag{B.54}$$

and

$$E^DE = V[WEV]^{-1}WVW = V[WV]^{-1}[WV]^{-1}WVW = V[WV]^{-1}W. \tag{B.55}$$

Therefore, the condition (1.26a) is satisfied. To check the condition (1.26b) we compute

$$\begin{aligned} E^DEE^D &= V[WEV]^{-1}WVWV[WEV]^{-1}W \\ &= V[WVWV]^{-1}WVWV[WEV]^{-1}W \\ &= V[WEV]^{-1}W = E^D. \end{aligned} \tag{B.56}$$

Therefore, the condition (1.26b) is also satisfied. Using (1.26c), (B.50), (B.52) and (B.53) we obtain

$$\begin{aligned} E^DE^{q+1} &= V[WEV]^{-1}W(VW)^{q+1} \\ &= V[WV]^{-1}[WV]^{-1}WVW(VW)^q \\ &= V[WV]^{-1}W(VW)^q \\ &= V[WV]^{-1}WVW(VW)^{q-1} \\ &= (VW)^q = E^q, \end{aligned} \tag{B.57}$$

where q is the index of E. Therefore, the condition (1.26c) is also satisfied. □

Example B.5. Compute the Drazin inverse of the matrix

$$E = \begin{bmatrix} 1 & 2 & 1 \\ 0 & 2 & 1 \\ 1 & 0 & 0 \end{bmatrix}. \tag{B.58}$$

Step 1. We find the pair of matrices

$$\begin{aligned} V &= \begin{bmatrix} -0.7427 & 0.3393 \\ -0.6652 & -0.4735 \\ -0.0775 & 0.8128 \end{bmatrix}, \\ W &= \begin{bmatrix} -0.8202 & -2.8157 & -1.4078 \\ 1.1521 & -0.2685 & -0.1342 \end{bmatrix}, \end{aligned} \tag{B.59}$$

which satisfy the conditions (B.50).
Step 2. Using (B.51) we compute

$$WEV = \begin{bmatrix} 6.7733 & -0.2679 \\ -1.9999 & 0.2267 \end{bmatrix}. \tag{B.60}$$

Step 3. The Drazin inverse of the matrix (B.58) has the form

$$\begin{aligned} E^D &= V[WEV]^{-1}W \\ &= \begin{bmatrix} -0.7427 & 0.3393 \\ -0.6652 & -0.4735 \\ -0.0775 & 0.8128 \end{bmatrix} \begin{bmatrix} 0.2267 & 0.2679 \\ 1.9999 & 6.7733 \end{bmatrix} \\ &\times \begin{bmatrix} -0.8202 & -2.8157 & -1.4078 \\ 1.1521 & -0.2685 & -0.1342 \end{bmatrix} \\ &= \begin{bmatrix} 2 & -2 & -1 \\ -3 & 4 & 2 \\ 5 & -6 & -3 \end{bmatrix}. \end{aligned} \tag{B.61}$$

Appendix C
Laplace Transforms of Continuous-Time Functions and $\mathcal{Z}$-Transforms of Discrete-Time Functions

C.1 Convolutions of Continuous-Time and Discrete-Time Functions and Their Transforms

Definition C.1. The Laplace transform of a continuous-time function $f(t)$ is defined by

$$\mathcal{L}[f(t)] = \int_0^\infty f(t)e^{-st}dt = F(s), \tag{C.1}$$

where $f(t) = 0$ for $t < 0$.

Definition C.2. The continuous-time function $f(t)$ defined by

$$f_1(t) * f_2(t) = \int_0^t f_1(t-\tau)f_2(\tau)d\tau \tag{C.2}$$

is called the convolution of the continuous-time functions $f_1(t)$ and $f_2(t)$.

Theorem C.1. If

$$F_1(s) = \mathcal{L}[f_1(t)], \;\; F_2(s) = \mathcal{L}[f_2(t)],$$

then

$$\mathcal{L}\left[\int_0^t f_1(t-\tau)f_2(\tau)d\tau\right] = F_1(s)F_2(s). \tag{C.3}$$

Proof. Taking into account that $f_1(t) = 0$, $f_2(t) = 0$ for $t < 0$ and using (C.1) we have

T. Kaczorek and K. Borawski, *Descriptor Systems of Integer and Fractional Orders*, Studies in Systems, Decision and Control 367,
https://doi.org/10.1007/978-3-030-72480-1

$$\mathcal{L}\left[\int_0^t f_1(t-\tau)f_2(\tau)d\tau\right] = \mathcal{L}\left[\int_0^\infty f_1(t-\tau)f_2(\tau)d\tau\right]$$
$$= \int_0^\infty\left[\int_0^\infty f_1(t-\tau)f_2(\tau)\right]e^{-st}d\tau dt$$
$$= \int_0^t f_1(u)e^{-su}du\int_0^t f_2(\tau)e^{-s\tau}d\tau = F_1(s)F_2(s). \tag{C.4}$$

□

Definition C.3. The $\mathcal{Z}$-transform of a discrete-time function f_i is defined by

$$\mathcal{Z}[f_i] = \sum_{k=0}^{\infty} f_k z^{-k} = F(z), \tag{C.5}$$

where $f_i = 0$ for $i < 0$.

Definition C.4. The discrete-time function defined by

$$f_{1,i} * f_{2,i} = \sum_{k=0}^{i} f_{1,i-k} f_{2,k} \tag{C.6}$$

is called the convolution of the discrete-time functions $f_{1,i}$, $f_{2,i}$.

Theorem C.2. If

$$F_1(z) = \mathcal{Z}[f_{1,i}],\ \ F_2(z) = \mathcal{Z}[f_{2,i}],$$

then

$$\mathcal{Z}\left[\sum_{k=0}^{i} f_{1,i-k} f_{2,k}\right] = F_1(z)F_2(z). \tag{C.7}$$

Proof. The proof is similar to the proof of Theorem C.1. □

C.2 Laplace Transforms of Derivative-Integrals

Theorem C.3. The Laplace transform of the function t^α has the form

$$\mathcal{L}[t^\alpha] = \frac{\Gamma(\alpha+1)}{s^{\alpha+1}} \tag{C.8}$$

for $\alpha \in \mathbb{R}, \alpha > -1$.

Proof. Using (C.1) and (2.1) for $\alpha > -1$ we obtain

$$\mathcal{L}[t^\alpha] = \int_0^\infty t^\alpha e^{-st} dt = \int_0^\infty \frac{x^\alpha}{s^{\alpha+1}} e^{-x} dx = \frac{1}{s^{\alpha+1}} \int_0^\infty x^\alpha e^{-x} dx = \frac{\Gamma(\alpha+1)}{s^{\alpha+1}}. \quad \text{(C.9)}$$

□

Theorem C.4. The Laplace transform of the first order derivative of the function $f(t)$ has the form

$$\mathcal{L}\left[\frac{d}{dt} f(t)\right] = sF(s) - f(0^+). \quad \text{(C.10)}$$

Proof. Using (C.1) we have

$$\begin{aligned} \mathcal{L}\left[\frac{d}{dt} f(t)\right] &= \int_0^\infty \frac{d}{dt} f(t) e^{-st} dt = \left[e^{-st} f\right]_0^\infty + s \int_0^\infty f(t) e^{-st} dt \\ &= sF(s) - f(0^+). \end{aligned} \quad \text{(C.11)}$$

□

Generalizing (C.10) for n-order derivative we obtain

$$\mathcal{L}\left[\frac{d^n}{dt^n} f(t)\right] = s^n F(s) - \sum_{k=1}^{n} s^{n-k} f^{(k-1)}(0^+), \quad \text{(C.12)}$$

where $f^{(k)}(0^+) = \left.\frac{df(t)}{dt}\right|_{t=0}$.

Theorem C.5. The Laplace transform of the fractional α-order integral has the form

$$\mathcal{L}[{}_0I_t^\alpha f(t)] = \mathcal{L}\left[\frac{1}{\Gamma(\alpha)} \int_0^t (t-\tau)^{\alpha-1} f(\tau) d\tau\right] = \frac{F(s)}{s^\alpha}. \quad \text{(C.13)}$$

Proof. Using (3.56), (C.8), Theorem C.1 and taking into account that $\alpha > 0$ we obtain

$$\begin{aligned} \mathcal{L}[{}_0I_t^\alpha f(t)] &= \mathcal{L}\left[\frac{1}{\Gamma(\alpha)} \int_0^t (t-\tau)^{\alpha-1} f(\tau) d\tau\right] \\ &= \frac{1}{\Gamma(\alpha)} \mathcal{L}\left[t^{\alpha-1}\right] \mathcal{L}[f(t)] = \frac{F(s)}{s^\alpha}. \end{aligned} \quad \text{(C.14)}$$

□

Theorem C.6. The Laplace transform of the fractional α-order integral of n-order derivative of the function $f(t)$ has the form

$$\mathcal{L}\left[{}_0I_t^\alpha f^{(n)}(t)\right] = s^{n-\alpha}F(s) - \sum_{k=1}^{n} s^{n-k-\alpha} f^{(k-1)}(0^+). \tag{C.15}$$

Proof. Using (3.56), (C.8) and Theorem C.1 we obtain

$$\begin{aligned}\mathcal{L}\left[{}_0I_t^\alpha f^{(n)}(t)\right] &= \mathcal{L}\left[\frac{1}{\Gamma(\alpha)}\int_0^t (t-\tau)^{\alpha-1} f^{(n)}(\tau)d\tau\right] \\ &= \frac{1}{\Gamma(\alpha)}\mathcal{L}\left[t^{\alpha-1}\right]\mathcal{L}\left[f^{(n)}(t)\right] \\ &= \frac{1}{\Gamma(\alpha)}\frac{\Gamma(\alpha)}{s^\alpha}\left[s^n F(s) - \sum_{k=1}^{n} s^{n-k} f^{(k-1)}(0^+)\right] \\ &= s^{n-\alpha}F(s) - \sum_{k=1}^{n} s^{n-k-\alpha} f^{(k-1)}(0^+).\end{aligned} \tag{C.16}$$

□

Theorem C.7. [90] The inverse Laplace transform of the expression $s^\alpha F(s)$ has the form

$$\mathcal{L}^{-1}[s^\alpha F(s)] = {}_0DI_t^\alpha f(t) + \sum_{k=0}^{n-1} \frac{t^{k-\alpha}}{\Gamma(k-\alpha+1)} f^{(k)}(0^+), \tag{C.17a}$$

where $n-1 < \alpha < n$, $n \in \mathbb{N}$ and

$${}_0DI_t^\alpha f(t) = \begin{cases} {}_0D_t^\alpha f(t) \text{ for } \alpha > 0, \\ {}_0I_t^\alpha f(t) \text{ for } \alpha < 0. \end{cases} \tag{C.17b}$$

C.3 $\mathcal{Z}$-Transforms of Discrete-Time Functions

Theorem C.8. If

$$\mathcal{Z}[x_i] = \sum_{i=0}^{\infty} x_i z^{-i}, \tag{C.18}$$

then

$$\mathcal{Z}[x_{i+1}] = zX(z) - zx_0, \tag{C.19a}$$

$$\mathcal{Z}[x_{i-p}] = z^{-p}X(z) + z^{-p}\sum_{j=-1}^{-p} x_j z^j. \tag{C.19b}$$

Proof. Using (C.18) we obtain

$$\begin{aligned}\mathcal{Z}[x_{i+1}] &= \sum_{i=0}^{\infty} x_{i+1}z^{-i} = \sum_{j=1}^{\infty} x_j z^{-(j-1)} = z\sum_{j=0}^{\infty} x_j z^{-j} - zx_0 \\ &= zX(z) - zx_0,\end{aligned}$$

$$\begin{aligned}\mathcal{Z}[x_{i-p}] &= \sum_{i=0}^{\infty} x_{i-p}z^{-i} = \sum_{j=-p}^{\infty} x_j z^{-(j+p)} = z^{-p}\sum_{j=-p}^{\infty} x_j z^{-j} \\ &= z^{-p}\sum_{j=0}^{\infty} x_j z^{-j} + z^{-p}\sum_{j=-1}^{-p} x_j z^{-j} \\ &= z^{-p}X(z) + z^{-p}\sum_{j=-1}^{-p} x_j z^{-j}.\end{aligned} \tag{C.20}$$

□

Appendix D
Nilpotent Matrices

Definition D.1. A real matrix $A \in \mathbb{R}^{n \times n}$ is called nilpotent if there exists a natural number $\mu \leq n$ such that $A^{\mu-1} = 0$ and $A^{\mu} = 0$. The natural number μ is called the nilpotency index of the matrix A.

Lemma D.1. Matrices of the form

$$A = \begin{bmatrix} 0 & a_{12} & \dots & a_{1,n-1} & 0 \\ 0 & 0 & \dots & 0 & 0 \\ \vdots & \vdots & \ddots & \vdots & \vdots \\ 0 & 0 & \dots & 0 & 0 \\ 0 & a_{n,2} & \dots & a_{n,n-1} & 0 \end{bmatrix} \in \mathbb{R}^{n \times n} \tag{D.1}$$

have the nilpotency index $\mu = 2$ for any values of the entries $a_{12}, \dots, a_{1,n-1}$, $a_{n,2}, \dots, a_{n,n-1}$ and the characteristic polynomials of the form

$$\det[\mathbb{I}_n \lambda - A] = \lambda^n. \tag{D.2}$$

Proof. Using (D.1) it is easy to check that $A^2 = 0$ and

$$\det[\mathbb{I}_n \lambda - A] = \begin{vmatrix} \lambda & -a_{12} & \dots & -a_{1,n-1} & 0 \\ 0 & \lambda & \dots & 0 & 0 \\ \vdots & \vdots & \ddots & \vdots & \vdots \\ 0 & 0 & \dots & \lambda & 0 \\ 0 & -a_{n,2} & \dots & -a_{n,n-1} & \lambda \end{vmatrix} = \lambda^n. \tag{D.3}$$

□

T. Kaczorek and K. Borawski, *Descriptor Systems of Integer and Fractional Orders*, Studies in Systems, Decision and Control 367,
https://doi.org/10.1007/978-3-030-72480-1

Lemma D.2. Matrices of the form

$$A = \begin{bmatrix} 0 & A_{12} \\ 0 & 0 \end{bmatrix} \in \mathbb{R}^{2n \times 2n} \tag{D.4}$$

have the nilpotency index $\mu = 2$ and the characteristic polynomials of the form

$$\det[\mathbb{I}_{2n}\lambda - A] = \lambda^{2n} \tag{D.5}$$

for any submatrices $A_{12} \in \mathbb{R}^{n \times n}$.

Proof. Using (D.4) it is easy to verify that $A^2 = 0$ and

$$\det[\mathbb{I}_{2n}\lambda - A] = \begin{vmatrix} \mathbb{I}_n\lambda & -A_{12} \\ 0 & \mathbb{I}_n \end{vmatrix} = \lambda^{2n}. \tag{D.6}$$

□

From the well-known property of the transposition of the matrix A, i.e. $\left(A^k\right)^T = \left(A^T\right)^k$ for $k = 1, 2, \ldots$ we have the following remark.

Remark D.1. The transpose matrix A^T has a nilpotency index μ if and only if the matrix A has the same nilpotency index μ.

Lemma D.3. A diagonal matrix A with at least one nonzero entry is not the nilpotent matrix.

Proof. This follows immediately from the relation

$$A^k = (\mathrm{diag}[a_1, \ldots, a_n])^k = \mathrm{diag}[a_1^k, \ldots, a_n^k] \neq 0 \tag{D.7}$$

for $k = 1, 2, \ldots$ if at least one from the entries $a_1, \ldots, a_n$ is nonzero. □

Lemma D.4. A nonnegative matrix $A \in \mathbb{R}^{n \times n}$ with at least one nonzero diagonal entry is not nilpotent matrix.

Proof. Let decompose the matrix A as the sum of the diagonal matrix D and the nonnegative matrix B with zero diagonal entries. Let assume that D and B are commuting matrices, i.e. $DB = DB$. Then

$$A^k = (D + B)^k = D^k + BD^{k-1} + \ldots + B^k \text{ for } k = 1, 2, \ldots . \tag{D.8}$$

If the matrix A has at least one nonzero diagonal entry, then $D \neq 0$ and by Lemma D.3 $D^k \neq 0$ for $k = 1, 2, \ldots$. From (D.8) we have $A^k \neq 0$ for $k = 1, 2, \ldots$ since $D^k \neq 0$ and the remaining entries are nonnegative. □

References

1. Ait-Rami, M., Tadeo, F.: Controller synthesis for positive linear systems with bounded controls. IEEE Trans. Circuits Syst. **54**(2), 151–155 (2007)
2. Antsaklis, P.J., Michel, A.N.: Linear Systems. Birkhauser, Boston (2006)
3. Ascher, U.M., Petzold, L.R.: Computer Methods for Ordinary Differential Equations and Differential-Algebraic Equations. SIAM, Philadelphia (1998)
4. Balakrishnan, A.V.: On superstability of semigroups. In: Proceedings 18th IFIP TC-7 Conference on System Modelling and Optimization, pp. 1–9, July 1997
5. Bartosiewicz, Z., Pawłuszewicz, E.: Linear control systems on time scale: unification of continuous and discrete. In: 10th IEEE International Conference on Methods and Models in Automation and Robotics, pp. 263–266, Międzyzdroje, Poland, August 30–2 September 2004
6. Benvenuti, L., Farina, L.: Positive and compartmental systems. IEEE Trans. Autom. Contr. **47**(2), 370–373 (2002)
7. Berger, T., Reis, T.: Controllability of linear differential-algebraic systems—a survey. In: Surveys in Differential-Algebraic Equations I. Differential-Algebraic Equations Forum. Springer, Heidelberg (2013)
8. Berman, A., Plemmons, R.J.: Nonnegative Matrices in the Mathematical Sciences. SIAM, Philadelphia (1994)
9. Blajer, W., Kołodziejczyk, K.: Control of under actuated mechanical systems with servo-constraints. Nonlinear Dynam. **50**, 781–791 (2007)
10. Bohner, M., Wintz, N.: The Kalman filter for linear systems on time scales. J. Math. Anal. Appl. **406**(2), 419–436 (2013)
11. Borawski, K.: Characteristic equations for descriptor linear electrical circuits. In: Proceedings of the Conference on MMAR 2017, Międzyzdroje, Poland, 28-31 August 2017
12. Borawski, K.: Modification of the stability and positivity of standard and descriptor linear electrical circuits by state feedbacks. Electr. Rev. **93**(11), 176–180 (2017)
13. Borawski, K.: Analysis of the positivity of descriptor continuous-time linear systems by the use of Drazin inverse matrix method. In: Szewczyk, R., Zieliński, C., Kaliczyńska, M. (eds.) Automation 2018: Advances in Automation, Robotics and Measurement Techniques (Advances in Intelligent Systems and Computing), Warsaw, Poland, 21–23 March 2018, pp. 172–182. Springer (2018)
14. Borawski, K.: Application of feedback control in linear descriptor dynamical systems. Ph.D. thesis, Białystok University of Technology (2019). (in Polish)
15. Borawski, K.: Superstabilization of descriptor continuous-time linear systems via state-feedback using Drazin inverse matrix method. Symmetry **12**(6), 940 (2020)

T. Kaczorek and K. Borawski, *Descriptor Systems of Integer and Fractional Orders*, Studies in Systems, Decision and Control 367,
https://doi.org/10.1007/978-3-030-72480-1

16. Brayton, R.K., Gustavson, F.G., Hachtel, G.D.: A new efficient algorithm for solving differential-algebraic systems using implicit backward differentiation formulas. Proc. IEEE **60**(1), 98–108 (1972)
17. Brenan, K.E.: Difference approximations for higher index differential-algebraic systems with applications in trajectory control. In: Proceedings of the 23rd IEEE Conference on Decision and Control, Las Vegas, USA, 12–14 December 1984, pp. 291–292 (1994)
18. Brenan, K.E., Campbell, S.L., Petzold, R.L.: Numerical Solution of Initial-Value Problems in Differential-Algebraic Equations. North-Holland, Amsterdam (1989)
19. Bru, R., Coll, C., Sanchez, E.: About positively discrete-time singular systems. In: System and Control: Theory and Applications, pp. 44–48, Athens. World Scientific and Engineering Society (2000)
20. Bunse-Gerstner, A., Mehrmann, V., Nichols, N.K.: Regularization of descriptor systems by derivative and proportional state feedback. SIAM J. Matrix Anal. Appl. **13**, (1992)
21. Bunse-Gerstner, A., Mehrmann, V., Nichols, N.K.: Regularization of descriptor systems by output feedback. IEEE Trans. Autom. Contr. **39**(8), 1742–1748 (1992)
22. Busłowicz, M.: Robust stability of convex combination of two fractional degree characteristic polynomial. Acta Mechanica et Automatica **2**(2), 5–10 (2008)
23. Busłowicz, M.: Simple stability conditions for linear positive discrete-time linear systems with delays. Bull. Pol. Acad. Sci. Tech. **56**(4), 325–328 (2008)
24. Busłowicz, M.: Stability analysis of continuous-time linear systems consisting of n subsystems with different fractional orders. Bull. Pol. Acad. Sci. Tech. **60**(2), 279–284 (2012)
25. Busłowicz, M., Ruszewski, A.: Necessary and sufficient conditions for stability of fractional discrete-time linear state-space systems. Bull. Pol. Acad. Sci. Tech. **61**(4), 779–786 (2013)
26. Campbell, S.L.: Singular systems of differential equations. Research Notes in Mathematics, 40, Pitman Advanced Publishing Program, Boston, Massachusetts (1980)
27. Campbell, S.L., Meyer, C.D., Rose, N.J.: Applications of the Drazin inverse to linear systems of differential equations with singular constant coefficients. SIAM J. Appl. Math. **31**(3), 411–425 (1976)
28. Chu, D., Ho, D.W.C.: Necessary and sufficient conditions for the output feedback regularization of descriptor systems. IEEE Trans. Autom. Contr. **44**(2), 405–412 (1999)
29. Chu, D., Ho, D.W.C.: Infinite eigenvalue assignment for singular systems. Linear Algebra Appl. **294**, 21–37 (2008)
30. Cobb, D.: Controllability, observability and duality in singular systems. IEEE Trans. Autom. Contr. **29**(12), 1076–1082 (1984)
31. Cuthrell, J.E., Biegler, L.T.: On the optimization of differential-algebraic process systems. AIChE J. **33**(8), 1257–1270 (1987)
32. Dai, L.: Singular Control Systems, vol. 118. Lecture Notes in Control and Information Sciences. Springer, Berlin (1989)
33. Daoutidis, P.: DAEs in chemical engineering: a survey. Surveys in Differential-Algebraic Equations I, vol. 2, Differential-Algebraic Equations Forum. Springer (2012)
34. Drazin, M.P.: Pseudo-inverses in associative rings and semigroups. Amer. Math. Monthly **65**, 506–514 (1958)
35. Dzieliński, A., Sierociuk, D.: Stability of discrete fractional state-space systems. In: II IFAC Workshop on Fractional Differentiation and its Applications, pp. 518–523, Porto, Portugal, 19–21 July 2006
36. Dzieliński, A., Sierociuk, D.: Stability of discrete fractional order state-space systems. J. Vib. Control **14**(9/10), 1543–1556 (2008)
37. Dzieliński, A., Sierociuk, D., Sarwas, G.: Ultracapacitor parameters identification based on fractional order model. In: Proceedings of the ECC 2009, Budapest, Hungary (2009)
38. Dziurla, B., Newcomb, R.W.: An example of the continuation method of solving semistate equations. In: Proceedings 23rd IEEE Conference on Decision and Control, pp. 274–279, Las Vegas, USA, 12–14, December 1984
39. Eich-Soellner, E., Führer, C.: Numerical Methods in Multibody Dynamics. Springer, Stuttgart (1998)

40. Farina, L., Rinaldi, S.: Positive Linear Systems: Theory and Applications. Wiley, New York (2000)
41. Führer, C., Leimkuhler, B.J.: Numerical solution of differential-algebraic equations for constrained mechanical motion. Numer. Math. **59**, 55–69 (1991)
42. Gantmacher, F.R.: The Theory of Matrices. Chelsea Publishing Company, New York (1959)
43. Gear, C.W.: Simultaneous numerical solution of differential-algebraic equations. IEEE Trans. Circuit Theory **18**(1), 89–95 (1971)
44. Guang-Ren, D.: Eigenstructure assignment and response analysis in descriptor linear systems with state feedback control. Int. J. Control **69**(5), 663–694 (1998)
45. Guang-Ren, D.: Analysis and Design of Descriptor Linear Systems. Springer, New York (2010)
46. Haug, E.J.: Computer-Aided Kinematics and Dynamics of Mechanical Systems. Allyn and Bacon, Boston (1989)
47. Kaczorek, T.: Theory of Control and Systems. PWN, Warsaw (1991). (in Polish)
48. Kaczorek, T.: Vectors and Matrices in Automation and Electrotechnics. WNT, Warsaw (1998)
49. Kaczorek, T.: Positive 1D and 2D Systems. Springer, London (2002)
50. Kaczorek, T.: The relationship between the infinite eigenvalue assignment for singular systems and the solvability of polynomial matrix equations. Int. J. Appl. Math. Comput. Sci. **14**(1), 161–167 (2003)
51. Kaczorek, T.: Infinite eigenvalue assignment by an output feedback for singular systems. Int. J. Appl. Math. Comput. Sci. **14**(1), 19–23 (2004)
52. Kaczorek, T.: Polynomial and Rational Matrices. Applications in Dynamics Systems Theory. Springer, London (2007)
53. Kaczorek, T.: Practical stability of positive fractional discrete-time linear systems. Bull. Pol. Acad. Sci. Tech. **56**(4), 313–317 (2008)
54. Kaczorek, T.: Stability of positive continuous-time linear systems with delays. Bull. Pol. Acad. Sci. Tech. **57**(4), 395–398 (2009)
55. Kaczorek, T.: Positive linear systems with different fractional orders. Bull. Pol. Acad. Sci. Tech. **58**(3), 453–458 (2010)
56. Kaczorek, T.: Positive linear systems consisting of n subsystems with different fractional orders. IEEE Trans. Circuits Syst. **58**(6), 1203–1210 (2011)
57. Kaczorek, T.: Selected Problems of Fractional Systems Theory. Springer, Berlin (2011)
58. Kaczorek, T.: Singular fractional linear systems and electrical circuits. Int. J. Appl. Math. Comput. Sci. **21**(2), 379–384 (2011)
59. Kaczorek, T.: Positive stable descriptor continuous-discrete 2D linear system. Electr. Rev. **88**(12a), 137–142 (2012)
60. Kaczorek, T.: Stability of descriptor positive linear systems. COMPEL **32**(1), 412–423 (2012)
61. Kaczorek, T.: Application of the Drazin inverse to the analysis of descriptor fractional discrete-time linear systems with regular pencils. Int. J. Appl. Math. Comput. Sci. **23**(1), 29–33 (2013)
62. Kaczorek, T.: Stability of positive fractional switched continuous-time linear systems. Bull. Pol. Acad. Sci. Tech. **61**(2), 349–352 (2013)
63. Kaczorek, T.: Drazin inverse matrix method for fractional descriptor continuous-time linear systems. Bull. Pol. Acad. Sci. Tech. **62**(3), 409–412 (2014)
64. Kaczorek, T.: Fractional descriptor observers for fractional descriptor continuous-time linear system. Arch. Control Sci. **24**(1), 27–37 (2014)
65. Kaczorek, T.: Polynomial approach to fractional descriptor electrical circuits. In: Proceedings of the International Conference on Intelligent Information and Engineering Systems, pp. 518–523, Rzeszów-Krynica, Poland, 16–20 September 2014 (2014)
66. Kaczorek, T.: Positive descriptor time-varying discrete-time linear systems and their asymptotic stability. TransNav: Int. J. Marine Navigat. Saf. Sea Transp. **9**(1), 83–89 (2015)
67. Kaczorek, T.: Cayley-Hamilton theorem for Drazin inverse matrix and standard inverse matrices. Bull. Pol. Acad. Sci. Tech. **64**(4), 793–797 (2016)
68. Kaczorek, T.: Drazin inverse matrix method for fractional descriptor discrete-time linear systems. Bull. Pol. Acad. Sci. Tech. **64**(2), 395–399 (2016)

69. Kaczorek, T.: Extension of Cayley-Hamilton theorem and a procedure for computation of the Drazin inverse matrices. In: Proceeding of the Conference MMAR 2017, Międzyzdroje, Poland, 28–31 August 2017
70. Kaczorek, T.: Superstabilization of positive linear electrical circuit by state-feedbacks. Bull. Pol. Acad. Sci. Tech. **65**(5), 703–708 (2017)
71. Kaczorek, T.: Characteristic equations of the standard and descriptor linear electrical circuits of integer and fractional orders. Acta Mechanica et Automatica **12**(3), 209–216 (2018)
72. Kaczorek, T.: Positivity and stability of descriptor linear systems with interval state matrices. Comput. Prob. Electr. Eng. **8**(1), 7–17 (2018)
73. Kaczorek, T.: Stability of interval positive continuous-time linear systems. Bull. Pol. Acad. Sci. Tech. **66**(1), 31–35 (2018)
74. Kaczorek, T.: Stability of interval positive fractional discrete-time linear systems. Int. J. Appl. Math. Comput. Sci. **28**(3), 451–456 (2018)
75. Kaczorek, T.: Absolute stability of a class of fractional positive nonlinear systems. Int. J. Appl. Math. Comput. Sci. **29**(1), 93–98 (2019)
76. Kaczorek, T.: Absolute stability of a class of nonlinear systems with nonpositive linear parts. Archives Control Sci. **29**(2), 247–258 (2019)
77. Kaczorek, T.: Absolute stability of a class of positive nonlinear continuous-time and discrete-time systems. Archives Control Sci. **29**(1), 157–167 (2019)
78. Kaczorek, T.: Global stability of nonlinear feedback systems with positive descriptor linear parts. Bull. Pol. Acad. Sci. Tech. **67**(1), 45–51 (2019)
79. Kaczorek, T.: Positivity of fractional descriptor linear continuous-time systems. Bull. Pol. Acad. Sci. Tech. **67**(2), 229–233 (2019)
80. Kaczorek, T.: Global stability of nonlinear feedback systems with fractional positive linear parts. Int. J. Appl. Math. Comput. Sci. **30**(3), 493–499 (2020)
81. Kaczorek, T.: Global stability of positive standard and fractional nonlinear feedback systems. Bull. Pol. Acad. Sci. Tech. **68**(2), 285–288 (2020)
82. Kaczorek, T., Borawski, K.: Pointwise completeness, pointwise degeneracy and stability of standard and positive linear systems after discretization. Measur. Autom. Monitor. **60**(6), 405–409 (2014)
83. Kaczorek, T., Borawski, K.: Positivity and stability of time-varying fractional discrete-time linear systems. Measur. Autom. Monitor. **61**(3), 84–87 (2015)
84. Kaczorek, T., Borawski, K.: Fractional descriptor continuous-time linear systems described by the Caputo-Fabrizio derivative. Int. J. Appl. Math. Comput. Sci. **26**(3), 533–541 (2016)
85. Kaczorek, T., Borawski, K.: Minimum energy control of descriptor discrete-time linear systems by the use of Weierstrass-Kronecker decomposition. Archives of Control Sciences **26**(2), 177–187 (2016)
86. Kaczorek, T., Borawski, K.: Eigenvalue assignment in fractional descriptor discrete-time linear systems. Archives Control Sci. **27**(1), 119–128 (2017)
87. Kaczorek, T., Borawski, K.: Stability of positive nonlinear systems. In: Proceedings of the Conference on MMAR 2017, Międzyzdroje, Poland, 28–31 August 2017
88. Kaczorek, T., Busłowicz, M.: Simple conditions for practical stability of positive fractional discrete-time linear systems. Int. J. Appl. Math. Comput. Sci. **19**(2), 263–269 (2009)
89. Kaczorek, T., Rogowski, K.: A computer algorithm for the solution of the state equations of descriptor fractional discrete-time linear systems. Measur. Autom. Monitor. **59**(7), 695–699 (2013)
90. Kaczorek, T., Rogowski, K.: Fractional Linear Systems and Electrical Circuits, Studies in Systems, Decision and Control, vol. 13. Springer, New York (2015)
91. Kaczorek, T., Ruszewski, A.: Global stability of discrete-time nonlinear systems with descriptor standard and fractional positive linear parts and scalar feedbacks. Archives Control Sci. **30**(4), 667–681 (2020)
92. Kaczorek, T., Sajewski, Ł.: Realization Problem for Positive and Fractional Systems. Printing House of Białystok University of Technology, Białystok (2013)

93. Kaczorek, T., Sajewski, Ł.: Global stability of positive different fractional orders nonlinear feedback systems. In: SIGNAL 2020, The Fifth International Conference on Advances in Signal, Image and Video Processing, pp. 6–10, Lisbon, Portugal (2020)
94. Kailath, T.: Linear Systems. Prentice-Hall, Englewood Cliffs (1980)
95. Kalman, R.E.: On the general theory of control systems. In: Proceedings of the 1st IFAC Congress Automatic Control, pp. 481–492 (1960)
96. Kalogeropoulos, G., Mitrouli, M.: On the computation of the Weierstrass canonical form of a regular matrix pencil. Control. Comput. **20**(3), 61–67 (1992)
97. Kalogeropoulos, G., Mitrouli, M., Pantelous, A., Triantafyllou, D.: The weierstrass canonical form of a regular matrix pencil: numerical issues and computational techniques. In: Proceedings of the 4th International Conference on Numerical Analysis and Its Applications, NAA 2008, Lozenetz, Bulgaria, 16–20 June 2008
98. Kharitonov, V.L.: Asymptotic stability of an equilibrium position of a family of systems of differential equations. Differentsialnye uravneniya **14**, 2086–2088 (1978)
99. Klamka, J.: Complete controllability of singular 2-D system. In: Proceedings of the 13th IMACS World Congress, pp. 1839–1840, Dublin, Ireland (1991)
100. Klamka, J.: Controllability of Dynamical Systems. Kluwer Academic Publishers, Dordrecht (1991)
101. Klamka, J., Kalinowski, J.: Reachability and controllablity of positive continuous-time linear systems with time variable coefficients. Bull. Pol. Acad. Sci. Tech. **49**(4), 633–641 (2001)
102. Kociszewski, R.: Full order observer synthesis for singular discrete-time fractional systems. Pomiary Automatyka Robotyka **20**(4), 9–13 (2016). (in Polish)
103. Kronecker, L.: Algebraische Reduction der Schaaren Bilinearer Formen. Sitzungsbreichte der Königlich Preußischen Akademie der Wissenschaften zu Berlin, pp. 1225–1237 (1890)
104. Kudrewicz, J.: Ustoicivost nieliniejnych sistem z obratnoj swjazju. Avtomatika i Telemechanika **25**(8), (1964)
105. Kunkel, P., Mehrmann, V.: Differential-Algebraic Equations. Analysis and Numerical Solution. EMS Publishing House, Zürich (2006)
106. Kunkel, P., Mehrmann, V., Rath, W.: Analysis and numerical solution of control problems in descriptor form. Math. Control Sign. Syst. **14**(1), 29–61 (2001)
107. Lamour, R., März, R., Tischendorf, C.: Differential Algebraic Equations: A Projector Based Analysis, Differential-Algebraic Equations Forum, vol. 1. Springer, Heidelberg (2012)
108. Leipholz, H.: Stability Theory. Academic Press, New York (1970)
109. Lewis, F.L.: Fundamental, reachability, and observability matrices for discrete descriptor systems. IEEE Trans. Autom. Contr. **AC–30**(5), 502–505 (1985)
110. Lewis, F.L.: A survey of linear singular systems. Circuits Syst. Signal Process. **5**(1), 3–36 (1986)
111. Lewis, F.L., Mertzios, B.G.: On the Analysis of Discrete Linear Time-Invariant Singular Systems. IEEE Trans. Autom. Contr. **35**(4), 506–511 (1990)
112. Losse, P.: The H-infinity optimal control problem for descriptor systems. Ph.D. thesis, Technische Universität Chemnitz, Germany (2011)
113. Lötstedt, P., Petzold, L.R.: Numerical solution of nonlinear differential equations with algebraic constraints i: convergence results for backward differentiation formulas. Math. Comput. **46**(174), 491–516 (1986)
114. Luenberger, D.G.: Time-invariant descriptor systems. Automatica **14**(5), 473–480 (1978)
115. Luenberger, D.G.: Introduction to Dynamic Systems: Theory. Models and Applications. Wiley, New York (1979)
116. Luenberger, D.G., Arbel, A.: Singular dynamic Leontief systems. Econometrica **45**(4), 991–995 (1977)
117. Manzo, V.E.: Estimation and control of descriptor systems. Ph.D. thesis, Université de Valenciennes et du Hainaut-Cambresis, France (2015)
118. Marszałek, W., Trzaska, Z.W.: DAE models of electrical power systems and their bifurcations around singularities. In: Proceedings of the 43rd IEEE Conference on Decision and Control, pp. 4833–4838, Nassau, Bahamas, 14–17 December 2004

119. McClamroch, N.H.: Singular systems of differential equations as dynamic models for constrained robot systems. In: Proceedings of the IEEE International Conference on Robotics and Automation, pp. 21–28, San Francisco, USA, 7–10 April 1986
120. Megawati, N.: A geometric approach to differential-algebraic systems: from bisimulation to control by interconnection. Ph.D thesis, University of Groningen, The Netherlands (2017)
121. Mertzios, B.G., Lewis, F.L.: Fundamental matrix of discrete singular systems. Circuits Syst. Signal Process. **8**(3), 341–355 (1989)
122. Miller, K.S., Ross, B.: An Introduction to the Fractional Calculus and Fractional Differential Equations. Wiley, New York (1993)
123. Mills, J.K., Goldenberg, A.A.: Force and Position Control of Manipulators During Constrained Motion Tasks. IEEE Trans. Robot. Autom. **5**(1), 30–46 (1989)
124. Mitkowski, W.: Stabilization of Dynamic Systems. WNT, Warsaw (1991). (in Polish)
125. Mitkowski, W.: Remarks on stability of positive linear systems. Control. Cybern. **29**(1), 295–304 (2000)
126. Mitkowski, W.: Dynamical properties of Metzler systems. Bull. Pol. Acad. Sci. Tech. **56**(4), 309–312 (2008)
127. Moysis, L., Mishra, V.K.: Existence of reachable and observable triples of linear discrete-time descriptor systems. Circuits Syst. Signal Process. **38**, 1086–1098 (2019)
128. Nashimoto, K.: Fractional Calculus. Descartes Press, Koriama (1984)
129. Newcomb, R.W., Dziurla, B.: Some circuits and systems applications of semistate theory. Circuits Syst. Signal Process. **8**(3), 235–260 (1989)
130. Oldham, K.B., Spanier, J.: The Fractional Calculus: Integrations and Differentiations of Arbitrary Order. Academic Press, New York (1974)
131. Ortigueira, M.D.: Fractional Calculus for Scientists and Engineers. Springer, Heidelberg (2011)
132. Ostalczyk, P.: Epitome of the Fractional Calculus: Theory and its Applications in Automatics. Publishing Department of Technical University of Łódź, Łódź (2008). (in Polish)
133. Ostalczyk, P.: Discrete Fractional Calculus: Applications in Control and Image Processing. World Scientific, River Edge (2016)
134. Phat, V.N., Sau, N.H.: On exponential stability of linear singular positive delayed systems. Appl. Math. Lett. **38**, 67–72 (2014)
135. Podlubny, I.: Fractional Differential Equations. Academic Press, San Diego (1999)
136. Podlubny, I., Dorcak, L., Misanek, J.: Application of fractional-order derivatives to calculation of heat load intensity change in blast furnace wall. Trans. Tech. Univ. Kosice **5**(5), 137–144 (1995)
137. Polyak, B.T., Shcherbakov, P.S.: Superstable linear control systems. I. Analysis. Autom. Remote Control **63**(8), 1239–1254 (2002)
138. Polyak, B.T., Shcherbakov, P.S.: Superstable linear control systems. II. Design. Autom. Remote Control **63**(11), 1745–1763 (2002)
139. Prasolov, V.: Problems and Theorems in Linear Algebra. American Mathematical Society, Rhode Island (1994)
140. Radwan, A.G., Soliman, A.M., Elwakil, A.S., Sedeek, A.: On the stability of linear systems with fractional-order elements. Chaos, Sol. Fractals **40**(5), 2317–2328 (2009)
141. Rami, M.A., Napp, D.: Characterization and stability of autonomous positive descriptor systems. IEEE Trans. Autom. Contr. **57**(10), 2668–2673 (2012)
142. Rami, M.A., Napp, D.: Positivity of discrete singular systems and their stability: an LP-based approach. Automatica **50**(1), 84–91 (2014)
143. Rantzer, A.: Scalable control of positive systems. Eur. J. Control. **24**, 72–80 (2015)
144. Raynaud, H.F., Zergainoh, A.: State-space representation for fractional order controllers. Automatica **36**, 1017–1021 (2000)
145. Reis, T.: An Infinite dimensional descriptor system model for electrical circuits with transmission lines. In: Proceedings of the Conference MTNS, Leuven, Belgium (2004)
146. Riaza, R.: Differential-Algebraic Systems: Analytical Aspects and Circuit Applications. World Scientific Publishing, Basel (1994)

147. Rose, N.J.: The Laurent expansion of a generalized resolvent with some applications. SIAM J. Math. Anal. **9**(4), 751–758 (1978)
148. Rosenbrock, H.H.: State-space and Multivariable Theory. Wiley, New York (1970)
149. Rumchev, V.G., Caccetta, L., Kostova, S.: Positive linear dynamic model of mobile source air pollution and problems of control. Syst. Sci. **29**(1), 31–46 (2003)
150. Runyi, Y., Dianhui, W.: Structural properties and poles assignability of LTI singular systems under output feedback. Automatica **39**(4), 685–692 (2003)
151. Ruszewski, A.: Practical and asymptotic stabilities for a class of delayed fractional discrete-time linear systems. Bull. Pol. Acad. Sci. Tech. **67**(3), 509–515 (2019)
152. Ruszewski, A.: Stability of discrete-time fractional linear systems with delays. Archives Control Sci **29**(3), 549–567 (2019)
153. Sajewski, Ł.: Decentralized stabilization of descriptor fractional positive continuous-time linear systems with delays. In: Proceedings of the Conference on MMAR 2017, pp. 482–487, Międzyzdroje, Poland, 28–31 August 2017
154. Sajewski, Ł.: Stabilization of positive descriptor fractional discrete-time linear system with two different fractional orders by decentralized controller. Bull. Pol. Acad. Sci. Tech. **65**(5), 709–714 (2017)
155. Samko, S.G., Kilbas, A.A., Marichev, O.I.: Fractional Integrals and Derivatives: Theory and Applications. Gordon and Breach Science Publishers, Amsterdam (1993)
156. Shafai B., Li, C.: Positive stabilization of singular systems by proportional derivative state feedback. In: IEEE Conference on Control Technology and Applications, pp. 1140–1146, Mauna Lani, USA, 27–30 August 2017
157. Simeon, B., Führer, C., Rentrop, P.: Differential-algebraic equations in vehicle system dynamics. Surv. Math. Ind. **1**, 1–37 (1991)
158. Sincovec, R., Erisman, A., Yip, E., Epton, M.: Analysis of descriptor systems using numerical algorithms. IEEE Trans. Autom. Contr. **26**(1), 139–147 (1981)
159. Solteiro Pires, E.J., Tenreiro Machado, J.A., Moura Oliveira, P.B.: Functional dynamics in genetic algorithms. In: Workshop on Fractional Differentiation and Its Applications, pp. 414–419, Porto, Portugal (2006)
160. Stanisławski, R., Latawiec, K.J.: Normalized finite fractional differences: computational and accuracy breakthroughs. Int. J. Appl. Math. Comput. Sci. **22**(4), 907–919 (2012)
161. Stanisławski, R., Latawiec, K.J.: Stability analysis for discrete-time fractional-order LTI state-space systems. Part I: new necessary and sufficient conditions for asymptotic stability. Bull. Pol. Acad. Sci. Tech. **61**(2), 353–361 (2013)
162. Stanisławski, R., Latawiec, K.J.: Stability analysis for discrete-time fractional-order LTI state-space systems. Part II: new stability criterion for FD-based systems. Bull. Pol. Acad. Sci. Tech. **61**(2), 363–370 (2013)
163. Stott, B.: Power system dynamic response calculations. Proc. IEEE **67**(2), 219–241 (1979)
164. Talagaev, Y.V.: Robust analysis and superstabilization of chaotic systems. In: Proceedings of the IEEE Conference on Control Applications, pp. 1431–1436, Juan Les Antibes, France, 8–10 October 2014
165. Trenn, S.: Distributional differential algebraic equations. Ph.D. thesis, Universitätsverlag Ilmenau, Germany (2009)
166. Uddin, M.M.: Computational Methods for Model Reduction of Large-Scale Sparse Structured Descriptor Systems. Ph.D. thesis, Otto-von-Guericke-Universität Magdeburg, Germany (2015)
167. Valcher, M.E.: Reachability properties of continuous-time positive systems. IEEE Trans. Autom. Contr. **54**(7), 1586–1590 (2009)
168. Van Dooren, P.: The computation of Kronecker's canonical form of a singular pencil. Linear Algebra Appl. **27**, 103–140 (1979)
169. Varga, A.: Numerical algorithms and software tools for analysis and modelling of descriptor systems. In: Proceedings of the 2nd IFAC Workshop on System Structure and Control, Prague, Czechoslovakia, 3–5 September 1992, pp. 392–395 (1992)

170. Verghese, G.C., Lévy, B.C., Kailath, T.: A generalized state-space for singular systems. IEEE Trans. Autom. Contr. **26**(4), 811–831 (1981)
171. Vinagre, B.M., Podlubny, I., Dorcák, L., Feliu, V.: On fractional PID controllers: a frequency domain approach. In: IFAC Workshop on Digital Control, pp. 53–58, Terrasa, Spain, April 2000
172. Virnik, E.: Stability analysis of positive descriptor systems. Linear Algebra Appl. **429**(10), 2640–2659 (2008)
173. Weierstrass. K.: Zur Theorie der Bilinearen und Quadratischen Formen Monatsh. Akad. Wiss. Berlin, 310–338 (1868)
174. Xudong, Z., Xingwen, L., Shen, Y., Hongyi, L.: Improved results on stability of continuous-time switched positive linear systems. Automatica **50**(2), 614–621 (2014)
175. Yago Sánchez, B.S.: Fractional-PID control for active reduction of vertical tail buffeting. Ph.D. thesis, Saint Louis University (1999)
176. Yip, E., Sincovec, R.: Solvability, controllability and observability of continuous descriptor systems. IEEE Trans. Autom. Contr. **26**(3), 702–707 (1981)
177. Younus, A., Javaid, I., Zehra, A.: On controllability and observability of fractional continuous-time linear systems with regular pencils. Bull. Pol. Acad. Sci. Tech. **65**(3), 297–304 (2017)
178. Zhang, Y., Zhang, Q., Tanaka, T., Yan, X.G.: Positivity of continuous-time descriptor systems with time delays. IEEE Trans. Autom. Contr. **59**(11), 3093–3097 (2014)

The manufacturer's authorised representative in the EU is Springer
Nature Customer Service Centre GmbH, Europaplatz 3, 69115 Heidelberg,
Germany. If you have any concerns regarding our products, please
contact ProductSafety@springernature.com

Printed and bound by CPI Group (UK) Ltd, Croydon, CR0 4YY
28/09/2026
02224088-0013